Genetic Regulation of Animal Reproduction

Genetic Regulation of Animal Reproduction

Editor

Bao Yuan

MDPI

Basel • Beijing • Wuhan • Barcelona • Belgrade • Novi Sad • Cluj • Manchester

Editor
Bao Yuan
Jilin University
Changchun
China

Editorial Office
MDPI
St. Alban-Anlage 66
4052 Basel, Switzerland

This is a reprint of articles from the Special Issue published online in the open access journal *Genes* (ISSN 2073-4425) (available at: https://www.mdpi.com/journal/genes/special_issues/FCM298DOM4).

For citation purposes, cite each article independently as indicated on the article page online and as indicated below:

Lastname, A.A.; Lastname, B.B. Article Title. *Journal Name* **Year**, *Volume Number*, Page Range.

ISBN 978-3-7258-0387-3 (Hbk)
ISBN 978-3-7258-0388-0 (PDF)
doi.org/10.3390/books978-3-7258-0388-0

Contents

About the Editor . . . vii

Xinyue Liu, Junyuan Zheng, Jialan Ding, Jiaxin Wu, Fuyuan Zuo and Gongwei Zhang
When Livestock Genomes Meet Third-Generation Sequencing Technology: From Opportunities to Applications
Reprinted from: *Genes* **2024**, *15*, 245, doi:10.3390/genes15020245 . . . 1

Jian-Bo Liu, Jia-Bao Zhang, Xiang-Min Yan, Peng-Gui Xie, Yao Fu, Xu-Huang Fu, et al.
DNA Double-Strand Break-Related Competitive Endogenous RNA Network of Noncoding RNA in Bovine Cumulus Cells
Reprinted from: *Genes* **2023**, *14*, 290, doi:10.3390/genes14020290 . . . 16

Tian Wang, Guokun Zhao, Song Yu, Yi Zheng, Haixiang Guo, Haoqi Wang, et al.
Sequencing of the Pituitary Transcriptome after GnRH Treatment Uncovers the Involvement of lncRNA-m23b/miR-23b-3p/CAMK2D in FSH Synthesis and Secretion
Reprinted from: *Genes* **2023**, *14*, 846, doi:10.3390/genes14040846 . . . 34

Fusheng Yao, Jia Hao, Zhaochen Wang, Meiqiang Chu, Jingyu Zhang, Guangyin Xi, et al.
WNT Co-Receptor LRP6 Is Critical for Zygotic Genome Activation and Embryonic Developmental Potential by Interacting with Oviductal Paracrine Ligand WNT2
Reprinted from: *Genes* **2023**, *14*, 891, doi:10.3390/genes14040891 . . . 55

Fupeng Wang, Yawen Tang, Yijie Cai, Ran Yang, Zongyu Wang, Xiaodong Wang, et al.
Intrafollicular Retinoic Acid Signaling Is Important for Luteinizing Hormone-Induced Oocyte Meiotic Resumption
Reprinted from: *Genes* **2023**, *14*, 946, doi:10.3390/genes14040946 . . . 69

Bo Zhai, Xu Li, Zhongli Zhao, Yang Cao, Xinxin Liu, Zheng Liu, et al.
Melatonin Protects the Apoptosis of Sheep Granulosa Cells by Suppressing Oxidative Stress via MAP3K8 and FOS Pathway
Reprinted from: *Genes* **2023**, *14*, 1067, doi:10.3390/genes14051067 . . . 84

Zijing Zhang, Qiaoting Shi, Xiaoting Zhu, Lei Jin, Limin Lang, Shijie Lyu, et al.
Identification and Functional Analysis of Transcriptome Profiles, Long Non-Coding RNAs, Single-Nucleotide Polymorphisms, and Alternative Splicing from the Oocyte to the Preimplantation Stage of Sheep by Single-Cell RNA Sequencing
Reprinted from: *Genes* **2023**, *14*, 1145, doi:10.3390/genes14061145 . . . 97

Yue Zou, Xiang Chen, Xingzhou Tian, Wei Guo, Yong Ruan, Wen Tang, et al.
Transcriptomic Analysis of the Developing Testis and Spermatogenesis in Qianbei Ma Goats
Reprinted from: *Genes* **2023**, *14*, 1334, doi:10.3390/genes14071334 . . . 118

Shuming Shi, Zhaojun Geng, Xianfeng Yu, Bing Hu, Liying Liu, Zhichao Chi, et al.
Salidroside Supplementation Affects In Vitro Maturation and Preimplantation Embryonic Development by Promoting Meiotic Resumption
Reprinted from: *Genes* **2023**, *14*, 1729, doi:10.3390/genes14091729 . . . 133

Jiage Dai, Jiabao Cai, Taipeng Zhang, Mingyue Pang, Xiaoling Xu, Jiahua Bai, et al.
Transcriptome and Metabolome Analyses Reveal the Mechanism of Corpus Luteum Cyst Formation in Pigs
Reprinted from: *Genes* **2023**, *14*, 1848, doi:10.3390/genes14101848 . . . 149

Jinlun Lu, Min Guo, Xiaodong Wang, Rui Wang, Guangyin Xi, Lei An, et al.
A Redesigned Method for CNP-Synchronized In Vitro Maturation Inhibits Oxidative Stress and Apoptosis in Cumulus-Oocyte Complexes and Improves the Developmental Potential of Porcine Oocytes
Reprinted from: *Genes* **2023**, *14*, 1885, doi:10.3390/genes14101885 . . . . **164**

Jianbo Liu, Jiabao Zhang, Yi Zheng, Guokun Zhao, Hao Jiang and Bao Yuan
miR-302d Targeting of CDKN1A Regulates DNA Damage and Steroid Hormone Secretion in Bovine Cumulus Cells
Reprinted from: *Genes* **2023**, *14*, 2195, doi:10.3390/genes14122195 . . . . **181**

Rihong Guo, Huili Wang, Chunhua Meng, Hongbing Gui, Yinxia Li, Fang Chen, et al.
Efficient and Specific Generation of *MSTN*-Edited Hu Sheep Using C-CRISPR
Reprinted from: *Genes* **2023**, *14*, 1216, doi:10.3390/genes14061216 . . . . **196**

About the Editor

Bao Yuan

Bao Yuan, professor and doctoral supervisor of the College of Animal Science, Jilin University. The main research interests are animal reproductive regulation and standardization of livestock laboratory animals. He has presided over 2 general projects of the National Natural Science Foundation of China and 3 provincial projects. In the past three years, he has published 22 SCI papers as the first or corresponding author and obtained 5 invention patents.

GCAT TACG GCAT *genes*

MDPI

Review

When Livestock Genomes Meet Third-Generation Sequencing Technology: From Opportunities to Applications

Xinyue Liu [1], Junyuan Zheng [1], Jialan Ding [1], Jiaxin Wu [1], Fuyuan Zuo [1,2] and Gongwei Zhang [1,2,*]

[1] College of Animal Science and Technology, Southwest University, Rongchang, Chongqing 402460, China; luckyliuxy@163.com (X.L.); 13896120486@163.com (J.Z.); djl20212023@163.com (J.D.); wjiaxin0225@163.com (J.W.); zfuyuan@163.com (F.Z.)

[2] Beef Cattle Engineering and Technology Research Center of Chongqing, Southwest University, Rongchang, Chongqing 402460, China

* Correspondence: zgw-vip@163.com

Abstract: Third-generation sequencing technology has found widespread application in the genomic, transcriptomic, and epigenetic research of both human and livestock genetics. This technology offers significant advantages in the sequencing of complex genomic regions, the identification of intricate structural variations, and the production of high-quality genomes. Its attributes, including long sequencing reads, obviation of PCR amplification, and direct determination of DNA/RNA, contribute to its efficacy. This review presents a comprehensive overview of third-generation sequencing technologies, exemplified by single-molecule real-time sequencing (SMRT) and Oxford Nanopore Technology (ONT). Emphasizing the research advancements in livestock genomics, the review delves into genome assembly, structural variation detection, transcriptome sequencing, and epigenetic investigations enabled by third-generation sequencing. A comprehensive analysis is conducted on the application and potential challenges of third-generation sequencing technology for genome detection in livestock. Beyond providing valuable insights into genome structure analysis and the identification of rare genes in livestock, the review ventures into an exploration of the genetic mechanisms underpinning exemplary traits. This review not only contributes to our understanding of the genomic landscape in livestock but also provides fresh perspectives for the advancement of research in this domain.

Keywords: third-generation sequencing; gene assembly; transcriptome; epigenetics

Citation: Liu, X.; Zheng, J.; Ding, J.; Wu, J.; Zuo, F.; Zhang, G. When Livestock Genomes Meet Third-Generation Sequencing Technology: From Opportunities to Applications. *Genes* **2024**, *15*, 245. https://doi.org/10.3390/genes15020245

Academic Editor: Penny K. Riggs

Received: 23 December 2023
Revised: 30 January 2024
Accepted: 10 February 2024
Published: 15 February 2024

1. Introduction

The utilization of high-throughput sequencing technologies in molecular genetics research has become increasingly prevalent. Next-generation sequencing (NGS) technology, represented by the Illumina sequencing platform, is known for its advantages in terms of accuracy and low cost. However, the short read lengths result in the inability to sequence certain repetitive genomic sequences, leading to assembly errors and gaps in the genome assembly. Additionally, it lacks the capability for direct DNA/RNA sequencing, rendering it unable to fully meet the evolving technological demands of modern biology. Third-generation sequencing (TGS) technologies, exemplified by single-molecule real-time (SMRT) sequencing from Pacific Biosciences (PacBio) and nanopore sequencing from Oxford Nanopore Technologies (ONT), have emerged as the leading methods in the genomic, transcriptomic, and epigenetic field due to their significant advantages, such as long read lengths, real-time base sequencing, direct sequencing, and shorter processing times [1,2]. SMRT utilizes two sequencing modes, continuous long reads (CLR) and circular consensus sequencing (CCS), displaying outstanding advantages in the sequencing of complex gene structures, the identification of SNP variations, the detection of gene structures, such as full-length transcriptomes, alternative splicing (AS) and fusion genes, and the discrimination between monozygotic twins [3,4]. ONT technology, driven by motor proteins, directly threads DNA/RNA strands through nanopores, representing a direct, real-time sequencing approach that eliminates the need for PCR amplification. This approach preserves base modification information and enables accurate quantitative analysis, significantly improving the accuracy, read length, and throughput of ONT sequencing [5]. TGS has addressed the shortcomings of NGS in genomic composition and transcript isoform analysis, establishing itself as the preferred technology for gene function studies [6].

2. Advances of TGS in the Genomic Research of Livestock

The TGS technology has found extensive applications in the genomics and molecular genetics research of livestock, serving as a powerful tool for the deciphering of genome structure, functionality, and evolution. Firstly, it efficiently conducts comprehensive sequencing of the livestock genome, providing a holistic genomic dataset that facilitates the discovery of genetic variations and the elucidation of gene functions [7]. Secondly, through long-read sequencing, TGS can more accurately detect structural variations within the genome, including insertions, deletions, and inversions, thereby offering more precise genomic information for the study of relevant traits [8]. Moreover, this technology proves advantageous in the superior assembly of complex genomes, particularly those in livestock characterized by highly repetitive sequences and chromosomal structures with intricate features [9]. Simultaneously, in epigenetic studies, the long reads generated by TGS enable a more comprehensive and accurate resolution of DNA methylation and RNA modification patterns [10], shedding light on the intricate network governing gene expression regulation in livestock. Finally, the application of TGS in genomic selection involves the comprehensive and accurate analysis of large-scale individual genome sequencing data, facilitating genomic selection strategies that accelerate genetic improvement in livestock breeds and enhance breeding efficiency [11]. This paper primarily summarizes the research progress of TGS technologies in livestock genetics, specifically in genome assembly, structural variation detection, transcriptome sequencing, and epigenetic analysis (Figure 1), providing novel insights for further genetic studies in livestock.

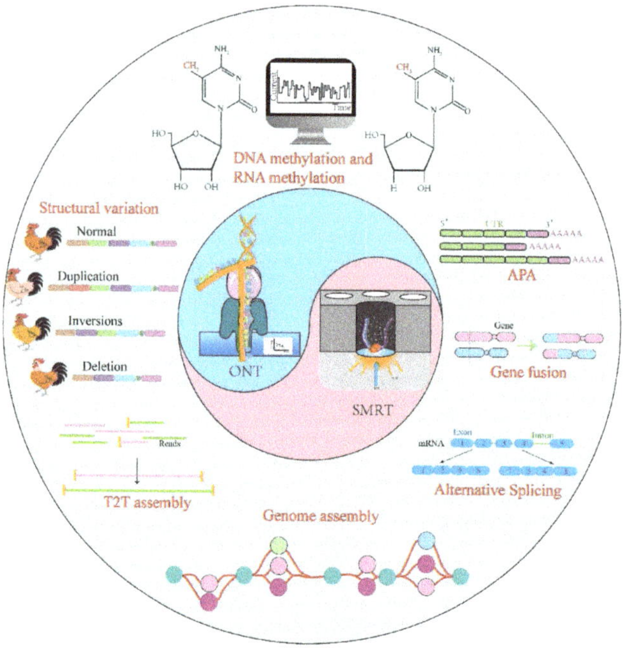

Figure 1. The application of TGS in the genomic research on livestock, including genome assembly, structural variation detection, transcriptome sequencing, and epigenetic analysis.

2.1. Progress in Genome Assembly Using TGS in Livestock

A complete genome is a prerequisite for the obtaining of accurate genetic information, the precise exploration of genetic details, and the deciphering of the mechanisms of genetic variation. It provides a theoretical foundation for the in-depth analysis of genetic features in germplasm resources and promotes the development of genomics and molecular breeding. Genome assembly involves the assembly of sequenced fragments into a complete genomic sequence; it uses two primary methods: de novo assembly and mapping-based assembly. Additionally, the assembly algorithms include overlap–layout–consensus (OLC) and the de Bruijn graph (DBG) [12]. Following the advent of NGS, the de novo sequencing of livestock species became feasible, typically by constructing a reference genome based on a representative breed [13]. TGS technologies demonstrate remarkable advantages and enormous potential in genome assembly research, and they are widely applied in various animals, such as cattle [14,15], gayals [16], yaks [17,18], buffalos [19,20], sheep [21,22], goats [23,24], pigs [25,26], chickens [27,28], ducks [29], and geese [30] (Table 1).

2.2. Progress in Pan-Genome Using TGS in Livestock

Due to the limited genetic diversity covered by a single reference genome, the concept of the pan-genome was introduced by researchers [31,32]. The pan-genome encompasses the sum of all genomic information within a species, capturing more genetic diversity. Presently, numerous species have developed pan-genomes based on NGS data, including the human [33], pig [34–36], goat [37], cattle [38,39], chicken [40], sheep [41] and goose species [30] (Figure 2A). The pan-genome comprises core genes, dispensable genes, and strain-specific genes [42]. Core genes are shared by all individuals of a species and are generally associated with biological functions and major phenotypic features. Dispensable genes exist in some but not all individuals, reflecting species-specific adaptations or unique biological traits [43]. Strain-specific genes are unique to specific individuals, indicating individual-specific traits (Figure 2B). The pan-genomic analyses conducted on these species have yielded novel insights. The pan-genome construction strategies include iterative assembly, de novo assembly, and graphical pan-genomes [44] (Figure 2C). These strategies utilize large-scale sequencing data, deep sequencing of a small number of individuals, and graph-based data structures to represent the gene sequences and structures of species.

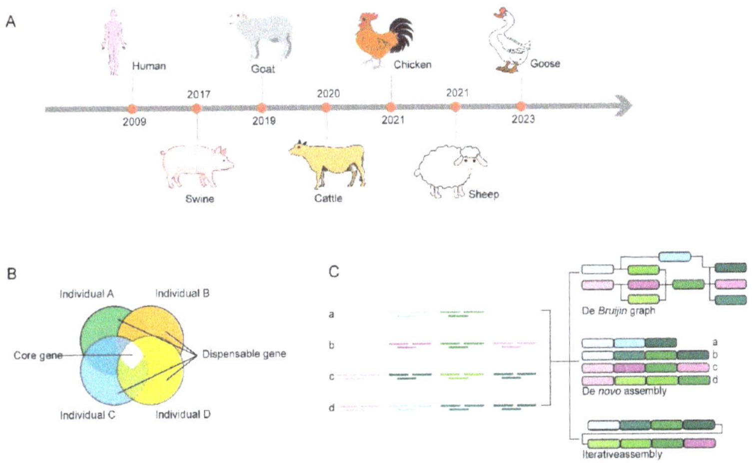

Figure 2. The development process and construction methods of pan-genomic research. (**A**) Numerous species have developed pan-genomes, including emiliania huxleyi. (**B**) Pan-genomes comprise core genes, dispensable genes, and strain-specific genes. (**C**) Pan-genome construction strategies include iterative assembly, de novo assembly, and graphical pan-genomes.

2.3. Progress in Telomere-to-Telomere Assembly Using TGS

Telomeres are specialized terminal regions of linear chromosomes in eukaryotes; they play crucial roles in cell division and chromosome replication. The need to understand the characteristics of repetitive sequences within genes for better genetic mechanism studies has made telomere assembly a research hotspot. With the rapid development of sequencing technologies, a new era of "Telomere-to-Telomere (T2T) assembly" is unfolding [45]. Through T2T assembly, the intricate structures and functions of complex regions in the genome, such as telomeres and centromeres, can be deeply explored. Achieving gapless chromosome assembly from telomere to telomere holds significant importance for the in-depth exploration of the complex region of centromeres in the genome and further uncovers crucial genetic variations [46]. Researchers have combined TGS with genome assembly technologies to successfully accomplish T2T assembly in various species, including humans [47,48] and chickens [28], obtaining high-quality genomes. Huang et al. pioneered the completion of a full-genome map for domestic chickens, using Huxu breed chickens as material and identifying the six missing chromosomes in the original genome assembly. The complete chicken chromosome models are useful in the reconstruction of the karyotype of the vertebrate ancestor [28].

Table 1. TGS techniques for genome assembly and pan-genome research in livestock.

Species	Feature	Breed	Sequencing Platform	Key Findings	Publication Year	Reference
Bos taurus	Cattle	1.OxO F1 2.NxB F1 3.GxP F	ONT	Constructed haplotype-resolved genomes for cattle and related species, established a pan-genome for cattle, and quantified structural diversity	2022	[14]
B. taurus	Cattle	Southern Yellow Cattle	PacBio-SMRT	Confirmed genetic diversity in the southern yellow cattle population in China, identified gene introgression events from five different wild cattle species	2023	[15]
B. taurus	Cattle	Hainan Cattle, Mongolian Cattle	ONT	Discovered significant structural variations influencing environmental adaptability in Chinese yellow cattle	2023	[49]
Bos frontalis	Cattle	Gayal	PacBio-SMRT	Conducted chromosome-level genome assembly for Dulong cattle	2023	[16]
Bos grunniens	Yak	Yak	ONT	Obtained high-quality chromosome-level genomes for wild and domestic yaks, a structural variation catalog for yaks, and a single-cell transcriptome atlas of lung tissues	2022	[17]
B. grunniens	Yak	White Yak	ONT	Revealed genetic introgression of unique structural variations in the color-sided yellow cattle, resulting in the creation of the color-sided Yak. Subsequent genetic variations gave rise to the white Yak	2023	[18]
Bubalus bubalis	Buffalo	Water Buffalo	PacBio-SMRT	Generated a detailed genomic map for water buffalo (2n = 50) and performed chromosome-level genome assembly	2019	[19]
B. bubalis	Buffalo	Swamp-type Water Buffalo, River-type Water Buffalo	PacBio-SMRT	Attained high-quality chromosome-level reference genomes for swamp-type water buffalo (2n = 48) and river-type water buffalo (2n = 50)	2020	[20]
Ovis aries	Sheep	Dorper Sheep	ONT	Revealed the genetic basis of allele-specific expression (ASE) genes and specific phenotypic traits in Dorper sheep	2022	[21]
O. aries	Sheep	15different breeds of sheep	PacBio-SMRT	Constructed high-quality pan-genome maps for different sheep breeds	2023	[22]
Capra hircus	Goat	Saanen Dairy Goat	PacBio-SMRT	Assembled the reference genome Saanen_v1 for Saanen dairy goats	2021	[23]
C. hircus	Goat	Tibetan Goat	PacBio-SMRT	Unveiled PAPSS2 as a key gene not only for high-altitude adaptation in goats but also a significant gene in genetic introgression analysis	2022	[24]

Table 1. *Cont.*

Species	Feature	Breed	Sequencing Platform	Key Findings	Publication Year	Reference
Sus scrofa	Pig	Tibetan Pig, Jinhua Pig, and 8 other breeds	ONT	Completed pan-genome maps for Anqing Liubai Pig, Laiwu Pig, Meishan Pig, Min Pig, Rongchang Pig, Wuzhishan Pig, Yorkshire Pig, European Wild Boar, etc.	2023	[25]
S. scrofa	Pig	Duroc	PacBio-SMRT	Assembled the reference genome Sscrofa11.1 for pigs from scratch	2020	[26]
Gallus gallus	Chicken	Huxu Chicken	ONT	First published complete genome atlas (T2T) for vertebrates; characterized the epigenetics of the W chromosome; elucidated the origin, sequence structure, and diversity of chicken centromeres	2023	[28]
G. gallus	Chicken	Chickens from Four Continents	PacBio-SMRT	Established the pan-genome of chickens, identified new coding genes, long non-coding RNAs, and new gene families; identified new gene clusters for studying collinearity	2022	[50]
G. gallus	Chicken	Wenshang Lu Hua Chicken	PacBio-SMRT	Obtained a high-quality chromosome-level reference genome for the Wenshang Lu Hua chicken	2023	[51]
Anas platyrhynchos	Duck	Peking Duck, Shaoxing Duck, and Mallard	PacBio-SMRT	Assembled chromosome-level high-quality genomes for Peking Duck, Shaoxing Duck, and mallard, refuting the "missing gene hypothesis" in birds	2021	[29]

O: Original Braunvieh cattle (Bos taurus taurus); N: Nellore (Bos taurus indicus); G: gaur (Bos gaurus) bull; B: Brown Swiss (Bos taurus taurus) cattle; P: Piedmontese (Bos taurus taurus) cow.

2.4. Understanding the Genetic Mechanisms of Livestock Traits Using TGS

The emergence of TGS technology has had a profound impact on the field of life sciences. Compared with traditional NGS, multiple aspects of TGS technology have shown significant advantages. The long read feature of TGS enables it to better cope with gene regions with complex structural variations (SVs) and repetitive sequences, particularly when exploring the genetic mechanism of animal traits; thus, it can more accurately reveal the genetic basis of animal traits. In addition, TGS technology has higher sequencing accuracy, which helps to more accurately identify mutation sites and mutations in the genome that may be related to animal traits and to provide strong support for in-depth research on the genetic mechanism of animal traits. Furthermore, TGS technology has improved sequencing speed, saved time costs, and enabled more efficient association analysis of large samples and genomic association research. This helps in the discovery of more genes and variation sites related to animal traits. In summary, TGS technology has significant advantages in the exploration of the genetic mechanisms of animal traits, providing a powerful tool for a deeper understanding of the genetic basis of animals.

SV is an important part of the pan-genome, including insertion, deletion, duplication, translocation, inversion, and so on. SVs are significant contributors to genetic variations in livestock [52]. TGS technologies can detect structural variations such as deletions, duplications, inversions, translocations, and more (greater than 50bp) [53]. In the detection of SV events, researchers have found that long-read sequencing is more convenient than short-read sequencing [54]. The state-of-the-art SV callers, i.e., cuteSV [55], NanoSV [56], NanoVar [57], Sniffles [58], SVIM [59], and PBSV (https://github.com/PacificBiosciences/pbsv (accessed on 9 February 2024)) can be used for TGS data.

Copy number variation (CNV) is a crucial component of SV; it describes the molecular phenomenon of genomic sequence repetition and plays a key role in promoting population diversity and both micro- and macroevolutionary processes in humans and animals [60]. Magini et al. found that nanopore sequencing, when compared with the current most advanced CNV detection techniques, can reduce the time required for CNV detection at the same resolution, and ONT has greater stability in identifying chimeric CNVs [61]. In this review, we focused on the application of TGS in livestock.

2.4.1. Understanding the Genetic Mechanisms of Traits in Ruminants

In cattle, researchers constructed a graphical pan-genome for 10 southern Chinese yellow cattle breeds, utilizing the high accuracy and continuity advantages of local directional haplotype genomes. They clarified five gene infiltration events from different wild cattle species in the population history of southern Chinese yellow cattle [15]. Liu et al. built a graphical pan-genome for yak and cattle species, discovering that nearly 90% of the domestic yak genome contains genes that have infiltrated from yellow cattle. They identified that the cause of the white yak coat color is due to the infiltration of an SV spanning the *KIT* gene [18]. Genetic infiltration from unique SVs in color-sided yellow cattle led to the formation of color-sided yaks; subsequently, the infiltrated structural variation produced a new genetic variation which resulted in the production of white yaks [18]. Meanwhile, Gao et al. constructed high-quality chromosome-level reference genomes for wild yaks and domestic yaks. Using a combination with a single-cell map of lung tissue, they indicated that the development of lung endothelial cells and their function in low-oxygen adaptation may be influenced by SV [17]. Concurrently, Xia et al. assembled high-quality chromosome-level genomes for Hainan cattle and Mongolian cattle. Utilizing various omics technologies, they identified significant SVs influencing the environmental adaptability of Chinese yellow cattle [15]. Researchers using the nanopore platform, analyzed alleles related to horns and found that the size of the inserted sequences at the Celtic locus ranged from 181 to 206 bp. Through alignment with ARS-UCD1.2, repetitive sequences were displayed on the Celtic locus, providing direct evidence for the presence of the Celtic locus in the Australian Brahman cattle population [62].

In sheep, Li et al. assembled high-quality genomes for 15 sheep breeds and constructed high-quality pan-genomic maps for different sheep breeds. They identified numerous divergent allele genes, complex multi-allele variations, and crucial candidate mutation sites related to tail length and thickness traits in sheep [22]. In goats, Li et al. constructed a goat pan-genome web interface for data visualization by comparing nine de novo assemblies from seven sibling species of domestic goats with ARS1 and by using resequencing and transcriptome data from goats for verification [37]. The researchers revealed the strongest high-altitude adaptation signature in Tibetan goats at the PAPSS2 locus by using data from 331 genomes and 104 transcriptomes, which provided evidence that interspecific introgression contributed to the high-altitude adaptability. This study expanded the gene repertoire of hypoxia adaptation in highland-dwelling mammals and provided new insights into their evolutionary origins [24].

2.4.2. Understanding the Genetic Mechanisms of Traits in Monogastric Animals

NGS technology has been used in the pan-genome assembly and gene structural variation detection in pigs; 12 de novo pig assemblies from Eurasia were compared to identify the missing sequences from the reference genome, and 72.5 Mb of non-redundant sequences (~3% of the genome) were found to be absent from the reference genome [36]. Furthermore, Li et al. systematically analyzed the presence/absence variation (PAV) of the coding sequences in 250 sequenced individuals from 32 pig breeds in Eurasia by constructing a pan-genome [35]. Jiang et al. utilized 11 pig breeds to construct a graphical pan-genome for pigs, revealing 206 Mb of new sequences and detecting 183,352 non-redundant structural variations. They explored the significant role of structural variations in high-altitude adaptation in Tibetan pigs [25].

In horses, Viluma et al. used PacBio sequencing technology to sequence bacterial artificial chromosome (BAC) clones spanning the major histocompatibility complex class II (MHC class II) of horses for the first time. They discovered numerous CNV sites, providing important resources for the association study of immune-mediated diseases in horses and the evolutionary analysis of genetic diversity in this region [63].

In chickens, 664 sample data were used to construct a pan-genome of chickens in order to gain a deeper understanding of the changes in the genome structure during evolution. *IGF2BP1*, as the causal variant of the chicken body size quantitative trait locus located at chromosome 27, was found for the first time. Therefore, the chicken pan-genome is a useful resource for biological discovery and breeding [40]. However, researchers performed a de novo assembly on 20 chicken individuals worldwide, constructing the first high-quality pan-genome based on de novo assembly for birds. They identified new coding genes, long non-coding RNAs, and novel gene clusters [50]. The genome sequences of a reference chicken genome (GRCg7b) and the resequencing reads of 15 Silkie chickens (8 males and 7 females) onto the Silkie genome were aligned to identify a total of 9,337,467 SNPs and 920,864 small insertions and deletions (indels, referring $\leq$50 bp) [27]. Using two different methods, the PanGenome Graph Builder (PGGB) and Minigraph-Cactus, pangenome references of the chicken genome were constructed. This new genome reference paradigm will better identify the mutations responsible for specific phenotypes, provide tools for in-depth research on the chicken genome structure and variation, and help in the understanding of chicken genetic characteristics and evolutionary processes; in the cultivation of chickens, it is necessary that new sustainability and disease resistance capabilities are met [64]. By performing nanopore sequencing on different chicken strains, they discovered that the high local variation rate of SV and the negative selection of harmful SV events drive the rapid evolution of piRNA [65].

In ducks, Zhu et al. assembled chromosome-level high-quality genomes for Peking ducks, Shaoxing ducks, and mallards, annotated thousands of new protein-coding genes, and refuted the hypothesis of missing genes in birds by confirming the presence of presumed "missing genes" in the genome [29]. The researchers conducted pan-genome estimations for gene PAV detection construction analysis by comparing various genome versions (Tianfu goose, Sichuan white goose, and Zhedong white goose) to investigate genomic sequences beyond the single-reference genome sequence. Although this strategy proved valuable in obtaining 612 Mb of a new sequence, 2,813 new genes, and a total of 20,503 genes across the pan-genome, the limitations, such as shorter contig lengths in the assembly of contigs using second-generation sequencing data, were not to be neglected [30].

3. Application of TGS in the Transcriptome of Livestock

Long-read RNA-sequencing technologies have now reached a mature stage, having already been used to study transcript structures, novel transcripts, and APA, as well as for early allele-specific analyses [4,66]. Applying TGS technology to sequence RNA from different tissues allows the analysis of alternative splicing, the identification of new transcripts (genes), and the optimization of gene structures, which reveal the hidden transcriptional complexity in humans [67] and in livestock [68,69]. The expression product of a gene is a protein, and alternative splicing may lead to protein changes, influencing biological phenotypes. Therefore, the study of alternative splicing is a focal point in current molecular genetic research in livestock [70]. The analysis of the TGS results for alternative splicing identification and regulatory network analysis across pigs and chickens has revealed that alternative splicing affects various biological processes [71,72]. Alternative splicing generates different transcript isoforms, allowing a single gene to be transcribed into multiple transcripts, which can be translated into different protein subtypes to regulate different phenotypes or exert distinct physiological functions [73,74].

3.1. Application of TGS in the Transcriptome of Ruminant Animals

In recent years, alternative splicing has been widely applied in ruminant animals, including cattle, sheep, and so on. Alternative splicing plays a crucial role in the transcriptional changes induced by environmental disturbances. Researchers sequenced RNA from different periods of bovine fat cells using ONT sequencing. The direct sequencing of full-length RNA accurately reflected the RNA modification status, detecting modifications at the single-base level and their potential roles in gene expression and selective splicing in fat cells, enhancing our understanding of the mechanisms underlying fat formation in cattle [75]. Additionally, studies on different mRNA modifications implied that TGS technology plays a crucial role in the discovery of new transcripts in animals (Table 2).

In cattle, researchers analyzed the full-length transcriptome using both SMRT and ONT technologies. Based on the full-length transcriptome, numerous alternative splicing events, alternative polyadenylation (APA) sites, novel isoforms, novel lncRNAs, and transcription factors provided a more comprehensive foundation for the exploration of the diversity of the cattle transcriptome. The full-length transcriptome was refined, revealing differentially expressed transcripts among various tissues [69,76]. Yuan et al. used SMRT technology to perform RNA long-read sequence analysis on muscle tissues from male sheep of different meat qualities and hybrid strains. They annotated the sheep genome and discovered a new isoform, ANKRD23, which is associated with tenderness and is potentially regulated by the CCCTC-binding factor (CTCF) [77].

3.2. *Application of TGS in the Transcriptome of Monogastric Animals*

Meanwhile, many studies have focused on the application of the transcriptome in monogastric animals such as pigs, chickens, and birds. In pigs, researchers sequenced the *Longissimus dorsi* muscle in pigs from different intramuscular fat using ONT sequencing, discovering new splicing bodies related to skeletal muscle development and fatty acid metabolism [71]. In poultry, Guan et al. sequenced different tissues of White Leghorn chickens using ONT sequencing, revealing tissue-specific transcripts. Two brain tissues (cerebellum and cortex) showed the highest number of expressed transcripts and sites, while the reproductive tissues (testes and ovaries) exhibited the most tissue-specific transcripts [78]. Genes related to follicle development were discovered through the analysis of the full-length transcriptome obtained from avian sequencing [72,79–81].

Table 2. Studies of TGS technology in the transcriptome.

Species	Feature	Breed	Sequencing Platform	Key Findings	Publication Year	References
B. taurus	Cattle	Hereford Cattle	ONT	Discovered tissue-specific transcripts in cattle, with the testes exhibiting the most complex transcriptome	2021	[69]
B. taurus	Cattle	Simmental Cattle	PacBio-SMRT	Analyzed the full-length transcriptome of Simmental cattle, providing a foundation for refining the cattle draft genome annotation, optimizing genome structure, and comprehensively characterizing the cattle transcriptome	2021	[76]
O. aries	Sheep	(Dorper × Hu) × Hu sheep; Dorper × (Dorper × Hu sheep)	PacBio-SMRT	Revealed the transcriptome complexity and identified many candidate transcripts in tail fat, which could enhance the understanding of molecular mechanisms behind tail fat deposition	2021	[82]
C. hircus	Goat	Chinese Cashmere Goat	PacBio-SMRT	Showed the superiority of full-length transcriptome data in gene annotation; more such data are required to improve the gene annotation for goat genome and that of other species	2023	[83]
S. scrofa	Pig	Large White Pig × Min Pig F2 Generation	ONT	Discovered differentially expressed mRNA isoforms involved in skeletal muscle development and fatty acid metabolism	2022	[71]
G. gallus	Chicken	White Leghorn Chicken	ONT	Identified the most tissue-specific transcripts in reproductive tissues (testes and ovaries) of chickens	2022	[78]
G. gallus	Chicken	Hy-Line Brown Chicken	ONT	Revealed mRNA and lncRNA expression differences between pre-GCs and post-GCs during chicken follicle selection; discovered significant estrogen-induced expression of three *DHCR7* isoforms	2023	[72,79]
Cairina moschata	Duck	Muscovy Duck	ONT	Obtained the full-length transcriptome of Muscovy duck follicles, providing structural and functional annotations for new transcripts	2021 2022	[80,81]

4. Advances of TGS in Epigenetic Studies of Livestock

4.1. *Application of TGS in DNA Methylation Modification*

DNA methylation is a common epigenetic modification found in prokaryotic and eukaryotic genomes. It plays a crucial role in the regulation of gene expression. Methylation in the gene promoter regions and at the transcription start sites can inhibit gene transcription, thereby exerting a significant impact on the regulation of biological activities [84].

Currently, the detection of epigenetic modifications can be achieved through both SMRT and ONT sequencing [85,86]. SMRT sequencing data often exhibit relatively weak signals for epigenetic modifications, requiring high coverage at specific sites to determine the presence of epigenetic gene modifications. This method is primarily applied in the analysis of smaller genomes, such as in single-cell epigenetic sequencing and the corresponding modification of the high-throughput detection of 5-methylcytosine (m5C) in bacterial genomes [87,88]. The duration of the fluorescence signal and the interval between two signals generated by SMRT correspond to the kinetics of DNA synthesis. The duration of the fluorescence signal at DNA methylation site 5-mC is much longer than that at non-methylation site C. Differences in duration and interval times allow the detection of various types of epigenetic modifications. The characteristics of SMRT sequencing, such as long read lengths and insensitivity to GC repeats, enable the complete detection of CpG island sites in high-GC repeat regions in epigenetics [89]. In the context of cloned cattle systems, the application of TGS to the study of differentially methylated regions (DMRs) revealed an increase in DNA methylation in multiple genes, including PEG1-DRM [90]. The application of nanopore sequencing in DNA methylation detection provides a new avenue for epigenetic research. Currently, ONT is used to evaluate the epigenetic characteristics of human cell line DNA, particularly CpG methylation and chromatin accessibility. CpG methylation and chromatin accessibility on long-stranded DNA are simultaneously evaluated by applying exogenous labeling of open chromatin with GpC methyltransferase [91]. In addition, ONT can also detect the copy number changes in circulating tumor DNA (ctDNA) and cancer-specific methylation changes in tumor patients [92].

TGS technology has significant advantages in gene methylation detection. Its long read length property enables researchers to detect DNA methylation more accurately; thus, the mystery of the epigenome is revealed more comprehensively. In livestock, ONT is expected to provide strong support for genetic improvement and disease prevention and control. Through the methylation detection of the animal genome, researchers can have a deep understanding of the genetic basis of animal growth and development, reproductive performance, disease resistance, and other traits and can provide key information for breeding work. In addition, ONT has important value in disease diagnosis and prevention and control, which helps in the development of more effective disease diagnosis methods and prevention and control strategies and to reduce the economic losses of animal husbandry. With further research and technological advancements, the study of DNA methylation is expected to make significant progress in the field of epigenetics.

4.2. Application of TGS in RNA Epigenetic Modifications

In recent years, research in epi-transcriptomics has provided a new direction for animal studies. The relationship between RNA epi-transcriptome and animal phenotypes and traits is of significant importance for understanding the essence of life and revealing evolutionary processes. Epigenetic studies are based on the structure and modifications of RNA molecules, including N6-methyladenosine (m6A), 5-methylcytosine (m5C), and the spatial structure of RNA. Currently, 150 types of RNA modifications have been discovered in various fields [93]. Epigenetics is closely related to the growth and development of animals, with methylation being one of the rich modifications in epigenetics. m6A is the most common methylation modification in mRNA; it dynamically and reversibly regulates various life activities, including gene expression, RNA metabolism, and protein translation [94]. The study of RNA epigenetic modifications relies on the advancements in sequencing technologies. ONT technology allows direct sequencing of RNA through the monitoring of the changes in the present caused by the passage of individual molecules through a membrane-embedded nanopore. This method enables the direct detection of RNA base modification sites, producing long reads that cover the entire transcript, making it a promising alternative for studying m6A [95].

While ONT sequencing technology has been widely applied in the study of RNA epigenetic modifications in plants and microorganisms, its application in livestock is relatively limited. However, the role of m6A epigenetics in livestock is substantial. Qin Jiang et al. found that m6A methylation plays a crucial regulatory role in the uniform deposition of fat and muscle in Jin Hua pigs and is associated with the decomposition metabolism of Landrace pigs. They further showed that MTCH2 promotes fat generation in muscles in an m6A-dependent manner [96]. m6A RNA methylation, through the regulation of metabolites by gut microbiota, such as folic acid and butyric acid, affects the nutrition, absorption, and metabolic mechanisms of livestock [97]. ONT sequencing technology is expected to provide targets for precise nutrition and targeted regulation in different livestock.

5. Prospects

With the rapid development of molecular biology technologies, TGS has gained widespread application in the research of superior breeding and genetic reproduction in livestock due to its advantages such as long reads, real-time base sequencing, and short turnaround time. The advantages of the low cost, high yield, and high accuracy of NGS technology have made it widely used in large-scale sequencing. However, compared with TGS technology, the short read length of NGS increases the difficulty and error rate of gene assembly. At the same time, the PCR technology used in NGS increases the error rate of sequencing. The long read length of TGS technology helps reduce the splicing cost in bioinformatics, providing convenience in subsequent data analysis and interpretation [98]. The use of PacBio Iso-Seq can detect transcripts that would have otherwise been missed by RNA-seq [99]. Further research found that the utility of Iso-Seq can uncover hidden mammalian transcriptional complexity, not seen by RNA-seq alone, which examined the correlation between RNA-seq and Iso-Seq estimations of relative transcript abundance and their predictions of differential gene expression [68]. However, several considerations need attention in the application of TGS: (1) Error Rates and Sequencing Artifacts: The issue of base error rates and sequencing artifacts remains a challenge in TGS applications. Researchers are addressing this by improving reagent purity, developing rapid detection kits, and enhancing the accuracy of original read lengths. The development of more efficient and precise sequencing methods and error correction modules provides possibilities for the addressing of these challenges [100,101]. (2) Nanopore Technology for Single-Molecule Sequencing: The accuracy, read length, and throughput of nanopore technology for sequencing single long DNA and RNA molecules have significantly improved. This calls for the development of new experimental techniques and bioinformatics methods to fully exploit nanopore long-read sequencing in the study of genomes, transcriptomes, epigenomes, and transcription [102]. (3) Expanding Applications in Livestock Genomics: While TGS has found broad applications in the genomic and transcriptomic analysis and sequencing of livestock, its exploration in the epigenetics of livestock breeding is relatively limited and not comprehensive. Therefore, future efforts should focus on leveraging TGS for epigenetic studies in livestock and explore the interactions among epigenetic modifications. This represents a promising direction for future research in livestock breeding [102].

In summary, TGS technology has been extensively applied in various domains related to livestock genomics, including genome assembly, detection of structural variations, transcriptome sequencing, and epigenetic analysis. As the costs of TGS continue to be controlled and sequencing functionalities are further optimized, it is anticipated that it will become a routine technology in livestock breeding research. Its role is expected to be pivotal in discovering rare genes, cultivating superior individuals, conducting population-based genetic breeding, and performing single-cell whole-transcriptome sequencing.

Author Contributions: X.L. undertook the primary responsibility for composing the review, encompassing literature collection, organization, and summarization. Additionally, G.Z. was accountable for establishing the article's structural framework and logical coherence. J.Z. meticulously reviewed and critically discussed key aspects of the relevant research. J.D. provided invaluable support in the development of the review content and played a pivotal role in chart creation and the editorial process.

J.W. contributed by meticulously editing and proofreading various sections of the review, ensuring linguistic accuracy and adherence to formatting standards. F.Z. and G.Z. took responsibility for the final review and approval of the review, ensuring the completeness and precision of the manuscript. Each author played a crucial role in the process of composing and revising the review and ensuring the quality and accuracy of the article; all authors collectively contributed to the completion of this review. All authors have read and agreed to the published version of the manuscript.

Funding: This work was supported by the National Training Program of Innovation and Entrepreneurship for Undergraduates (grant no. 202310635023) and the Technological Innovation and Application Development Project of Chongqing (grant no. cstc2021jscx-gksbX0012).

Conflicts of Interest: The authors declare that the research was conducted in the absence of any financial or non-financial relationships that could be construed as potential conflicts of interest.

References

1. van Dijk, E.L.; Jaszczyszyn, Y.; Naquin, D.; Thermes, C. The Third Revolution in Sequencing Technology. *Trends Genet.* **2018**, *34*, 666–681. [CrossRef] [PubMed]
2. Shendure, J.; Balasubramanian, S.; Church, G.M.; Gilbert, W.; Rogers, J.; Schloss, J.A.; Waterston, R.H. DNA sequencing at 40: Past, present and future. *Nature* **2017**, *550*, 345–353. [CrossRef] [PubMed]
3. Chen, L.; Wang, J.; Tan, L.; Lu, C.; Fu, G.; Fu, L.; Zhang, X.J.; Wang, Q.; Ma, C.L.; Cong, B.; et al. Highly accurate mtGenome haplotypes from long-read SMRT sequencing can distinguish between monozygotic twins. *Forensic Sci. Int. Genet.* **2020**, *47*, 102306. [CrossRef] [PubMed]
4. Byrne, A.; Cole, C.; Volden, R.; Vollmers, C. Realizing the potential of full-length transcriptome sequencing. *Philos. Trans. R. Soc. London. Ser. B Biol. Sci.* **2019**, *374*, 20190097. [CrossRef] [PubMed]
5. Delahaye, C.; Nicolas, J. Sequencing DNA with nanopores: Troubles and biases. *PLoS ONE* **2021**, *16*, e0257521. [CrossRef] [PubMed]
6. Logsdon, G.A.; Vollger, M.R.; Eichler, E.E. Long-read human genome sequencing and its applications. *Nat. Rev. Genet.* **2020**, *21*, 597–614. [CrossRef] [PubMed]
7. van Dijk, E.L.; Naquin, D.; Gorrichon, K.; Jaszczyszyn, Y.; Ouazahrou, R.; Thermes, C.; Hernandez, C. Genomics in the long-read sequencing era. *Trends Genet.* **2023**, *39*, 649–671. [CrossRef]
8. Hu, T.; Li, J.; Long, M.; Wu, J.; Zhang, Z.; Xie, F.; Zhao, J.; Yang, H.P.; Song, Q.Q.; Lian, S.; et al. Detection of Structural Variations and Fusion Genes in Breast Cancer Samples Using Third-Generation Sequencing. *Front. Cell Dev. Biol.* **2022**, *10*, 854640. [CrossRef]
9. Nakano, K.; Shiroma, A.; Shimoji, M.; Tamotsu, H.; Ashimine, N.; Ohki, S.; Shinzato, M.; Minami, M.; Nakanishi, T.; Teruya, K.; et al. Advantages of genome sequencing by long-read sequencer using SMRT technology in medical area. *Hum. Cell* **2017**, *30*, 149–161. [CrossRef]
10. Liu, L.; Zhang, Y.; Jiang, D.; Du, S.; Deng, Z.; Wang, L.; Chen, S. Recent Advances in the Genomic Profiling of Bacterial Epigenetic Modifications. *Biotechnol. J.* **2019**, *14*, e1800001. [CrossRef]
11. Gurgul, A.; Jasielczuk, I.; Szmatoła, T.; Sawicki, S.; Semik-Gurgul, E.; Długosz, B.; Bugno-Poniewierska, M. Application of Nanopore Sequencing for High Throughput Genotyping in Horses. *Animals* **2023**, *13*, 2227. [CrossRef]
12. Yang, C.; Chowdhury, D.; Zhang, Z.; Cheung, W.K.; Lu, A.; Bian, Z.; Zhang, L. A review of computational tools for generating metagenome-assembled genomes from metagenomic sequencing data. *Comput. Struct. Biotechnol. J.* **2021**, *19*, 6301–6314. [CrossRef]
13. Sohn, J.-I.; Nam, J.-W. The present and future ofde novowhole-genome assembly. *Brief. Bioinform.* **2016**, *19*, 23–40. [CrossRef]
14. Leonard, A.S.; Crysnanto, D.; Fang, Z.H.; Heaton, M.P.; Vander Ley, B.L.; Herrera, C.; Bollwein, H.; Bickhart, D.M.; Kuhn, K.L.; Smith, T.P.L.; et al. Structural variant-based pangenome construction has low sensitivity to variability of haplotype-resolved bovine assemblies. *Nat. Commun.* **2022**, *13*, 3012. [CrossRef]
15. Dai, X.; Bian, P.; Hu, D.; Luo, F.; Huang, Y.; Jiao, S.; Wang, X.H.; Gong, M.; Li, R.; Cai, Y.D.; et al. A Chinese indicine pangenome reveals a wealth of novel structural variants introgressed from other Bos species. *Genome Res.* **2023**, *33*, 1284–1298. [CrossRef] [PubMed]
16. Li, Y.; Wang, S.; Zhang, Z.; Luo, J.; Lin, G.L.; Deng, W.D.; Guo, Z.F.; Han, F.M.; Wang, L.L.; Li, J.; et al. Large-Scale Chromosomal Changes Lead to Genome-Level Expression Alterations, Environmental Adaptation, and Speciation in the Gayal (*Bos frontalis*). *Mol. Biol. Evol.* **2023**, *40*, msad006. [CrossRef]
17. Gao, X.; Wang, S.; Wang, Y.F.; Li, S.; Wu, S.X.; Yan, R.G.; Zhang, Y.W.; Wan, R.D.; He, Z.; Song, R.D.; et al. Long read genome assemblies complemented by single cell RNA-sequencing reveal genetic and cellular mechanisms underlying the adaptive evolution of yak. *Nat. Commun.* **2022**, *13*, 4887. [CrossRef] [PubMed]
18. Liu, X.; Liu, W.; Lenstra, J.A.; Zheng, Z.; Wu, X.; Yang, J.; Li, B.W.; Yang, Y.Z.; Qiu, Q.; Liu, H.Y.; et al. Evolutionary origin of genomic structural variations in domestic yaks. *Nat. Commun.* **2023**, *14*, 5617. [CrossRef]
19. Low, W.Y.; Tearle, R.; Bickhart, D.M.; Rosen, B.D.; Kingan, S.B.; Swale, T.; Thibaud-Nissen, F.; Murphy, T.D.; Young, R.; Lefevre, L.; et al. Chromosome-level assembly of the water buffalo genome surpasses human and goat genomes in sequence contiguity. *Nat. Commun.* **2019**, *10*, 260. [CrossRef]

20. Luo, X.; Zhou, Y.; Zhang, B.; Zhang, Y.; Wang, X.; Feng, T.; Li, Z.P.; Cui, K.Q.; Wang, Z.Q.; Luo, C.; et al. Understanding divergent domestication traits from the whole-genome sequencing of swamp- and river-buffalo populations. *Natl. Sci. Rev.* **2020**, *7*, 686–701. [CrossRef] [PubMed]
21. Qiao, G.; Xu, P.; Guo, T.; Wu, Y.; Lu, X.; Zhang, Q.; He, X.; Zhu, S.H.; Zhao, H.C.; Lei, Z.H.; et al. Genetic Basis of Dorper Sheep (*Ovis aries*) Revealed by Long-Read De Novo Genome Assembly. *Front. Genet.* **2022**, *13*, 846449. [CrossRef] [PubMed]
22. Li, R.; Gong, M.; Zhang, X.; Wang, F.; Liu, Z.; Zhang, L.; Yang, Q.M.; Xu, Y.; Xu, M.S.; Zhang, H.H.; et al. A sheep pangenome reveals the spectrum of structural variations and their effects on tail phenotypes. *Genome Res.* **2023**, *33*, 463–477. [CrossRef] [PubMed]
23. Li, R.; Yang, P.; Dai, X.; Asadollahpour Nanaei, H.; Fang, W.; Yang, Z.; Cai, Y.D.; Zheng, Z.Q.; Wang, X.H.; Jiang, Y. A near complete genome for goat genetic and genomic research. *Genet. Sel. Evol.* **2021**, *53*, 74. [CrossRef] [PubMed]
24. Li, C.; Wu, Y.; Chen, B.; Cai, Y.; Guo, J.; Leonard, A.S.; Kalds, P.; Zhou, S.W.; Zhang, J.C.; Ping, Z.; et al. Markhor-derived Introgression of a Genomic Region Encompassing PAPSS2 Confers High-altitude Adaptability in Tibetan Goats. *Mol. Biol. Evol.* **2022**, *39*, msac253. [CrossRef] [PubMed]
25. Jiang, Y.F.; Wang, S.; Wang, C.L.; Xu, R.H.; Wang, W.W.; Jiang, Y.; Wang, M.S.; Jiang, L.; Dai, L.H.; Wang, J.R.; et al. Pangenome obtained by long-read sequencing of 11 genomes reveal hidden functional structural variants in pigs. *iScience* **2023**, *26*, 106119. [CrossRef] [PubMed]
26. Warr, A.; Affara, N.; Aken, B.; Beiki, H.; Bickhart, D.M.; Billis, K.; Chow, W.; Eory, L.; Finlayson, H.A.; Flicek, P.; et al. An improved pig reference genome sequence to enable pig genetics and genomics research. *Gigascience* **2020**, *9*, giaa051. [CrossRef] [PubMed]
27. Zhu, F.; Yin, Z.-T.; Zhao, Q.-S.; Sun, Y.-X.; Jie, Y.-C.; Smith, J.; Yang, Y.Z.; Burt, D.W.; Hincke, M.; Zhang, Z.D.; et al. A chromosome-level genome assembly for the Silkie chicken resolves complete sequences for key chicken metabolic, reproductive, and immunity genes. *Commun. Biol.* **2023**, *6*, 1233. [CrossRef] [PubMed]
28. Huang, Z.; Xu, Z.; Bai, H.; Huang, Y.; Kang, N.; Ding, X.; Liu, J.; Luo, H.R.; Yang, C.T.; Chen, W.J.; et al. Evolutionary analysis of a complete chicken genome. *Proc. Natl. Acad. Sci. USA* **2023**, *120*, e2216641120. [CrossRef]
29. Zhu, F.; Yin, Z.T.; Wang, Z.; Smith, J.; Zhang, F.; Martin, F.; Ogeh, D.; Hincke, M.; Lin, F.B.; Burt, D.W.; et al. Three chromosome-level duck genome assemblies provide insights into genomic variation during domestication. *Nat. Commun.* **2021**, *12*, 5932. [CrossRef]
30. Gao, G.; Zhang, H.; Ni, J.; Zhao, X.; Zhang, K.; Wang, J.; Kong, X.D.; Wang, Q.G. Insights into genetic diversity and phenotypic variations in domestic geese through comprehensive population and pan-genome analysis. *J. Anim. Sci. Biotechnol.* **2023**, *14*, 150. [CrossRef]
31. Tettelin, H.; Masignani, V.; Cieslewicz, M.J.; Donati, C.; Medini, D.; Ward, N.L.; Angiuoli, S.V.; Crabtree, J.; Jones, A.L.; Durkin, A.S.; et al. Genome analysis of multiple pathogenic isolates of *Streptococcus agalactiae*: Implications for the microbial "pan-genome". *Proc. Natl. Acad. Sci. USA* **2005**, *102*, 13950–13955. [CrossRef]
32. Smith, T.P.L.; Bickhart, D.M.; Boichard, D.; Chamberlain, A.J.; Djikeng, A.; Jiang, Y.; Low, W.Y.; Pausch, H.; Demyda-Peyras, S.; Prendergast, J.; et al. The Bovine Pangenome Consortium: Democratizing production and accessibility of genome assemblies for global cattle breeds and other bovine species. *Genome Biol.* **2023**, *24*, 139. [CrossRef] [PubMed]
33. Li, R.; Li, Y.; Zheng, H.; Luo, R.; Zhu, H.; Li, Q.; Qian, W.B.; Ren, Y.Y.; Tian, G.; Li, J.X.; et al. Building the sequence map of the human pan-genome. *Nat. Biotechnol.* **2010**, *28*, 57–63. [CrossRef]
34. Li, M.; Chen, L.; Tian, S.; Lin, Y.; Tang, Q.; Zhou, X.; Li, D.Y.; Yeung, C.K.L.; Che, T.D.; Jin, L.; et al. Comprehensive variation discovery and recovery of missing sequence in the pig genome using multiple de novo assemblies. *Genome Res.* **2017**, *27*, 865–874. [CrossRef]
35. Li, Z.; Liu, X.; Wang, C.; Li, Z.; Jiang, B.; Zhang, R.; Tong, L.; Qu, Y.; He, S.; Chen, H.; et al. The pig pangenome provides insights into the roles of coding structural variations in genetic diversity and adaptation. *Genome Res.* **2023**, *33*, 1833–1847. [CrossRef] [PubMed]
36. Tian, X.; Li, R.; Fu, W.; Li, Y.; Wang, X.; Li, M.; Du, D.; Tang, Q.Z.; Cai, Y.D.; Long, Y.M.; et al. Building a sequence map of the pig pan-genome from multiple de novo assemblies and Hi-C data. *Sci. China Life Sci.* **2019**, *63*, 750–763. [CrossRef]
37. Li, R.; Fu, W.; Su, R.; Tian, X.; Du, D.; Zhao, Y.; Zheng, Z.Q.; Chen, Q.M.; Gao, S.; Cai, Y.D.; et al. Towards the Complete Goat Pan-Genome by Recovering Missing Genomic Segments from the Reference Genome. *Front. Genet.* **2019**, *10*, 1169. [CrossRef]
38. Crysnanto, D.; Pausch, H. Bovine breed-specific augmented reference graphs facilitate accurate sequence read mapping and unbiased variant discovery. *Genome Biol.* **2020**, *21*, 184. [CrossRef]
39. Zhou, Y.; Yang, L.; Han, X.; Han, J.; Hu, Y.; Li, F.; Xia, H.; Peng, L.W.; Boschiero, C.; Rosen, B.D.; et al. Assembly of a pangenome for global cattle reveals missing sequences and novel structural variations, providing new insights into their diversity and evolutionary history. *Genome Res.* **2022**, *32*, 1585–1601. [CrossRef]
40. Wang, K.; Hu, H.; Tian, Y.; Li, J.; Scheben, A.; Zhang, C.; Li, Y.Y.; Wu, J.F.; Yang, L.; Fan, X.W.; et al. The Chicken Pan-Genome Reveals Gene Content Variation and a Promoter Region Deletion in IGF2BP1 Affecting Body Size. *Mol. Biol. Evol.* **2021**, *38*, 5066–5081. [CrossRef]
41. Li, R.; Yang, P.; Li, M.; Fang, W.; Yue, X.; Nanaei, H.A.; Gan, S.Q.; Du, D.; Cai, Y.D.; Dai, X.L.; et al. A Hu sheep genome with the first ovine Y chromosome reveal introgression history after sheep domestication. *Sci. China Life Sci.* **2021**, *64*, 1116–1130. [CrossRef]

42. Golicz, A.A.; Bayer, P.E.; Barker, G.C.; Edger, P.P.; Kim, H.; Martinez, P.A.; Chan, C.K.K.; Severn-Ellis, A.; McCombie, W.R.; Parkin, I.A.P.; et al. The pangenome of an agronomically important crop plant *Brassica oleracea*. *Nat. Commun.* **2016**, *7*, 13390. [CrossRef]
43. Rosconi, F.; Rudmann, E.; Li, J.; Surujon, D.; Anthony, J.; Frank, M.; Jones, D.S.; Rock, C.; Rosch, J.W.; Johnston, C.D.; et al. A bacterial pan-genome makes gene essentiality strain-dependent and evolvable. *Nat. Microbiol.* **2022**, *7*, 1580–1592. [CrossRef]
44. Golicz, A.A.; Batley, J.; Edwards, D. Towards plant pangenomics. *Plant Biotechnol. J.* **2016**, *14*, 1099–1105. [CrossRef]
45. Kille, B.; Balaji, A.; Sedlazeck, F.J.; Nute, M.; Treangen, T.J. Multiple genome alignment in the telomere-to-telomere assembly era. *Genome Biol.* **2022**, *23*, 182. [CrossRef] [PubMed]
46. Chen, J.; Wang, Z.; Tan, K.; Huang, W.; Shi, J.; Li, T.; Hu, J.; Wang, K.; Wang, C.; Xin, B.B.; et al. A complete telomere-to-telomere assembly of the maize genome. *Nat. Genet.* **2023**, *55*, 1221–1231. [CrossRef] [PubMed]
47. Logsdon, G.A.; Vollger, M.R.; Hsieh, P.; Mao, Y.; Liskovykh, M.A.; Koren, S.; Nurk, S.; Mercuri, L.; Dishuck, P.C.; Rhie, A.; et al. The structure, function and evolution of a complete human chromosome 8. *Nature* **2021**, *593*, 101–107. [CrossRef] [PubMed]
48. Miga, K.H.; Koren, S.; Rhie, A.; Vollger, M.R.; Gershman, A.; Bzikadze, A.; Brooks, S.; Howe, E.; Porubsky, D.; Logsdon, G.A.; et al. Telomere-to-telomere assembly of a complete human X chromosome. *Nature* **2020**, *585*, 79–84. [CrossRef] [PubMed]
49. Xia, X.; Zhang, F.; Li, S.; Luo, X.; Peng, L.; Dong, Z.; Pausch, H.; Leonard, A.S.; Crysnanto, D.; Wang, S.K.; et al. Structural variation and introgression from wild populations in East Asian cattle genomes confer adaptation to local environment. *Genome Biol.* **2023**, *24*, 211. [CrossRef] [PubMed]
50. Li, M.; Sun, C.; Xu, N.; Bian, P.; Tian, X.; Wang, X.; Wang, Y.Z.; Jia, X.Z.; Heller, R.; Wang, M.S.; et al. De Novo Assembly of 20 Chicken Genomes Reveals the Undetectable Phenomenon for Thousands of Core Genes on Microchromosomes and Subtelomeric Regions. *Mol. Biol. Evol.* **2022**, *39*, msac066. [CrossRef] [PubMed]
51. Xue, Q.; Xing, W.J.; Li, G.H.; Zhou, C.H.; Zhang, H.Y.; Yin, J.Z.; Jiang, Y.X.; Zhu, Y.F.; Han, W. Assembly of high-quality Wenshang barred chickens genome based on PacBio third-generation sequencing. *China Anim. Husb. Vet. Med.* **2023**, *50*, 3869–3881.
52. Bickhart, D.M.; Liu, G.E. The challenges and importance of structural variation detection in livestock. *Front. Genet.* **2014**, *5*, 37. [CrossRef] [PubMed]
53. Chang, L.; Deng, E.; Wang, J.; Zhou, W.; Ao, J.; Liu, R.; Su, D.; Fan, X.Y. Single-cell third-generation sequencing-based multi-omics uncovers gene expression changes governed by ecDNA and structural variants in cancer cells. *Clin. Transl. Med.* **2023**, *13*, e1351. [CrossRef] [PubMed]
54. Gao, Y.; Ma, L.; Liu, G.E. Initial Analysis of Structural Variation Detections in Cattle Using Long-Read Sequencing Methods. *Genes* **2022**, *13*, 828. [CrossRef] [PubMed]
55. Jiang, T.; Liu, Y.Z.; Jiang, Y.; Li, J.Y.; Gao, Y.; Cui, Z.; Liu, Y.D.; Liu, B.; Wang, Y.D. Long-read-based human genomic structural variation detection with cuteSV. *Genome Biol.* **2020**, *21*, 189. [CrossRef] [PubMed]
56. Stancu, M.C.; van Roosmalen, M.J.; Renkens, I.; Nieboer, M.M.; Middelkamp, S.; de Ligt, J.; Pregno, G.; Giachino, D.; Mandrile, G.; Valle-Inclan, J.E.; et al. Mapping and phasing of structural variation in patient genomes using nanopore sequencing. *Nat. Commun.* **2017**, *8*, 1326. [CrossRef] [PubMed]
57. Tham, C.Y.; Tirado-Magallanes, R.; Goh, Y.; Fullwood, M.J.; Koh, B.T.H.; Wang, W.; Ng, C.H.; Chng, W.J.; Thiery, A.; Tenen, D.G.; et al. NanoVar: Accurate characterization of patients' genomic structural variants using low-depth nanopore sequencing. *Genome Biol.* **2020**, *21*, 56. [CrossRef] [PubMed]
58. Sedlazeck, F.J.; Rescheneder, P.; Smolka, M.; Fang, H.; Nattestad, M.; von Haeseler, A.; Schatz, M.C. Accurate detection of complex structural variations using single-molecule sequencing. *Nat. Methods* **2018**, *15*, 461–468. [CrossRef] [PubMed]
59. Heller, D.; Vingron, M. SVIM: Structural variant identification using mapped long reads. *Bioinformatics* **2019**, *35*, 2907–2915. [CrossRef]
60. Pös, O.; Radvanszky, J.; Buglyó, G.; Pös, Z.; Rusnakova, D.; Nagy, B.; Szemes, T. DNA copy number variation: Main characteristics, evolutionary significance, and pathological aspects. *Biomed. J.* **2021**, *44*, 548–559. [CrossRef]
61. Magini, P.; Mingrino, A.; Gega, B.; Mattei, G.; Semeraro, R.; Bolognini, D.; Mongelli, P.; Desiderio, L.; Pittalis, M.C.; Pippucci, T.; et al. Third-Generation Cytogenetic Analysis: Diagnostic Application of Long-Read Sequencing. *J. Mol. Diagn.* **2022**, *24*, 711–718. [CrossRef] [PubMed]
62. Lamb, H.J.; Ross, E.M.; Nguyen, L.T.; Lyons, R.E.; Moore, S.S.; Hayes, B.J. Characterization of the poll allele in Brahman cattle using long-read Oxford Nanopore sequencing. *J. Anim. Sci.* **2020**, *98*, skaa127. [CrossRef]
63. Viluma, A.; Mikko, S.; Hahn, D.; Skow, L.; Andersson, G.; Bergstrom, T.F. Genomic structure of the horse major histocompatibility complex class II region resolved using PacBio long-read sequencing technology. *Sci. Rep.* **2017**, *7*, 45518. [CrossRef]
64. Rice, E.S.; Alberdi, A.; Alfieri, J.; Athrey, G.; Balacco, J.R.; Bardou, P.; Blackmon, H.; Charles, M.; Cheng, H.H.; Fedrigo, O.; et al. A pangenome graph reference of 30 chicken genomes allows genotyping of large and complex structural variants. *BMC Biol.* **2023**, *21*, 267. [CrossRef]
65. Sun, Y.H.; Cui, H.; Song, C.; Shen, J.T.; Zhuo, X.; Wang, R.H.; Yu, X.H.; Ndamba, R.; Mu, Q.; Gu, H.W.; et al. Amniotes co-opt intrinsic genetic instability to protect germ-line genome integrity. *Nat. Commun.* **2023**, *14*, 812. [CrossRef] [PubMed]
66. Tilgner, H.; Grubert, F.; Sharon, D.; Snyder, M.P. Defining a personal, allele-specific, and single-molecule long-read transcriptome. *Proc. Natl. Acad. Sci. USA* **2014**, *111*, 9869–9874. [CrossRef] [PubMed]
67. Glinos, D.A.; Garborcauskas, G.; Hoffman, P.; Ehsan, N.; Jiang, L.H.; Gokden, A.; Dai, X.G.; Aguet, F.; Brown, K.L.; Garimella, K.; et al. Transcriptome variation in human tissues revealed by long-read sequencing. *Nature* **2022**, *608*, 353–359. [CrossRef]

68. Ren, Y.; Tseng, E.; Smith, T.P.L.; Hiendleder, S.; Williams, J.L.; Low, W.Y. Long read isoform sequencing reveals hidden transcriptional complexity between cattle subspecies. *BMC Genom.* **2023**, *24*, 108. [CrossRef]
69. Halstead, M.M.; Islas-Trejo, A.; Goszczynski, D.E.; Medrano, J.F.; Zhou, H.; Ross, P.J. Large-Scale Multiplexing Permits Full-Length Transcriptome Annotation of 32 Bovine Tissues From a Single Nanopore Flow Cell. *Front. Genet.* **2021**, *12*, 664260. [CrossRef]
70. Liu, J.; Lang, K.; Tan, S.; Jie, W.; Zhu, Y.; Huang, S.; Huang, W. A web-based database server using 43,710 public RNA-seq samples for the analysis of gene expression and alternative splicing in livestock animals. *BMC Genom.* **2022**, *23*, 706. [CrossRef]
71. Shu, Z.; Wang, L.; Wang, J.; Zhang, L.; Hou, X.; Yan, H.; Wang, L. Integrative Analysis of Nanopore and Illumina Sequencing Reveals Alternative Splicing Complexity in Pig Longissimus Dorsi Muscle. *Front. Genet.* **2022**, *13*, 877646. [CrossRef]
72. Li, D.; Zhong, C.; Sun, Y.; Kang, L.; Jiang, Y. Identification of genes involved in chicken follicle selection by ONT sequencing on granulosa cells. *Front. Genet.* **2022**, *13*, 1090603. [CrossRef]
73. Jin, Y.; Dong, H.; Shi, Y.; Bian, L. Mutually exclusive alternative splicing of pre-mRNAs. *Wiley Interdiscip. Rev. RNA* **2018**, *9*, e1468. [CrossRef]
74. Lopez, A.J. Alternative splicing of pre-mRNA: Developmental consequences and mechanisms of regulation. *Annu. Rev. Genet.* **1998**, *32*, 279–305. [CrossRef]
75. Peng, L.; Zhang, X.; Du, Y.; Li, F.; Han, J.; Liu, O.; Dai, S.L.; Zhang, X.; Liu, G.E.; Yang, L.G.; et al. New insights into transcriptome variation during cattle adipocyte adipogenesis by direct RNA sequencing. *iScience* **2023**, *26*, 107753. [CrossRef]
76. Chang, T.; An, B.; Liang, M.; Duan, X.; Du, L.; Cai, W.; Zhu, B.; Gao, X.; Chen, Y.; Xu, L.Y.; et al. PacBio Single-Molecule Long-Read Sequencing Provides New Light on the Complexity of Full-Length Transcripts in Cattle. *Front. Genet.* **2021**, *12*, 664974. [CrossRef] [PubMed]
77. Yuan, Z.; Ge, L.; Zhang, W.; Lv, X.; Wang, S.; Cao, X.; Sun, W. Preliminary Results about Lamb Meat Tenderness Based on the Study of Novel Isoforms and Alternative Splicing Regulation Pathways Using Iso-seq, RNA-seq and CTCF ChIP-seq Data. *Foods* **2022**, *11*, 1068. [CrossRef] [PubMed]
78. Guan, D.; Halstead, M.M.; Islas-Trejo, A.D.; Goszczynski, D.E.; Cheng, H.H.; Ross, P.J.; Zhou, H. Prediction of transcript isoforms in 19 chicken tissues by Oxford Nanopore long-read sequencing. *Front. Genet.* **2022**, *13*, 997460. [CrossRef] [PubMed]
79. Zhong, C.; Liu, Z.; Li, D.; Kang, L.; Jiang, Y. Long-read sequencing reveals the effect of follicle-stimulating hormone on the mRNA profile of chicken granulosa cells from prehierarchical follicles. *Poult. Sci.* **2023**, *102*, 102600. [CrossRef] [PubMed]
80. Lin, J.; Guan, L.; Ge, L.; Liu, G.; Bai, Y.; Liu, X. Nanopore-based full-length transcriptome sequencing of Muscovy duck (*Cairina moschata*) ovary. *Poult. Sci.* **2021**, *100*, 101246. [CrossRef] [PubMed]
81. Lin, J.; Ge, L.; Mei, X.; Niu, Y.; Chen, C.; Hou, S.; Liu, X. Integrated ONT Full-Length Transcriptome and Metabolism Reveal the Mechanism Affecting Ovulation in Muscovy Duck (*Cairina moschata*). *Front. Vet. Sci.* **2022**, *9*, 890979. [CrossRef]
82. Yuan, Z.; Ge, L.; Sun, J.; Zhang, W.; Wang, S.; Cao, X.; Sun, W. Integrative analysis of Iso-Seq and RNA-seq data reveals transcriptome complexity and differentially expressed transcripts in sheep tail fat. *PeerJ* **2021**, *9*, e12454. [CrossRef]
83. Zhang, H.; Liang, Y.; Chen, S.; Xuan, Z.; Jiang, Y.; Li, R.; Cao, Y. Full-length transcriptome sequencing reveals extreme incomplete annotation of the goat genome. *Anim. Genet.* **2023**, *54*, 421–424. [CrossRef] [PubMed]
84. Luo, C.; Hajkova, P.; Ecker, J.R. Dynamic DNA methylation: In the right place at the right time. *Science* **2018**, *361*, 1336–1340. [CrossRef] [PubMed]
85. Flusberg, B.A.; Webster, D.R.; Lee, J.H.; Travers, K.J.; Olivares, E.C.; Clark, T.A.; Korlach, J.; Turner, S.W. Direct detection of DNA methylation during single-molecule, real-time sequencing. *Nat. Methods* **2010**, *7*, 461–465. [CrossRef]
86. Rand, A.C.; Jain, M.; Eizenga, J.M.; Musselman-Brown, A.; Olsen, H.E.; Akeson, M.; Paten, B. Mapping DNA methylation with high-throughput nanopore sequencing. *Nat. Methods* **2017**, *14*, 411–413. [CrossRef] [PubMed]
87. Zhuo, R.; Xu, M.; Wang, X.; Zhou, B.; Wu, X.; Leone, V.; Chang, E.B.; Zhong, X. The regulatory role of N(6) -methyladenosine modification in the interaction between host and microbes. *Wiley Interdiscip. Rev. RNA* **2022**, *13*, e1725. [CrossRef] [PubMed]
88. Fang, G.; Munera, D.; Friedman, D.I.; Mandlik, A.; Chao, M.C.; Banerjee, O.; Feng, Z.X.; Losic, B.; Mahajan, M.C.; Jabado, O.J.; et al. Genome-wide mapping of methylated adenine residues in pathogenic Escherichia coli using single-molecule real-time sequencing. *Nat. Biotechnol.* **2012**, *30*, 1232–1239. [CrossRef] [PubMed]
89. Rhoads, A.; Au, K.F. PacBio Sequencing and Its Applications. *Genom. Proteom. Bioinform.* **2015**, *13*, 278–289. [CrossRef]
90. Mangiavacchi, P.M.; Caldas-Bussiere, M.C.; Mendonca, M.D.S.; Rumpf, R.; Lemos Junior, P.E.S.; Alves, C.S.; Carneiro, W.S.; Dias, A.J.B.; Riso, A.F.L. Multi-locus DNA methylation analysis of imprinted genes in cattle from somatic cell nuclear transfer. *Theriogenology* **2022**, *186*, 95–107. [CrossRef]
91. Lee, I.; Razaghi, R.; Gilpatrick, T.; Molnar, M.; Gershman, A.; Sadowski, N.; Sedlazeck, F.J.; Hansen, K.D.; Simpson, J.T. Timp, W. Simultaneous profiling of chromatin accessibility and methylation on human cell lines with nanopore sequencing. *Nat. Methods* **2020**, *17*, 1191–1199. [CrossRef]
92. Katsman, E.; Orlanski, S.; Martignano, F.; Fox-Fisher, I.; Shemer, R.; Dor, Y.; Zick, A.; Eden, A.; Petrini, L.; Conticello, S.G.; et al. Detecting cell-of-origin and cancer-specific methylation features of cell-free DNA from Nanopore sequencing. *Genome Biol.* **2022**, *23*, 158. [CrossRef]
93. Dunin-Horkawicz, S.; Czerwoniec, A.; Gajda, M.J.; Feder, M.; Grosjean, H.; Bujnicki, J.M. MODOMICS: A database of RNA modification pathways. *Nucleic Acids Res.* **2006**, *34*, D145–D149. [CrossRef]
94. Hengwei, Y.; Raza, S.H.A.; Wenzhen, Z.; Xinran, Y.; Almohaimeed, H.M.; Alshanwani, A.R.; Assiri, R.; Aggad, W.S.; Zan, L.S. Research progress of m(6)A regulation during animal growth and development. *Mol. Cell. Probes* **2022**, *65*, 101851. [CrossRef]

95. Jenjaroenpun, P.; Wongsurawat, T.; Wadley, T.D.; Wassenaar, T.M.; Liu, J.; Dai, Q.; Wanchai, V.; Akel, N.S.; Jamshidi-Parsian, A.; Franco, A.T.; et al. Decoding the epitranscriptional landscape from native RNA sequences. *Nucleic Acids Res.* **2021**, *49*, e7. [CrossRef] [PubMed]
96. Jiang, Q.; Sun, B.; Liu, Q.; Cai, M.; Wu, R.; Wang, F.; Yao, Y.X.; Wang, Y.Z.; Wang, X.X. MTCH2 promotes adipogenesis in intramuscular preadipocytes via an m(6)A-YTHDF1-dependent mechanism. *FASEB J.* **2019**, *33*, 2971–2981. [CrossRef] [PubMed]
97. Ibeagha-Awemu, E.M.; Kiefer, H.; McKay, S.; Liu, G.E. Editorial: Epigenetic Variation Influences on Livestock Production and Disease Traits. *Front. Genet.* **2022**, *13*, 942747. [CrossRef] [PubMed]
98. Kumar, K.; Cowley, M.; Wetterstrand, K.; Watson, J.; Crick, F.; Sanger, F.; Nicklen, S.; Coulson, A.; Metzker, M.; Schadt, E.; et al. Next-Generation Sequencing and Emerging Technologies. *Semin. Thromb. Hemost.* **2019**, *45*, 661–673. [PubMed]
99. Low, W.Y.; Tearle, R.; Liu, R.; Koren, S.; Rhie, A.; Bickhart, D.M.; Rosen, B.D.; Kronenberg, Z.N.; Kingan, S.B.; Tseng, E.; et al. Haplotype-resolved genomes provide insights into structural variation and gene content in Angus and Brahman cattle. *Nat. Commun.* **2020**, *11*, 2071. [CrossRef] [PubMed]
100. Amarasinghe, S.L.; Su, S.; Dong, X.; Zappia, L.; Ritchie, M.E.; Gouil, Q. Opportunities and challenges in long-read sequencing data analysis. *Genome Biol.* **2020**, *21*, 30. [CrossRef] [PubMed]
101. Garalde, D.R.; Snell, E.A.; Jachimowicz, D.; Sipos, B.; Lloyd, J.H.; Bruce, M.; Pantic, N.; Admasu, T.; James, P.; Warland, A.; et al. Highly parallel direct RNA sequencing on an array of nanopores. *Nat. Methods* **2018**, *15*, 201–206. [CrossRef] [PubMed]
102. Wang, Y.; Zhao, Y.; Bollas, A.; Wang, Y.; Au, K.F. Nanopore sequencing technology, bioinformatics and applications. *Nat. Biotechnol.* **2021**, *39*, 1348–1365. [CrossRef] [PubMed]

genes

MDPI

Article

DNA Double-Strand Break-Related Competitive Endogenous RNA Network of Noncoding RNA in Bovine Cumulus Cells

Jian-Bo Liu [1,2,†], Jia-Bao Zhang [1,†], Xiang-Min Yan [1,3], Peng-Gui Xie [4], Yao Fu [1], Xu-Huang Fu [1], Xu-Lei Sun [1], Dong-Xu Han [1], Sheng-Peng Li [1], Yi Zheng [1], Yan Gao [1], Nam-Hyung Kim [1,5], Bao Yuan [1] and Hao Jiang [1,*]

1 Department of Laboratory Animals, College of Animal Sciences, Jilin University, Changchun 130062, China
2 Experimental Testing Center, Jilin Agricultural Science and Technology University, Jilin 132101, China
3 Institute of Animal Husbandry, Xinjiang Academy of Animal Husbandry, Urumqi 830057, China
4 Yili Vocational and Technical College, Yili 835000, China
5 Department of Animal Science, Chungbuk National University, Cheongju 361-763, Republic of Korea
* Correspondence: jhhaojiang@jlu.edu.cn
† These authors contributed equally to this work.

Abstract: (1) Background: DNA double strand breaks (DSBs) are the most serious form of DNA damage that affects oocyte maturation and the physiological state of follicles and ovaries. Non-coding RNAs (ncRNAs) play a crucial role in DNA damage and repair. This study aims to analyze and establish the network of ncRNAs when DSB occurs and provide new ideas for next research on the mechanism of cumulus DSB. (2) Methods: Bovine cumulus cells (CCs) were treated with bleomycin (BLM) to construct a DSB model. We detected the changes of the cell cycle, cell viability, and apoptosis to determine the effect of DSBs on cell biology, and further evaluated the relationship between the transcriptome and competitive endogenous RNA (ceRNA) network and DSBs. (3) Results: BLM increased γH2AX positivity in CCs, disrupted the G1/S phase, and decreased cell viability. Totals of 848 mRNAs, 75 long noncoding RNAs (lncRNAs), 68 circular RNAs (circRNAs), and 71 microRNAs (miRNAs) in 78 groups of lncRNA–miRNA–mRNA regulatory networks, 275 groups of circRNA–miRNA–mRNA regulatory networks, and five groups of lncRNA/circRNA–miRNA–mRNA co-expression regulatory networks were related to DSBs. Most differentially expressed ncRNAs were annotated to cell cycle, p53, PI3K-AKT, and WNT signaling pathways. (4) Conclusions: The ceRNA network helps to understand the effects of DNA DSBs activation and remission on the biological function of CCs.

Keywords: DNA double-strand breaks; DDR; ncRNA; ceRNA; cumulus cells; bovine

Citation: Liu, J.-B.; Zhang, J.-B.; Yan, X.-M.; Xie, P.-G.; Fu, Y.; Fu, X.-H.; Sun, X.-L.; Han, D.-X.; Li, S.-P.; Zheng, Y.; et al. DNA Double-Strand Break-Related Competitive Endogenous RNA Network of Noncoding RNA in Bovine Cumulus Cells. *Genes* **2023**, *14*, 290. https://doi.org/10.3390/genes14020290

Academic Editor: Benjamin N. Sacks

Received: 26 December 2022
Revised: 14 January 2023
Accepted: 18 January 2023
Published: 22 January 2023

1. Introduction

The follicle is a basic component of the mammalian ovary, and its development is coordinated by a variety of factors. Growing follicles develop from primordial follicle, as well as preantral and antral follicle. With the development of the follicle, the oocyte volume gradually increases and, at the same time, accumulates a large amount of genetic material and nutrients. There are multiple layers of cumulus cells (CCs) around it, and the cumulus cell–oocyte complexes (COCs) are close and orderly, which plays an important role in oocyte growth, meiosis, maturation, and ovulation [1,2]. However, most follicles undergo atresia degeneration before ovulation and cannot successfully ovulate in monotocous animals.

COCs are considered functional units of female mammalian germ cells [3]. During follicle development, maintaining genomic integrity is fundamental to ensuring reproductive success. However, long-term disturbances of in vitro and in vivo factors can lead to the accumulation of various lesions, including double-strand breaks (DSBs). DSBs are the most severe form of DNA damage caused by the simultaneous cleavage of double-stranded DNA at opposite or nearby positions by exogenous or endogenous factors [4,5]. γH2AX, formed by phosphorylation at DSB sites, plays an important role in extending the signaling cascade during the DNA damage stress response (DDR), which occurs early after injury and is considered the most effective biomarker for the detection of DSBs [6].

After DSBs occur, if the DDR cannot be properly activated, chromatin remodeling [7], cell cycle arrest [8], apoptosis or other forms of cell death may occur [9]. All of these may cause the quality of oocytes to decline [10], which in turn leads to the failure of oocytes to mature and become fertilized normally, resulting in disease or mutation in the offspring [11–13]. Studies have shown that DSBs affect the transition of porcine oocytes from MI to MII, inhibit first polar body excretion and oocyte maturation, and result in MI spindle defects [14,15]. The occurrence of DSBs in bovine oocytes affects their maturation rate [16]. DSBs disrupt bovine COC communication and cause oocytes to exit meiosis [17], and the DNA damage of CCs inhibits oocytes from resuming meiosis and affects blastocyst yield, hatchability, and embryo quality, while bovine embryonic developmental delay is positively correlated with the degree of DNA damage [18–20]. CCs control DSB-induced DDR in oocytes [21]. DSBs and premature ovarian failure are associated with the BRCA-AT pathway and the PTEN/PI3K/Akt pathways, whereas the CHK1/CHK2 to p53 and p63 pathways can eliminate oocyte DSBs [22–24]. However, research on the reproductive system of female animals mostly focuses on the regulation of coding genes, and research on the regulation of coding genes at the noncoding RNA (ncRNA) level, especially the interaction between ncRNAs, is scarce.

Only approximately 2% of genes in the genome encode proteins, and most of the others encode transcripts that ncRNAs, including long noncoding RNAs (lncRNAs), circular RNAs (circRNAs), and small RNAs (sRNAs) [25,26]. As research has increased, it has been found that ncRNAs are involved in many biological functions and biological processes, including development, proliferation, apoptosis, survival, differentiation, and carcinogenesis [27–30]. Previous studies have shown that microRNAs (miRNAs) have important roles in bovine oocytes and early embryos. miRNAs also affect blastocyst activation and implantation in rats. The overexpression of miR-155 and miR-224 in mice inhibited CC expansion and oocyte maturation [31–34]. The knockdown of lncRNA–ROSE results in abnormal oocyte cytokinesis and impaired preimplantation embryo development in mice, and the inhibition of circRNA ARMC4 affects oocyte meiosis and early embryonic development in pigs [35,36]. It is known that the binding of miRNAs to mRNA leads to gene silencing, and competitive endogenous RNAs (ceRNAs) can regulate gene expression by affecting miRNAs through competitive binding to miRNA response elements (MREs), indicating that there are important biologically significant RNA–miRNA regulatory pathways [37]. For example, the lncRNA Lnc-RI, which is a ceRNA, affects DSBs by competitively binding with miR-193a-3p and releasing the expression of RAD51, [38]. circRNA DB activates the USP7/Cyclin A2 signaling pathway through miR-34a, which promotes the increase in USP7 expression and the decrease in DNA damage [39], indicating that the ceRNA regulatory network can affect DSBs in cells. Although research in this field is rapidly advancing, little is known about the impact of ceRNA regulatory networks on DSBs or the DDR. In particular, there is a lack of a systematic analysis of the relationships between DSBs/DDR and ncRNAs in bovine CCs.

Bleomycin (BLM), the fermentation product of Streptomyces verticillus, is a commonly used inducer of cellular DSBs and drug for the treatment of cancer [40]. BLM also affects the cell cycle, proliferation, and apoptosis [41,42]. In this study, bovine CCs were first treated with 200 μM BLM [43] to construct a DSB model. Subsequently, the effects of DSBs on the cell cycle, cell viability, and apoptosis of CCs were detected, and the differential expression profiles of lncRNAs, circRNAs, miRNAs, and mRNAs during the DNA-DSB process of bovine CCs were determined by high-throughput sequencing technology. Finally, we predicted the lncRNA/circRNA–miRNA–RNA regulatory network based on the ceRNA mechanism. The results of this study are helpful for elucidating the mechanism underlying the interaction between ncRNAs and mRNAs in bovine CCs during DSBs/the DDR and provide a meaningful new direction for preventing and improving the effect of DSB-induced oocyte quality reduction on fertility in animals.

2. Materials and Methods

2.1. CCs Collection, Culture and BLM Treatment

CCs were harvested as we described before [44]. In brief, the ovaries were obtained from healthy cows of Changchun Haoyue Halal Meat Products Co., Ltd., in Jilin Province and transferred to the laboratory within 1 h. Follicular fluid with COC in the follicles 3–8 mm in diameter was obtained with a 10 mL syringe. After washing three times with HEPES (Gibco, Paisley, Scotland, United Kingdom), COCs with more than five tightly wrapped CC layers were collected under a microscope. Subsequently, the CCs were separated from the oocyte with 0.1% hyaluronidase. Then, CCs were harvested and centrifuged for 5 min, washed twice with phosphate-buffered saline (PBS), cultured in DMEM/F12 (Gibco, Grand Island, NY, USA) supplemented with 1% penicillin, streptomycin (HyClone, Logan, UT, USA), and 10% fetal bovine serum (Biological Industries, Kibbutz Beit Haemek, Israel), and placed in an incubator at 38.5 °C under 5% CO_2. For BLM (Zeocin, Thermo Fisher, Carlsbad, CA, USA, diluted with culture media) treatments, 5×10^5 cells/well CCs were seeded into 6-well cell culture plates and cultured for 6 h. Then, solutions of 0 µM (Negative Control(NC)group) and 200 µM BLM (BLM group) were applied to the cells for 3 h.

2.2. γH2AX Detection

After BLM treatments, CCs were dissociated with 500 µL of trypsin for 3 min. After 500× *g* centrifugation and washing three times with PBS, the CCs were fixed in 100 µL of 4% paraformaldehyde solution for 10 min. Then, 900 µL of prechilled methanol was added, after which the mixture was gently homogenized and then incubated on ice for 30 min. Next, the cells were centrifuged at 500× *g* for 5 min, transferred to 1 mL of incubation solution (500 g bovine serum albumin in 100 mL PBS) and mixed well after 500× *g* centrifugation. CCs were treated with 100 µL of incubation solution-diluted (1:50) Alexa Fluor 488-labeled γH2AX antibody (Cell Signaling Technology, Danvers, MA, USA) for 1 h at room temperature. Subsequently, the cells were washed three times with incubation solution and then centrifuged at 500× *g* for 5 min. Finally, the cells were resuspended in 200 µL of PBS and analyzed using flow cytometry (Beckman Coulter, Brea, CA, USA).

2.3. Cell Cycle Analysis

The cell cycle was determined in strict accordance with the instructions of the cell cycle kit (Beyotime, Shanghai, China). CCs (5×10^5 cells/well) were seeded in 6-well plates. After the cells adhered, they were treated with 0 µM BLM-NC or 200 µM BLM for 3 h, trypsinized, and collected into 1.5 mL centrifuge tubes. Cells were resuspended in 100 µL of PBS after centrifugation at 500× *g* for 5 min. The cell cycle distribution was detected using a flow cytometer (Beckman Coulter, Brea, CA, USA), and the data were processed using MODFIT software (Verity Software House, Topsham, ME, USA).

2.4. Cell Proliferation Assays

The proliferation of bovine CCs was measured using a CCK-8 kit (Dojindo, Kyushu, Japan). In brief, each well of a 96-well plate contained 1×10^4 CCs in 100 µL of media and was cultured for 12 h at 38.5 °C and 5% CO_2. BLM (0 µM) was added to the NC group, or BLM (200 µM) was added to the BLM group. The normal media was replaced immediately after 3 h of treatment. After 0, 24, 48, and 72 h of incubation, 10 µL of CCK-8 solution was added to each well. The plates were incubated in a 38.5 °C incubator for 3 h, and the absorbance at 450 nm was measured using a microplate reader (BioTek Instruments, Winooski, VT, USA).

2.5. Apoptosis Analysis

The method of detecting apoptosis was performed strictly in accordance with the apoptosis assay kit instructions (BD, CA, USA). Bovine CCs (5×10^5 cells/well) were seeded into 6-well plates. Cells were treated with BLM (0 or 200 µM) for 3 h. Then, 100 µL of PBS was added to each centrifuge tube, followed by the addition of 5 μL of FITC solution and 5 μL of propidium iodide (20 μg/mL). The cells were subsequently incubated for 15 min at room temperature in the dark. The samples were measured within 2 h using a flow cytometer (Beckman Coulter, Brea, CA, USA). Apoptotic cells and dead cells were distinguished by staining with propidium iodide and FITC.

2.6. RNA Extraction

After treatment, total RNA was extracted using TriPure (Roche, Basel, Switzerland), according to the manufacturer's recommended protocol. All the RNA samples were checked using a NanoDrop ND-2000 Spectrophotometer (Thermo, Waltham, MA, USA), a Qubit 2.0 instrument (Thermo, Waltham, MA, USA), and an Agilent 2100 device (Agilent, Karlsruhe, Baden-Württemberg, Germany) and via electrophoresis and were then stored at −80 °C for further experiments.

2.7. LncRNA, circRNA, and mRNA Library Construction

An Epicenter Ribo-ZeroTM kit (Epicenter, Madison, WI, USA) was used to remove rRNA from the sample, and then fragmentation buffer was added to randomly interrupt rRNA-depleted RNA. First-strand cDNA was synthesized with six-base random primers, and then buffer, dATP, dUTP, dCTP, dGTP, RNase H, and DNA polymerase I were added to synthesize doubled-stranded cDNA. This cDNA was purified using AMPure XP magnetic beads, the purified double-stranded cDNA was end repaired, A was added, and the sequencing adapter was ligated. Then, AMPure XP magnetic beads were used for fragment size selection. Finally, the U chain was degraded, and the cDNA library was obtained via PCR enrichment. LncRNA and circRNA libraries were sequenced by the Illumina HiSeq platform (Biomarker Technologies, Beijing, China).

2.8. miRNA Library Construction

Six microliters of RNase-free ddH2O solution, including 1.5 μg of RNA, was used to construct a library by using a Small RNA Sample Pre Kit (Thermo, Waltham, MA, USA). T4 RNA Ligase 1 and T4 RNA Ligase 2 (truncated) were attached to the 3′ and 5′ ends of the small RNA due to the phosphate group at the 5′ end and the hydroxyl group at the 3′ end, respectively. cDNA, PCR amplification, and gel separation techniques were used to screen the target fragments, and the fragments obtained by gel extraction were considered small RNA libraries.

After the library was constructed, a Qubit 2.0 instrument (Thermo, Waltham, MA, USA) was used to test the library concentration, the library concentration was diluted to 1 ng/μL, and an Agilent 2100 Bioanalyzer (Agilent, German) was used to detect the insert size. qPCR was used to accurately quantify the effective concentration of the library to ensure the quality of the library. The miRNA library was sequenced using HiSeq 2500 high-throughput sequencing (Biomarker Technologies, Beijing, China).

2.9. Bioinformatic Analysis of LncRNA

Transcriptomes were assembled using StringTie based on data mapped to the reference genome. Assembled transcripts were annotated using the gffcompare program (The Center for Computational Biology, USA). The CPC/CNCI/CPA/Pfam programs were combined to predict nonprotein-coding RNAs from unknown transcripts [45–48]. Transcripts that were more than 200 nt in length and with more than two exons were selected as lncRNA candidates, and CPC/CNCI/CPAT/Pfam with the ability to distinguish protein-coding genes from noncoding genes was used to further screen different types of lncRNAs, including lincRNAs, intronic lncRNAs, antisense lncRNAs, and sense lncRNAs.

2.10. Bioinformatic Analysis of circRNAs

CIRI and find_circ software [49] were used to predict circRNAs. First, CIRI software was used to predict circRNA sequences. Then, the find_circ software first took 20 bp from both ends of the reads on the genome alignment as anchor points and then used the anchor points as independent reads on the genome to align and identify the only matching site. If the alignment positions of the two anchors are reversed in the linear direction, the reads of the anchors are extended until the joint position of the circular RNA is found. A GT/AG splicing signal is considered to indicate circRNA.

2.11. Bioinformatic Analysis of miRNA

We obtained clean data from the raw data by removing reads containing adapters, reads containing poly-N sequences, and reads of low quality, trimmed and cleaned the data by removing sequences smaller than 18 nt or larger than 30 nt, and calculated the Q20 value, Q30 value, GC content, and sequence repeat levels. By the use of Bowtie tools [50], the clean reads were compared with the content in the Silva database, GtRNAdb database, Rfam database, and Repbase database to filter ribosomal RNA (rRNA), transfer RNA (tRNA), small nuclear RNA (snRNA), small nucleolar RNA (snoRNA), and other ncRNAs and repeats. Next, the remaining reads were used to detect known miRNAs, and newly discovered miRNAs were obtained by comparison with miRBase. Novel miRNA secondary structures were predicted using MTide tools [51].

2.12. Expression Analysis

The DESeq R package (1.10.1) [52] was used to analyze the expression of RNAs in each sample, and a fold change (fold change) greater than or equal to 1.5 and an FDR less than 0.05 were used as the criteria for screening differentially expressed RNAs. A fold change represents the ratio of expression levels between the two groups of samples.

2.13. Functional Enrichment Analysis

We used DAVID software to analyze the Gene Ontology (GO) of differentially expressed RNAs [53,54]. The Kyoto encyclopedia of genes and genomes (KEGG) was used to analyze differentially expressed RNAs with the KOBAS software [55]. GO and KEGG pathways identified as having a p value < 0.05 were considered significantly enriched.

2.14. RT–qPCR

The differential expression levels of ncRNAs and mRNAs in the NC group or BLM group were measured by RT–qPCR. To determine the resistance of ncRNAs to RNase R digestion, total RNA was treated with RNase R (Epicenter, WI, USA) prior to cDNA synthesis. Fluorescence quantitative PCR was performed using SYBR Green (Tiangen, Beijing, China) according to the manufacturer's protocol. Each 20 μL of the RT–qPCR mixture consisted of 8 μL of deionized water, 10 μL of SuperReal PreMix Plus (Tiangen, Beijing, China), 1 μL of cDNA, and 0.5 μL each of forward and reverse primers (10 mM). The RT–qPCR conditions included denaturation at 95 °C for 180 s and 40 cycles of 95 °C for 15 s, 60 °C for 20 s, and 72 °C for 15 s. Gene expression was quantified using a Mastercycler ep realplex (Eppendorf, Hamburg, Germany) and the $2^{-\Delta\Delta ct}$ method with GAPDH or miRNA-U6 as the standard. The sequences of the primers used are shown in Supplementary Table S1.

2.15. CeRNA Regulatory Network Analysis

The miRanda, RNAhybrid, and TargetScan databases were used to predict the miRNAs targeting mRNAs, lncRNAs, and circRNAs. The input files were miRNA, mRNA, lncRNA, and circRNA base sequence files that we obtained from the RNA-seq. Then, the miRNA, mRNA, lncRNA, and circRNA with different expressions were selected to construct the ceRNA network. Overlapping the same miRNAs to construct a lncRNA/circRNA–miRNA–mRNA regulatory network and the interaction between the networks were analyzed using Cytoscape software [56]. The sequences of the ncRNAs used are shown in Supplementary Table S2.

2.16. Statistical Analysis

The experimental results are expressed as the means ± SDs of three independent experiments. The data were analyzed using the ANOVA program of SPSS 23.0 (IBM, Armonk, NY, USA), and $p < 0.05$ was considered significant.

3. Results

3.1. BLM Treatment Induced DSBs, Disrupted the Cell Cycle, and Inhibited Cell Proliferation in Bovine CCs

As shown in Figure 1a, the γH2AX-positive rate of CCs in the BLM group was higher than that in the NC group ($p < 0.01$). The results (Figure 1b) showed that the relative cell cycle level in the G1 phase of the BLM group was lower than that of the NC group ($p < 0.01$), while that of the S phase was higher than that of the NC group ($p < 0.01$). In the G2 phase, the relative cell cycle levels of the two groups were not different ($p > 0.05$). The cell cycle of bovine CCs treated with BLM was altered, the retention time of the G1 phase in the BLM group was significantly lower than that of the control group, and the retention time of the S phase was significantly higher than that of the control group.

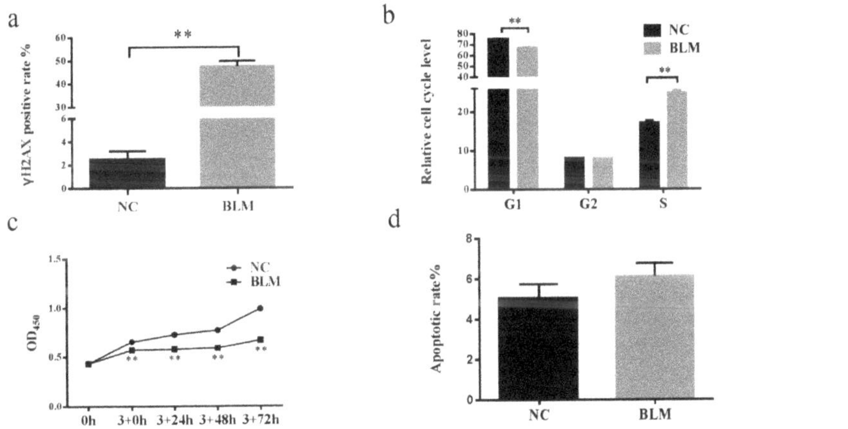

Figure 1. Effect of BLM on bovine CCs. Bovine CCs were treated with BLM (200 μM) for 3 h. (**a**) BLM treatment enhanced the γH2AX-positive rate. (**b**) The percentage of S phase was increased, and G1 phase was decreased significantly. (**c**) The value of OD450 was undermined after treatment with BLM for 3 h and measured for 72 h at intervals of 24 h. (**d**) BLM did not affect the cell apoptosis rate. Significant differences are represented with ** ($p < 0.01$).

CCK8-related data (Figure 1c) showed that the cell viability increased in both the NC and the BLM groups, but the cell viability in the BLM group was lower than that in the NC group at 3 + 0 h, 3 + 24 h, 3 + 48 h, and 3 + 72 h ($p < 0.01$).

In addition, the relative apoptosis of CCs induced by BLM was higher than that of the NC group (Figure 1d), but the difference was not significant. This indicated that most CCs treated with BLM did not begin the apoptosis process.

3.2. Overview of RNA Sequencing

We constructed two cDNA libraries, each with three replicates, to identify the mRNA, lncRNA, and circRNA characteristics and abundances of CCs in the NC group and the BLM group. We obtained a total of 124.06 Gb of clean data for mRNA, lncRNA, and circRNA analysis, and the clean data per sample reached 17.75 Gb. After removing junction and low-quality reads, 143,882,865 and 132,777,966 clean reads were obtained from the NC and BLM groups, respectively. The average GC content in the NC group and BLM group was 56.58% and 51.74%, respectively. With reference to the bovine genome sequence (NCBI: txid9913), we obtained 95.54% and 83.79% of the mapped reads of the clean data from the

NC group and the BLM group, respectively. The sequencing depth of the NC group was 7.4345 ± 0.3171, and that of the BLM group was 8.3795 ± 0.5218. Supplementary Figure S1a shows the distribution of mapped reads covering depth on the reference genome. The Q30 value of the NC group was 94.39% ± 0.1955%, and that of the BLM group was 94.52% ± 0.1169%.

We also obtained a clean data of 147.98 M for miRNA analysis. We also performed some processing of the clean read, such as removing low-quality sequences, reads with unknown base content greater than or equal to 10%, reads without 3′ linker sequences, 3′ linker sequences, and sequences that were less than 18 or more than 30 nucleotides in length; we ultimately obtained 25,735,570 and 17,420,398 clean reads from the NC and BLM groups, respectively. We referenced the bovine genome sequence (NCBI: txid9913) and obtained 70.91% and 69.20% mapped reads from the NC and BLM groups, respectively. The distribution of the mapped reads coverage depth across the reference genome is shown in Figure S1b. The Q30 values of the NC group and BLM group were 98.91% ± 0.0698%, and 98.88 ± 0.0458, respectively.

3.3. Differentially Expressed mRNA Identification

After the bovine genome reference sequences were compared, 1750 (3275 in total) functional mRNAs were annotated. The length of mRNA was mainly concentrated at 400 bp or more than 3000 bp (Figure 2a). From the mRNA expression profile data, 848 mRNAs were differentially expressed (fold change ≥ 1.5, $p < 0.05$), of which 396 mRNAs were upregulated and 452 were downregulated and showed hierarchical clustering (Supplementary Table S3). The mRNAs listed in Supplementary Table S7 have been shown to be associated with DSBs. Next, three upregulated and three downregulated mRNAs were randomly selected to verify the sequencing data by RT–qPCR. The results showed that they have the same expression pattern as that in the sequencing data (Figure 2b).

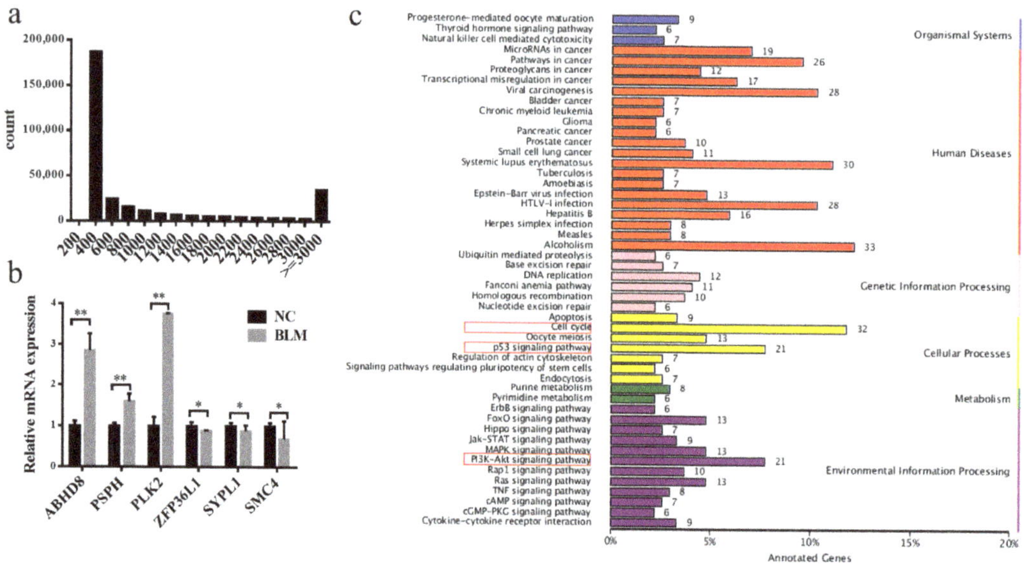

Figure 2. Identification of differentially expressed mRNAs. (**a**) MRNA length analysis. (**b**) Expression of up- and downregulated mRNAs in the NC or BLM group, as determined by RT–PCR. (**c**) KEGG pathway analysis of differentially expressed mRNAs. Red box is a relevant path for highlight DSBs or DDR. Significant differences are represented with * ($p < 0.05$) and ** ($p < 0.01$).

Subsequently, we performed GO term and KEGG pathway analyses on all differentially expressed mRNAs. The results showed that GO terms (Supplementary Figure S2) mainly focused on cellular processes, biological regulation, metabolic processes, and responses to stimuli, which were associated with DSBs and the DDR. The KEGG pathway (Figure 2c) analysis revealed pathways related to the cell cycle, p53 signaling, and PI3K-Akt signaling. This indicated that BLM-induced DSBs in CCs would cause the differential expression of mRNA and lead to changes in the biological state of cells.

3.4. *Differentially Expressed lncRNA Identification*

The analysis of CNCI, CPC, Pfam, and CPAT data revealed 6436 lncRNAs: 3388 lincRNAs (52.6%), 886 antisense lncRNAs (13.8%), 1720 intron lncRNAs (26.7%), and 442 lncRNAs (6.9%) (Figure 3a,b). The predominant length of differentially expressed lncRNAs was between 200 and 800 bp (Figure 3c). Furthermore, the expression levels of protein-coding genes were slightly higher than those of lncRNA transcripts (Figure 3d).

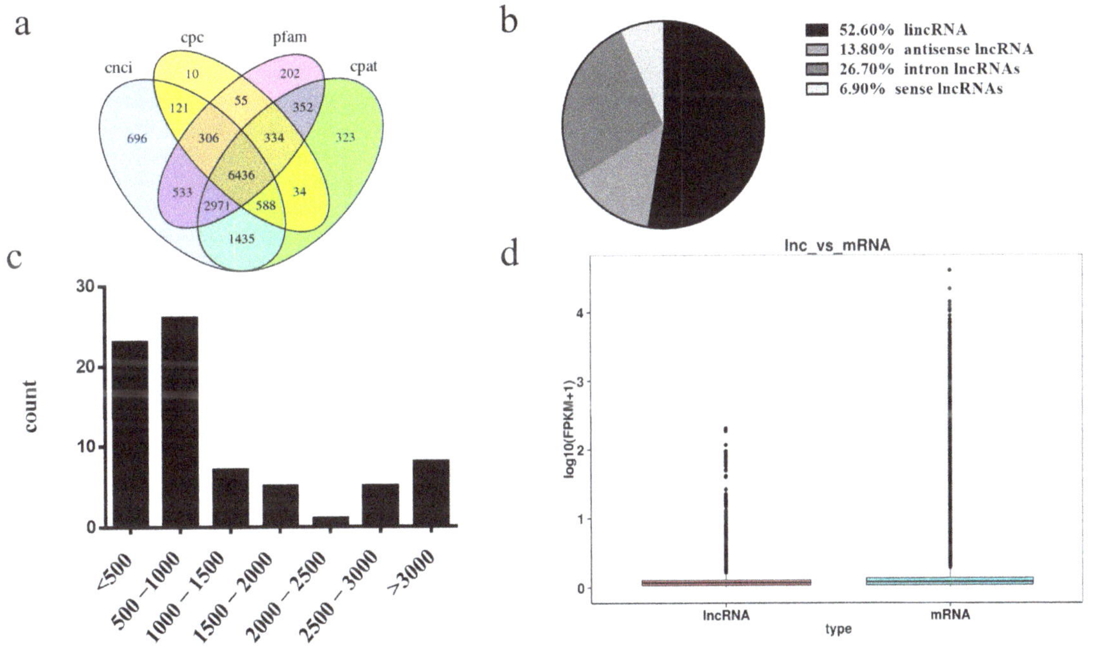

Figure 3. Analysis of lncRNAs. (**a**) Cross-analysis of CNC, CPC, Pfam, and CPAT data revealing lncRNAs. (**b**) Classification of 6436 lncRNAs, including lincRNAs, antisense lncRNAs, intronic lncRNAs, and sense lncRNAs. (**c**) Distributions of the sequence length of lncRNAs. (**d**) Boxplots of the expression levels (log10(FPKM)) of lncRNAs and protein-coding genes.

Among the 75 differentially expressed lncRNAs, 38 were upregulated, while 37 were downregulated (Supplementary Table S4). Three upregulated and three downregulated lncRNAs were randomly selected for RT–qPCR, and their expression patterns (Figure 4a) were consistent with the sequencing data.

Next, the identified GO terms and KEGG pathways were used to explore the potential functions of lncRNAs by searching 100 kb protein-coding gene sequences upstream and downstream of all identified lncRNAs to predict potential cis-regulatory targets of lncRNAs. A GO enrichment analysis showed that these genes were mainly related to GO terms involving cellular processes, cellular components, and binding (Supplementary Figure S3). As shown in Figure 4b, the pathways significantly enriched by KEGG mainly included the Hippo signaling pathway, the Wnt signaling pathway, and pathways related to the adherens junction and tight junction.

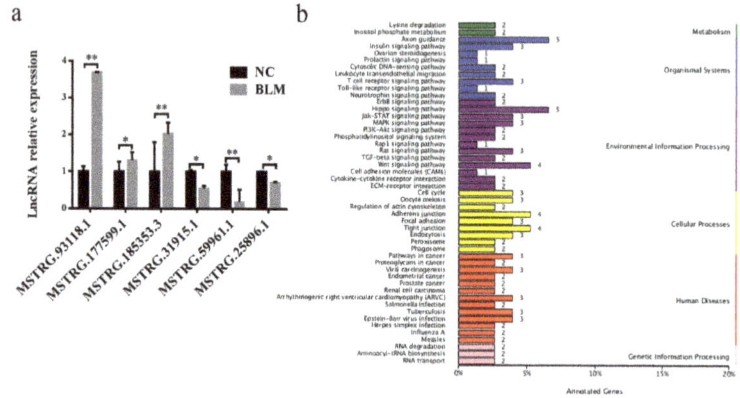

Figure 4. Identification of differentially expressed lncRNAs. (**a**) Expression of up- and downregulated lncRNAs in the NC and BLM groups, as determined by RT–PCR. (**b**) KEGG pathway analysis of differentially expressed lncRNAs. Significant differences are represented with * ($p < 0.05$) and ** ($p < 0.01$).

3.5. Differentially Expressed circRNA Identification

A total of 594 circRNAs, including exons, intergenic regions, and introns, which were mainly 400–800 bp in length, were identified (Figure 5a). Among them, the number of intronic circRNAs was the largest, followed by intergenic regions, and exons were the least abundant (Figure 5a). Thirty-two were upregulated, while 34 were downregulated (Supplementary Table S5) (fold change ≥ 1.5, $p < 0.05$). Three upregulated and three downregulated circRNAs were randomly selected for RT–qPCR (Figure 5b), and the results were consistent with the HiSeq data.

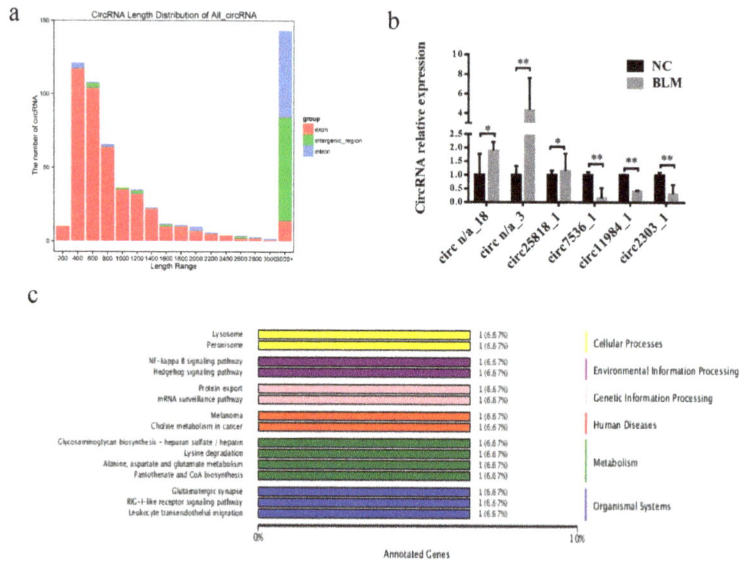

Figure 5. Identification of differentially expressed circRNAs. (**a**) CircRNA length and derivation analysis. (**b**) Expression of up- and downregulated circRNAs in the NC and BLM groups, as determined by RT–PCR. (**c**) KEGG pathway of differentially expressed circRNAs. Significant differences are represented with * ($p < 0.05$) and ** ($p < 0.01$).

The GO analysis revealed that many host genes were closely related to biological processes, cellular components, and molecular functions (Supplementary Figure S4), including cellular processes, cells, cellular parts, binding, and catalytic activities. Different gene products coordinate with each other to perform biological functions, and the pathway annotation of circRNA host genes can better help to understand the functions of these genes. KEGG analysis showed that 15 pathways, including those related to NF-κB, mRNA surveillance, and lysine degradation, were associated with potential CC functions (Figure 5c).

3.6. *Differentially Expressed miRNA Identification*

In this study, 1836 miRNAs, of which 467 were known and 1369 were newly predicted, were identified. Most miRNAs were mainly between 20 and 24 bp in length (Figure 6a). The results showed (Supplementary Table S6) 40 upregulated miRNAs and 31 downregulated miRNAs (fold change ≥ 1.5, $p < 0.05$). Among the 15 known differentially expressed miRNAs, three are involved in DSBs, including miR-486, miR-451, and miR-145 [57–59] (Supplementary Table S7). Then, six randomly selected miRNAs were verified by RT–qPCR (Figure 6b), and their levels were consistent with the sequencing data. We analyzed the nucleotide bias at each position of the miRNA (Figure 6c).

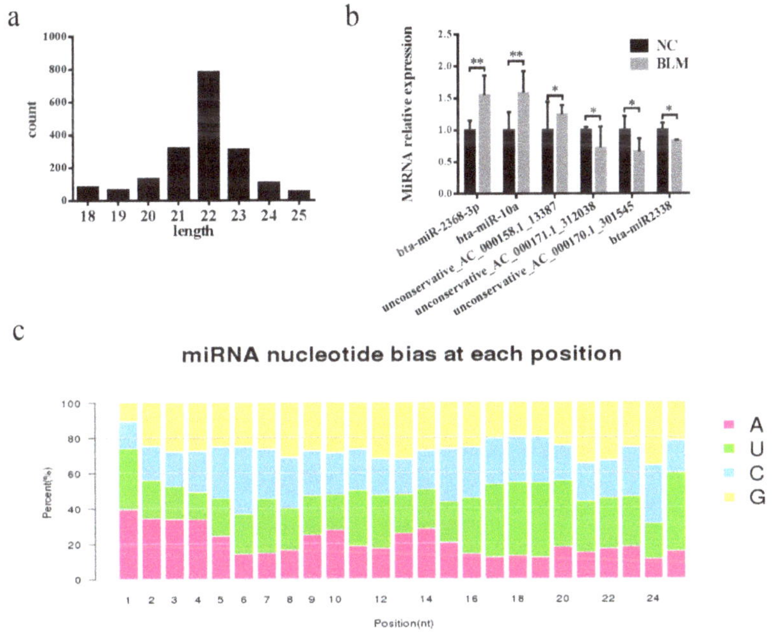

Figure 6. Identification of differentially expressed miRNAs. (**a**) miRNA length analysis. (**b**) Expression of up- and downregulated miRNAs in the NC and BLM groups, as determined by RT–PCR. (**c**) miRNA nucleotide bias at each position. Significant differences are represented with * ($p < 0.05$) and ** ($p < 0.01$).

Next, we predicted the functions of differentially expressed miRNAs by targeting relationships between miRNAs and mRNAs, as well as performing further GO (Supplementary Figure S5) and KEGG (Supplementary Figure S6) analyses. Cellular processes, single biological processes, metabolic regulation, cellular fractions, cells, binding, etc., were the main components identified via the GO analysis. The KEGG functional enrichment pathway analysis mainly revealed pathways involved in endocytosis, lysosome, protein processing, axon guidance, transcriptional dysregulation in cancer, dilated cardiomyopathy, neuroligand–receptor interaction, etc.

3.7. Construction of the lncRNA–miRNA–mRNA Regulatory Network

We predicted the ceRNA regulatory network of lncRNA–miRNA–mRNA using Cytoscape software (Figure 7). Figure 7a shows 52 groups of lncRNA–miRNA–mRNA regulatory networks, including lnc-MSTRG.197583.1, which acts as a molecular sponge to adsorb miR-unconservative-AC-00017.1-311957 to promote the expression of mRNAs. lnc-MSTRG.69684.1, 155681.4, and 24637.1 interact with miR-unconservative-AC-000166.1-251601, resulting in abnormal mRNA expression, and lnc-MSTRG.177599.1 affects miR-unconservative-AC-000173.1-360722 and 000162.1-112908 to target the same mRNA. The MiRNA-mediated regulation of NOTCH1 and CEP170B by lnc-MSTRG.197583.1, 69684.1, 155681.4, and 24637.1, lnc-MSTRG.177599.1, 69684.1, 155681.4, and 24637.1 regulate CEP170B and SPTBN5.

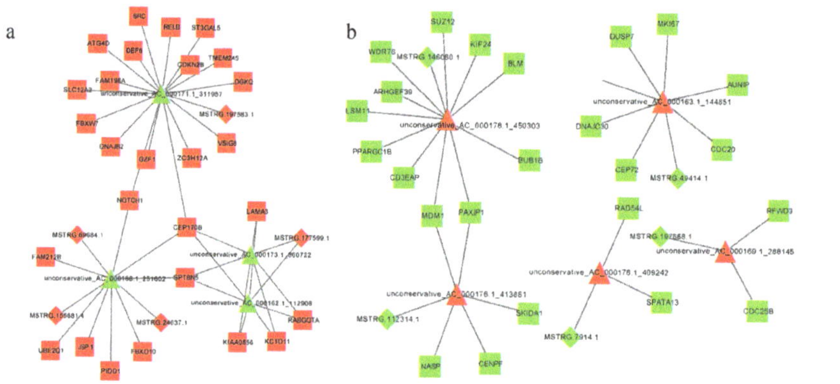

Figure 7. Construction of the lncRNA–miRNA–mRNA ceRNA regulatory network. (**a**,**b**) lncRNA–miRNA–mRNA ceRNA regulatory network. The diamonds represent lncRNAs, the triangles represent miRNAs, the squares represent mRNAs, red indicates upregulation, and green indicates downregulation.

In addition, five repressed lncRNAs promoted the expression of five miRNAs and decreased the expression of 24 mRNAs consisting of 26 ceRNAs (Figure 7b), such as lnc-MSTRG.146080.1, 49414.1, 112314.1, 7914.1, and 16, which promote miR-unconservative-AC-000178.1-450303, 000176.1-413881, 000176.1-409242, 000169.1-288145, and 000163.1, respectively. Lnc-MSTRG.146080.1 and 49414.1 allowed the expression of targeted miRNAs and corepressed MDM1 and PAXIP1.

3.8. Construction of the circRNA–miRNA–mRNA Regulatory Network

We predicted a ceRNA regulatory network involving 276 circRNA–miRNA–mRNA pairs using Cytoscape software, of which 124 pairs promoted the upregulation of mRNA expression (Figure 8a) and 151 pairs inhibited mRNA expression (Figure 8b). When circRNA-AC_000176.1: 39742140 | 39768649, AC_000175.1: 46155528 | 46202870, and AC_000158.1: 107948998 | 107975652 jointly suppressed miR-unconservative_AC_000159.1_32276, unconservative_AC_000162.1_112908, AC_000171.1_312038, and AC_000173.1_360722 released 18 mRNAs. CircRNA-AC_000159.1:20992753 | 21047706 acted as a molecular sponge to adsorb miR-unconservative_AC_000162.1_112908 and AC_000173.1_360722 to increase mRNA expression.

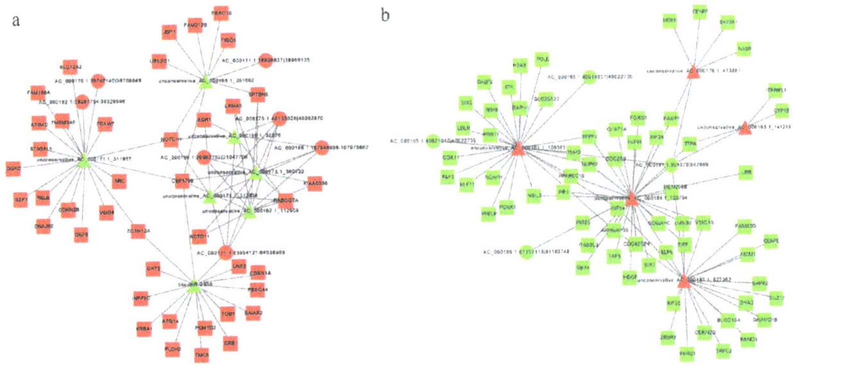

Figure 8. Construction of the circRNA–miRNA–mRNA ceRNA regulatory network. (**a**,**b**) circles represent circRNAs, triangles represent miRNAs, squares represent mRNAs, red indicates upregulation, and green indicates downregulation.

3.9. Construction of the LncRNA/CircRNA–miRNA–mRNA Regulatory Network

We predicted five groups of lncRNA/circRNA–miRNA–mRNA regulatory networks based on the same miRNA-binding sites (Supplementary Figure S7a). For example, the joint regulation of mRNA by lncRNA MSTRG.197583.1 and circRNA AC_000176.1:39742140 | 39768649 and AC_000182.1:38287754 | 38329596 via miR-unconservative_AC_000171.1_311957, miR-unconservative_AC_000162.1_112908, and AC_000173.1_360722 mediated lncRNA MSTRG.177599.1 and circRNA AC_000158.1: 107948998 | 107975652, AC_000159.1: 20992753 | 21047706, AC_000171.1: 63954121 | 64038908, and AC_000175.1: 46155528 | 46202870 promoted mRNA expression; miR-unconservative_AC_000168.1_251602 was inhibited by lnc-MSTRG.155681.4, MSTRG.24637.1, MSTRG.69684.1, and circ-AC_000171.1 Furthermore, the inhibition of lnc-MSTRG.112314.1 and circ-AC_000181.1:934078 | 947669 released the expression of miR-unconservative_AC_000176.1_413881 to silence the mRNA (Supplementary Figure S7b). These results showed that the treatment of CCs with BLM resulted in changes in the expression levels of various components, and a series of multiple regulatory events took place along with them.

4. Discussion

BLM has a dose/time-dependent effect on the induction of γH2AX, which is formed by DSBs through the generation of free radicals [15,60]. The formation of γH2AX functions not only in DSBs, but also as a recruitment factor for the DDR [61]. CCs regulate oocyte development through gap links, and DSBs cause bovine COC communication disorder, ultimately causing oocytes to exit meiosis [17] and control the DSB-induced oocyte DDR process [21], this leads to DNA injury and inhibits the oocyte resumption of meiosis, which generates DNA damage that inhibits the recovery of oocyte meiosis [18]. DNA damage to cells leads to cell cycle arrest [62], inhibiting cell viability [63], but cells complete DNA replication and histone synthesis in S phase to lay the foundation for cell division [64]. These results indicated that after BLM induced DSBs in CCs, a large amount of γH2AX was formed, which altered the cell cycle and cell viability and then affected the development of oocytes. Our study showed (Figure 1) that after bovine CCs were treated with 200 μM BLM for 3 h, the level of γH2AX in CCs increased, which promoted the rapid passage of cells through the G1 phase, but a large number of cells were arrested in the S phase, which in turn affected cell proliferation but did not significantly affect apoptosis. These results suggest that DSBs participate in oocyte development by regulating the biological functions of CCs. However, the molecular mechanism is still unclear, especially the relationship between ncRNA and DSB. Therefore, we performed high-throughput sequencing of the DSB model of CCs to investigate the potential functions of mRNA, lncRNA, circRNA, and miRNA during DSB/DDR processes in CCs.

During double strand breaking, cells are bound to detect damage and temporarily block cell cycle progression to allow time to repair or exit the cell cycle [65]. The rest of the cell cycle is required to engage the DDR through the cell cycle checkpoint pathway in response to DSBs [66]. Our KEGG pathway analysis (Figures 2c and 4b) showed that the most enriched cellular process was the cell cycle. This suggests that cell cycle arrest caused by DSBs in CCs may promote DDR to maintain the biological function of CCs and ensure the connection between CCs and oocytes. Our analysis also showed that KEGG also enriched the p53 and PI3K-AKT signaling pathways (Figures 2c and 4b), which play very important roles in DSB and DDR processes. On the one hand, DNA damage induces ATM lysine acetylation to activate ATM. The kinase activity, in turn, activates the ATM-chk2-p53 pathway and the PI3K-AKT pathway [67,68]. The p53 pathway can eliminate DSBs in oocytes [24], and the PI3K-AKT pathway promotes DSB repair [69]. Therefore, our experiments show that after the occurrence of DSBs in bovine CCs, the main purpose of the cell cycle, p53 signaling pathway, and PI3K-AKT pathway is to activate the DDR and eliminate DSBs in cells to stabilize the genome and maintain the normal function of CCs and oocytes. However, there are still many pathways related to CCs or oocytes, DSBs/the DDR that have not been investigated. We will investigate this in future research. In addition, we need to explore whether DSBs/the DDR is caused by pathway alteration or whether DSBs activate the pathway.

Currently, the analysis of DSBs is mainly related to genes encoding proteins, while little is known about ncRNAs. Recent reports have demonstrated that ncRNAs are involved in the regulation of multiple biological functions and pathological processes by regulating gene expression at the transcriptional and posttranscriptional levels [70], such as aging [71]. In addition, the competitive binding of lncRNA RI to miR-193a-3p affects homologous recombination(HR) repair of DSBs [38]. CircRNA circ-DB affects DNA damage through the regulation of USP7 by miR-34a [39]. LncRNAs regulate the cell cycle and proliferation through p53 [72,73]. miRNAs affect the cell cycle or apoptosis through p53 [74–78], and the p53 pathway is closely related to oocyte DSBs [24]. This finding indicates that ncRNA can affect DSBs of oocytes through the p53 signaling pathway. We will investigate this in future research. In this study, we discovered 6436 lncRNAs, 594 circRNAs, and 1836 miRNAs from the DSB model of bovine CCs using the Illumina HiSeq Xten platform. It is well known that CCs play an important role in the maturation of oocytes. To our knowledge, this study is the first to determine the expression pattern of ncRNAs following the induction of DSBs in bovine CCs.

Although less conservative and expressed at lower levels than protein-coding genes, ncRNAs are often regulated by transcription factors and expressed specifically in cells [29]. It has been reported that lncRNAs can act as miRNA sponges to regulate target genes [38]. FBXW7 can be rapidly recruited to DNA damage sites, and its phosphorylation mediated by ATM promotes its retention at DNA damage sites for subsequent NHEJ repair [79]. lncRNA–MIF inhibits tumorigenesis through miR-586 and attenuates the inhibitory effect of miR-586 on FBXW7 [80]. The lncRNA TINCR, as a ceRNA, cleaves miR-544a from its target gene FBXW7 to regulate the proliferation and invasion of lung cancer cells [81]. DNA damage activates PIDD1 and NF-κB to promote cell survival [82], and the loss of PAXIP1 promotes DDR [83]. RFWD3-mediated ubiquitination promotes the HR [84]. In this study, 38 lncRNAs were upregulated and 37 lncRNAs were downregulated, and some lncRNAs were predicted to be involved in the ceRNA regulatory network, suggesting that lncRNAs may be involved in the DSB/DDR regulatory process. Based on previous studies, we speculate that lnc-MSTRG.197583.1 may act as a sponge to adsorb miRNAs to release FBXW7 to promote the DDR. After DSBs occur, miR-unconservative-AC-000166.1-251601 may be inhibited by lnc-MSTRG.69684.1, 155681.4, and 24637.1 to promote PIDD1 mRNA expression, thereby maintaining cell stability. The deletion of lnc-MSTRG.146080.1 releases miRNA, which inhibits the expression of PAXIP1 indirectly, possibly promoting the DDR. lnc-MST·RG.197668.1 may regulate the efficiency of RFWD3 HR mediated by miRNA. These findings indicate that BLM induces DSBs in CCs, resulting in changes in the expression of a large number of ncRNAs. They affect the γH2AX, cell cycle, and cell viability of CCs through regulatory networks, thereby affecting the development and biological functions of oocytes, and even follicles. In bovine CCs, the relationship between most ceRNA regulatory networks and DSBs/the DDR is unclear, warranting experimental validation based on the predicted results.

DDR is an important defense mechanism against genomic instability [85]. The presence of DSBs in cells activates the DDR, which is an extensive signaling network that includes DNA repair, cell cycle checkpoint activation, cellular senescence, and apoptosis [86]. DDR involves several pathways including HR and NHEJ to repair DSBs [87]. Loss of function and mutation of key DDR genes lead to premature ovarian failure [88]. Non-coding RNA (ncRNA) has recently emerged as a vital component of the DNA damage response (DDR), which was previously believed to be solely regulated by proteins. Many species of ncRNA can directly or indirectly influence DDR and enhance DNA repair, particularly in response to double-strand DNA breaks. The abnormal expression of miRNA can affect the DDR [89]. miR-34a regulates DDR through FOXP1 [90]. lncRNAs control the DNA damage response through interaction with DDRNAs at DSBs [91]. LncRNA DDSR1 has both an early role by modulating repair pathway choices, and a later function when it regulates gene expression [92]. Circ_0057504 promotes DDR via the NONO-SFPQ complex [93]. The LncRNA OTUD6B-AS1/miR-26a-5p/MTDH pathway affects the stability of DDR genes [94]. By overlapping the same miRNAs, we predicted the lncRNA/circRNA–miRNA–mRNA regulatory network (Supplementary Figure S6) and found a total of five groups. In these regulatory networks, the differential expression of lncRNA/circRNA and mRNA was negatively correlated with miRNAs, which may be because the abnormal expression of lncRNA/circRNA caused by DSBs affects the role of miRNA-induced mRNAs in cell survival or DDRs, or the abnormal expression of these RNAs results in DSBs/the DDR. These changes will have effects on CCs, which in turn affect oocyte development. However, these are only guesses based on previous research reports combined with our findings, and further experimental verification is needed.

5. Conclusions

In conclusion, we systematically investigated the potential interactions between ncRNAs and mRNAs in the DSB/DDR processes of bovine CCs. For example, lnc-MSTRG.197583.1 may act as a sponge to adsorb miRNAs to release FBXW7, miR-unconservative-AC-000166.1-251601 may be inhibited by lnc-MSTRG.69684.1, 155681.4, and 24637.1 to promote PIDD1 mRNA expression, the deletion of lnc-MSTRG.146080.1 releases miRNA, which inhibits the expression of PAXIP1 indirectly, and lnc-MST·RG.197668.1 may regulate RFWD3 mediated by miRNAs. These findings will provide insight into the DNA-DSB process and its ability to mediate follicular development and atresia.

Supplementary Materials: The following supporting information can be downloaded at: https://www.mdpi.com/article/10.3390/genes14020290/s1, Figure S1: The distribution of mapped reads covering depth on the reference genome; Figure S2: The GO annotation of differentially expressed mRNAs; Figure S3: The GO annotation of differentially expressed lncRNAs; Figure S4: The GO annotation of differentially expressed circRNAs; Figure S5: The GO annotation of miRNAs; Figure S6: KEGG pathway of differentially expressed miRNAs Figure S7: Construction of the LncRNA/CircRNA–miRNA–mRNA Regulatory Network; Table S1: S1.Primers used for qRT-PCR; Table S2: Sequences of ncRNAs; Table S3: Differentially expressed mRNAs; Table S4: Differentially expressed lncRNAs; Table S5: Differentially expressed circRNAs; Table S6: Differentially expressed miRNAs; Table S7: The relationship between differentially expressed RNAs and DSB/DDR.

Author Contributions: Conceptualization: H.J. and J.-B.Z. Methodology: H.J., N.-H.K. and B.Y. Validation: J.-B.Z. Formal analysis: J.-B.L., X.-H.F., S.-P.L. and Y.G. Investigation: J.-B.L., X.-M.Y., P.-G.X., Y.F., X.-L.S. and Y.Z. Resources: S.-P.L., Y.Z. and Y.G. Data curation: J.-B.L. and D.-X.H. Writing—original draft preparation: J.-B.L. and J.-B.Z. Writing—review and editing: H.J. and J.-B.Z. Visualization: J.-B.L. and Y.Z. Supervision: H.J. and J.-B.Z. Project administration: B.Y. and J.-B.Z. Funding acquisition: J.-B.Z. All authors have read and agreed to the published version of the manuscript.

Funding: This study was supported by the National Natural Science Foundation of China (31972570 and U20A2053), China Agriculture Research System of MOF and MARA (CARS-37), and Research and Demonstration of Integrated Breeding Technology for New Kerqin Beef Cattle (KJXM2020002).

Institutional Review Board Statement: The experiments were strictly performed according to the guidelines of the Guide for the Care and Use of Laboratory Animals of Jilin University. In addition, all experimental protocols were approved by the Institutional Animal Care and Use Committee of Jilin University (Permit Number: 201801026).

Informed Consent Statement: Not applicable.

Data Availability Statement: The data used to support the findings of this study are available from the corresponding author upon request.

Acknowledgments: The Haoyue Halal Meat Products Company in Changchun, Jilin province, China, provided the ovaries of bovine. RNAs sequencing was provided by Beijing Biomarker Technologies Co., Ltd., China.

Conflicts of Interest: The authors declare that they have no competing interests.

References

1. Buccione, R.; Vanderhyden, B.C.; Caron, P.J.; Eppig, J.J. FSH-induced expansion of the mouse cumulus oophorus in vitro is dependent upon a specific factor(s) secreted by the oocyte. *Dev. Biol.* **1990**, *138*, 16–25. [CrossRef] [PubMed]
2. Liu, Y.X. Interaction and signal transduction between oocyte and samatic cells in the ovary. *Front. Biosci.* **2007**, *12*, 2782–2796. [CrossRef] [PubMed]
3. Su, Y.Q.; Sugiura, K.; Eppig, J.J. Mouse oocyte control of granulosa cell development and function: Paracrine regulation of cumulus cell metabolism. *Semin. Reprod. Med.* **2009**, *27*, 32–42. [CrossRef]
4. Neal, J.A.; Meek, K. Choosing the right path: Does DNA-PK help make the decision? *Mutat. Res.* **2011**, *711*, 73–86. [CrossRef] [PubMed]
5. Davis, A.J.; Chen, B.P.; Chen, D.J. DNA-PK: A dynamic enzyme in a versatile DSB repair pathway. *DNA Repair* **2014**, *17*, 21–29. [CrossRef]
6. Tu, W.Z.; Li, B.; Huang, B.; Wang, Y.; Liu, X.D.; Guan, H.; Zhang, S.M.; Tang, Y.; Rang, W.Q.; Zhou, P.K. gammaH2AX foci formation in the absence of DNA damage: Mitotic H2AX phosphorylation is mediated by the DNA-PKcs/CHK2 pathway. *FEBS Lett.* **2013**, *587*, 3437–3443. [CrossRef]
7. Pandita, T.K.; Richardson, C. Chromatin remodeling finds its place in the DNA double-strand break response. *Nucleic Acids Res.* **2009**, *37*, 1363–1377. [CrossRef]
8. van Gent, D.C.; Hoeijmakers, J.H.; Kanaar, R. Chromosomal stability and the DNA double-stranded break connection. *Nat. Rev. Genet.* **2001**, *2*, 196–206. [CrossRef]
9. Bohgaki, T.; Bohgaki, M.; Hakem, R. DNA double-strand break signaling and human disorders. *Genome Integr.* **2010**, *1*, 15. [CrossRef]
10. Raman, R.S.; Chan, P.J.; Corselli, J.U.; Patton, W.C.; Jacobson, J.D.; Chan, S.R.; King, A. Comet assay of cumulus cell DNA status and the relationship to oocyte fertilization via intracytoplasmic sperm injection. *Hum. Reprod.* **2001**, *16*, 831–835. [CrossRef]
11. Tuppi, M.; Kehrloesser, S.; Coutandin, D.W.; Rossi, V.; Luh, L.M.; Strubel, A.; Hotte, K.; Hoffmeister, M.; Schafer, B.; De Oliveira, T.; et al. Oocyte DNA damage quality control requires consecutive interplay of CHK2 and CK1 to activate p63. *Nat. Struct. Mol. Biol.* **2018**, *25*, 261–269. [CrossRef] [PubMed]
12. Horta, F.; Catt, S.; Ramachandran, P.; Vollenhoven, B.; Temple-Smith, P. Female ageing affects the DNA repair capacity of oocytes in IVF using a controlled model of sperm DNA damage in mice. *Hum. Reprod.* **2020**, *35*, 529–544. [CrossRef] [PubMed]
13. Marangos, P.; Carroll, J. Oocytes progress beyond prophase in the presence of DNA damage. *Curr. Biol.* **2012**, *22*, 989–994. [CrossRef] [PubMed]
14. Wang, H.; Luo, Y.; Zhao, M.H.; Lin, Z.; Kwon, J.; Cui, X.S.; Kim, N.H. DNA double-strand breaks disrupted the spindle assembly in porcine oocytes. *Mol. Reprod. Dev.* **2016**, *83*, 132–143. [CrossRef]
15. Zhang, T.; Zhang, G.L.; Ma, J.Y.; Qi, S.T.; Wang, Z.B.; Wang, Z.W.; Luo, Y.B.; Jiang, Z.Z.; Schatten, H.; Sun, Q.Y. Effects of DNA damage and short-term spindle disruption on oocyte meiotic maturation. *Histochem. Cell Biol.* **2014**, *142*, 185–194. [CrossRef]
16. Wang, L.; Xu, X.; Teng, M.; Zhao, G.; Lei, A. Coping with DNA Double-Strand Breaks via ATM Signaling Pathway in Bovine Oocytes. *Int. J. Mol. Sci.* **2020**, *21*, 8892. [CrossRef]
17. Yamamoto, T.; Iwata, H.; Goto, H.; Shiratuki, S.; Tanaka, H.; Monji, Y.; Kuwayama, T. Effect of maternal age on the developmental competence and progression of nuclear maturation in bovine oocytes. *Mol. Reprod. Dev.* **2010**, *77*, 595–604. [CrossRef]
18. Sun, M.H.; Zheng, J.; Xie, F.Y.; Shen, W.; Yin, S.; Ma, J.Y. Cumulus Cells Block Oocyte Meiotic Resumption via Gap Junctions in Cumulus Oocyte Complexes Subjected to DNA Double-Strand Breaks. *PLoS ONE* **2015**, *10*, e0143223. [CrossRef]

19. Sirini, M.A.; Anchordoquy, J.M.; Anchordoquy, J.P.; Pascua, A.M.; Nikoloff, N.; Carranza, A.; Relling, A.E.; Furnus, C.C. The presence of acylated ghrelin during in vitro maturation of bovine oocytes induces cumulus cell DNA damage and apoptosis, and impairs early embryo development. *Zygote* **2017**, *25*, 601–611. [CrossRef]
20. Takahashi, M.; Keicho, K.; Takahashi, H.; Ogawa, H.; Schultz, R.M.; Okano, A. Effect of oxidative stress on development and DNA damage in in-vitro cultured bovine embryos by comet assay. *Theriogenology* **2000**, *54*, 137–145. [CrossRef]
21. Sun, M.H.; Yang, M.; Xie, F.Y.; Wang, W.; Zhang, L.; Shen, W.; Yin, S.; Ma, J.Y. DNA Double-Strand Breaks Induce the Nuclear Actin Filaments Formation in Cumulus-Enclosed Oocytes but Not in Denuded Oocytes. *PLoS ONE* **2017**, *12*, e0170308. [CrossRef] [PubMed]
22. Turan, V.; Oktay, K. BRCA-related ATM-mediated DNA double-strand break repair and ovarian aging. *Hum. Reprod. Update* **2020**, *26*, 43–57. [CrossRef] [PubMed]
23. Maidarti, M.; Anderson, R.A.; Telfer, E.E. Crosstalk between PTEN/PI3K/Akt Signalling and DNA Damage in the Oocyte: Implications for Primordial Follicle Activation, Oocyte Quality and Ageing. *Cells* **2020**, *9*, 200. [CrossRef] [PubMed]
24. Rinaldi, V.D.; Bloom, J.C.; Schimenti, J.C. Oocyte Elimination through DNA Damage Signaling from CHK1/CHK2 to p53 and p63. *Genetics* **2020**, *215*, 373–378. [CrossRef]
25. Consortium, E.P. An integrated encyclopedia of DNA elements in the human genome. *Nature* **2012**, *489*, 57–74. [CrossRef]
26. Guil, S.; Esteller, M. RNA-RNA interactions in gene regulation: The coding and noncoding players. *Trends Biochem. Sci.* **2015**, *40*, 248–256. [CrossRef]
27. Lian, Y.; Yan, C.; Xu, H.; Yang, J.; Yu, Y.; Zhou, J.; Shi, Y.; Ren, J.; Ji, G.; Wang, K. A Novel lncRNA, LINC00460, Affects Cell Proliferation and Apoptosis by Regulating KLF2 and CUL4A Expression in Colorectal Cancer. *Mol. Nucleic Acids* **2018**, *12*, 684–697. [CrossRef]
28. Wang, X.; Guo, B.; Li, Q.; Peng, J.; Yang, Z.; Wang, A.; Li, D.; Hou, Z.; Lv, K.; Kan, G.; et al. miR-214 targets ATF4 to inhibit bone formation. *Nat. Med.* **2013**, *19*, 93–100. [CrossRef]
29. Zhang, K.; Han, X.; Zhang, Z.; Zheng, L.; Hu, Z.; Yao, Q.; Cui, H.; Shu, G.; Si, M.; Li, C.; et al. The liver-enriched lnc-LFAR1 promotes liver fibrosis by activating TGFbeta and Notch pathways. *Nat. Commun.* **2017**, *8*, 144. [CrossRef]
30. Shi, Z.; Chen, T.; Yao, Q.; Zheng, L.; Zhang, Z.; Wang, J.; Hu, Z.; Cui, H.; Han, Y.; Han, X.; et al. The circular RNA ciRS-7 promotes APP and BACE1 degradation in an NF-kappaB-dependent manner. *FEBS J.* **2017**, *284*, 1096–1109. [CrossRef]
31. Gilchrist, G.C.; Tscherner, A.; Nalpathamkalam, T.; Merico, D.; LaMarre, J. MicroRNA Expression during Bovine Oocyte Maturation and Fertilization. *Int. J. Mol. Sci.* **2016**, *17*, 396. [CrossRef] [PubMed]
32. Liu, W.; Niu, Z.; Li, Q.; Pang, R.T.; Chiu, P.C.; Yeung, W.S. MicroRNA and Embryo Implantation. *Am. J. Reprod. Immunol.* **2016**, *75*, 263–271. [CrossRef] [PubMed]
33. Dehghan, Z.; Mohammadi-Yeganeh, S.; Salehi, M. MiRNA-155 regulates cumulus cells function, oocyte maturation, and blastocyst formation. *Biol. Reprod.* **2020**, *103*, 548–559. [CrossRef] [PubMed]
34. Li, X.; Wang, H.; Sheng, Y.; Wang, Z. MicroRNA-224 delays oocyte maturation through targeting Ptx3 in cumulus cells. *Mech. Dev.* **2017**, *143*, 20–25. [CrossRef] [PubMed]
35. Iyyappan, R.; Aleshkina, D.; Zhu, L.; Jiang, Z.; Kinterova, V.; Susor, A. Oocyte specific lncRNA variant Rose influences oocyte and embryo development. *Noncoding RNA Res.* **2021**, *6*, 107–113. [CrossRef] [PubMed]
36. Cao, Z.; Gao, D.; Xu, T.; Zhang, L.; Tong, X.; Zhang, D.; Wang, Y.; Ning, W.; Qi, X.; Ma, Y.; et al. Circular RNA profiling in the oocyte and cumulus cells reveals that circARMC4 is essential for porcine oocyte maturation. *Aging* **2019**, *11*, 8015–8034. [CrossRef]
37. Salmena, L.; Poliseno, L.; Tay, Y.; Kats, L.; Pandolfi, P.P. A ceRNA hypothesis: The Rosetta Stone of a hidden RNA language? *Cell* **2011**, *146*, 353–358. [CrossRef]
38. Shen, L.; Wang, Q.; Liu, R.; Chen, Z.; Zhang, X.; Zhou, P.; Wang, Z. LncRNA lnc-RI regulates homologous recombination repair of DNA double-strand breaks by stabilizing RAD51 mRNA as a competitive endogenous RNA. *Nucleic Acids Res.* **2018**, *46*, 717–729. [CrossRef]
39. Zhang, H.; Deng, T.; Ge, S.; Liu, Y.; Bai, M.; Zhu, K.; Fan, Q.; Li, J.; Ning, T.; Tian, F.; et al. Exosome circRNA secreted from adipocytes promotes the growth of hepatocellular carcinoma by targeting deubiquitination-related USP7. *Oncogene* **2018**, *38*, 2844–2859. [CrossRef]
40. Chen, J.; Stubbe, J. Bleomycins: Towards better therapeutics. *Nat. Rev. Cancer* **2005**, *5*, 102–112. [CrossRef]
41. Cloos, J.; Temmink, O.; Ceelen, M.; Snel, M.H.; Leemans, C.R.; Braakhuis, B.J. Involvement of cell cycle control in bleomycin-induced mutagen sensitivity. *Environ. Mol. Mutagen.* **2002**, *40*, 79–84. [CrossRef] [PubMed]
42. Wang, K.; Zhang, T.; Lei, Y.; Li, X.; Jiang, J.; Lan, J.; Liu, Y.; Chen, H.; Gao, W.; Xie, N.; et al. Identification of ANXA2 (annexin A2) as a specific bleomycin target to induce pulmonary fibrosis by impeding TFEB-mediated autophagic flux. *Autophagy* **2018**, *14*, 269–282. [CrossRef] [PubMed]
43. Groselj, A.; Krzan, M.; Kosjek, T.; Bosnjak, M.; Sersa, G.; Cemazar, M. Bleomycin pharmacokinetics of bolus bleomycin dose in elderly cancer patients treated with electrochemotherapy. *Cancer Chemother. Pharmacol.* **2016**, *77*, 939–947. [CrossRef] [PubMed]
44. Chen, H.; Liu, C.; Jiang, H.; Gao, Y.; Xu, M.; Wang, J.; Liu, S.; Fu, Y.; Sun, X.; Xu, J.; et al. Regulatory Role of miRNA-375 in Expression of BMP15/GDF9 Receptors and its Effect on Proliferation and Apoptosis of Bovine Cumulus Cells. *Cell. Physiol. Biochem.* **2017**, *41*, 439–450. [CrossRef]
45. Kong, L.; Zhang, Y.; Ye, Z.Q.; Liu, X.Q.; Zhao, S.Q.; Wei, L.; Gao, G. CPC: Assess the protein-coding potential of transcripts using sequence features and support vector machine. *Nucleic Acids Res.* **2007**, *35*, W345–W349. [CrossRef]

46. Sun, L.; Luo, H.; Bu, D.; Zhao, G.; Yu, K.; Zhang, C.; Liu, Y.; Chen, R.; Zhao, Y. Utilizing sequence intrinsic composition to classify protein-coding and long non-coding transcripts. *Nucleic Acids Res.* **2013**, *41*, e166. [CrossRef]
47. Wang, L.; Park, H.J.; Dasari, S.; Wang, S.; Kocher, J.P.; Li, W. CPAT: Coding-Potential Assessment Tool using an alignment-free logistic regression model. *Nucleic Acids Res.* **2013**, *41*, e74. [CrossRef]
48. Finn, R.D.; Bateman, A.; Clements, J.; Coggill, P.; Eberhardt, R.Y.; Eddy, S.R.; Heger, A.; Hetherington, K.; Holm, L.; Mistry, J.; et al. Pfam: The protein families database. *Nucleic Acids Res.* **2014**, *42*, D222–D230. [CrossRef]
49. Chen, L.; Wang, C.; Sun, H.; Wang, J.; Liang, Y.; Wang, Y.; Wong, G. The bioinformatics toolbox for circRNA discovery and analysis. *Brief. Bioinform.* **2021**, *22*, 1706–1728. [CrossRef]
50. Langmead, B.; Trapnell, C.; Pop, M.; Salzberg, S.L. Ultrafast and memory-efficient alignment of short DNA sequences to the human genome. *Genome Biol.* **2009**, *10*, R25. [CrossRef]
51. Zhang, Z.; Jiang, L.; Wang, J.; Gu, P.; Chen, M. MTide: An integrated tool for the identification of miRNA-target interaction in plants. *Bioinformatics* **2015**, *31*, 290–291. [CrossRef]
52. Love, M.I.; Huber, W.; Anders, S. Moderated estimation of fold change and dispersion for RNA-seq data with DESeq2. *Genome Biol.* **2014**, *15*, 550. [CrossRef] [PubMed]
53. Huang, W.; Sherman, B.T.; Lempicki, R.A. Systematic and integrative analysis of large gene lists using DAVID bioinformatics resources. *Nat. Protoc.* **2009**, *4*, 44–57. [CrossRef] [PubMed]
54. Huang, W.; Sherman, B.T.; Lempicki, R.A. Bioinformatics enrichment tools: Paths toward the comprehensive functional analysis of large gene lists. *Nucleic Acids Res.* **2009**, *37*, e16. [CrossRef] [PubMed]
55. Mao, X.; Cai, T.; Olyarchuk, J.G.; Wei, L. Automated genome annotation and pathway identification using the KEGG Orthology (KO) as a controlled vocabulary. *Bioinformatics* **2005**, *21*, 3787–3793. [CrossRef]
56. Shannon, P.; Markiel, A.; Ozier, O.; Baliga, N.S.; Wang, J.T.; Ramage, D.; Amin, N.; Schwikowski, B.; Ideker, T. Cytoscape: A software environment for integrated models of biomolecular interaction networks. *Genome Res.* **2003**, *13*, 2498–2504. [CrossRef] [PubMed]
57. Wei, S.; Xue, J.; Sun, B.; Zou, Z.; Chen, C.; Liu, Q.; Zhang, A. miR-145 via targeting ERCC2 is involved in arsenite-induced DNA damage in human hepatic cells. *Toxicol. Lett.* **2018**, *295*, 220–228. [CrossRef]
58. Mocavini, I.; Pippa, S.; Licursi, V.; Paci, P.; Trisciuoglio, D.; Mannironi, C.; Presutti, C.; Negri, R. JARID1B expression and its function in DNA damage repair are tightly regulated by miRNAs in breast cancer. *Cancer Sci.* **2018**, *110*, 1232–1243. [CrossRef]
59. Zhu, H.; Zhang, L.; Xu, J.; Zhu, C.; Zhao, H.; Zhu, Y.; Lv, G. AntogomiR-451 protects human gastric epithelial cells from ethanol via activating AMPK signaling. *Biochem. Biophys. Res. Commun.* **2018**, *497*, 339–346. [CrossRef]
60. Banath, J.P.; Olive, P.L. Expression of phosphorylated histone H2AX as a surrogate of cell killing by drugs that create DNA double-strand breaks. *Cancer Res.* **2003**, *63*, 4347–4350.
61. Stucki, M.; Jackson, S.P. gammaH2AX and MDC1: Anchoring the DNA-damage-response machinery to broken chromosomes. *DNA Repair* **2006**, *5*, 534–543. [CrossRef] [PubMed]
62. Oosthuysen, W.F.; Mueller, T.; Dittrich, M.T.; Schubert-Unkmeir, A. Neisseria meningitidis causes cell cycle arrest of human brain microvascular endothelial cells at S phase via p21 and cyclin G2. *Cell. Microbiol.* **2016**, *18*, 46–65. [CrossRef] [PubMed]
63. Mahaddalkar, T.; Mehta, S.; Cheriyamundath, S.; Muthurajan, H.; Lopus, M. Tryptone-stabilized gold nanoparticles target tubulin and inhibit cell viability by inducing an unusual form of cell cycle arrest. *Exp. Cell Res.* **2017**, *360*, 163–170. [CrossRef]
64. Stillman, B. Cell cycle control of DNA replication. *Science* **1996**, *274*, 1659–1664. [CrossRef]
65. Shaltiel, I.A.; Krenning, L.; Bruinsma, W.; Medema, R.H. The same, only different—DNA damage checkpoints and their reversal throughout the cell cycle. *J. Cell Sci.* **2015**, *128*, 607–620. [CrossRef] [PubMed]
66. Huang, M.; Miao, Z.H.; Zhu, H.; Cai, Y.J.; Lu, W.; Ding, J. Chk1 and Chk2 are differentially involved in homologous recombination repair and cell cycle arrest in response to DNA double-strand breaks induced by camptothecins. *Mol. Cancer Ther.* **2008**, *7*, 1440–1449. [CrossRef]
67. Sun, Y.; Xu, Y.; Roy, K.; Price, B.D. DNA damage-induced acetylation of lysine 3016 of ATM activates ATM kinase activity. *Mol. Cell. Biol.* **2007**, *27*, 8502–8509. [CrossRef]
68. Tang, S.; Hou, Y.; Zhang, H.; Tu, G.; Yang, L.; Sun, Y.; Lang, L.; Tang, X.; Du, Y.E.; Zhou, M.; et al. Oxidized ATM promotes abnormal proliferation of breast CAFs through maintaining intracellular redox homeostasis and activating the PI3K-AKT, MEK-ERK, and Wnt-beta-catenin signaling pathways. *Cell Cycle* **2015**, *14*, 1908–1924. [CrossRef]
69. Schotz, U.; Balzer, V.; Brandt, F.W.; Ziemann, F.; Subtil, F.S.B.; Rieckmann, T.; Kocher, S.; Engenhart-Cabillic, R.; Dikomey, E.; Wittig, A.; et al. Dual PI3K/mTOR Inhibitor NVP-BEZ235 Enhances Radiosensitivity of Head and Neck Squamous Cell Carcinoma (HNSCC) Cell Lines Due to Suppressed Double-Strand Break (DSB) Repair by Non-Homologous End Joining. *Cancers* **2020**, *12*, 467. [CrossRef]
70. Beermann, J.; Piccoli, M.T.; Viereck, J.; Thum, T. Non-coding RNAs in Development and Disease: Background, Mechanisms, and Therapeutic Approaches. *Physiol. Rev.* **2016**, *96*, 1297–1325. [CrossRef]
71. Olivieri, F.; Albertini, M.C.; Orciani, M.; Ceka, A.; Cricca, M.; Procopio, A.D.; Bonafe, M. DNA damage response (DDR) and senescence: Shuttled inflamma-miRNAs on the stage of inflamm-aging. *Oncotarget* **2015**, *6*, 35509–35521. [CrossRef] [PubMed]
72. Schmitt, A.M.; Garcia, J.T.; Hung, T.; Flynn, R.A.; Shen, Y.; Qu, K.; Payumo, A.Y.; Peres-da-Silva, A.; Broz, D.K.; Baum, R.; et al. An inducible long noncoding RNA amplifies DNA damage signaling. *Nat. Genet.* **2016**, *48*, 1370–1376. [CrossRef] [PubMed]

73. Shihabudeen Haider Ali, M.S.; Cheng, X.; Moran, M.; Haemmig, S.; Naldrett, M.J.; Alvarez, S.; Feinberg, M.W.; Sun, X. LncRNA Meg3 protects endothelial function by regulating the DNA damage response. *Nucleic Acids Res.* **2018**, *47*, 1505–1522. [CrossRef] [PubMed]
74. Galluzzi, L.; Morselli, E.; Vitale, I.; Kepp, O.; Senovilla, L.; Criollo, A.; Servant, N.; Paccard, C.; Hupe, P.; Robert, T.; et al. miR-181a and miR-630 regulate cisplatin-induced cancer cell death. *Cancer Res.* **2010**, *70*, 1793–1803. [CrossRef]
75. He, L.; He, X.; Lim, L.P.; de Stanchina, E.; Xuan, Z.; Liang, Y.; Xue, W.; Zender, L.; Magnus, J.; Ridzon, D.; et al. A microRNA component of the p53 tumour suppressor network. *Nature* **2007**, *447*, 1130–1134. [CrossRef]
76. Park, S.Y.; Lee, J.H.; Ha, M.; Nam, J.W.; Kim, V.N. miR-29 miRNAs activate p53 by targeting p85 alpha and CDC42. *Nat. Struct. Mol. Biol.* **2009**, *16*, 23–29. [CrossRef]
77. Le, M.T.; Teh, C.; Shyh-Chang, N.; Xie, H.; Zhou, B.; Korzh, V.; Lodish, H.F.; Lim, B. MicroRNA-125b is a novel negative regulator of p53. *Genes Dev.* **2009**, *23*, 862–876. [CrossRef]
78. Xie, S.Y.; Li, Y.J.; Wang, P.Y.; Jiao, F.; Zhang, S.; Zhang, W.J. miRNA-regulated expression of oncogenes and tumor suppressor genes in the cisplatin-inhibited growth of K562 cells. *Oncol. Rep.* **2010**, *23*, 1693–1700. [CrossRef]
79. Zhang, Q.; Mady, A.S.A.; Ma, Y.; Ryan, C.; Lawrence, T.S.; Nikolovska-Coleska, Z.; Sun, Y.; Morgan, M.A. The WD40 domain of FBXW7 is a poly(ADP-ribose)-binding domain that mediates the early DNA damage response. *Nucleic Acids Res.* **2019**, *47*, 4039–4053. [CrossRef]
80. Zhang, P.; Cao, L.; Fan, P.; Mei, Y.; Wu, M. LncRNA-MIF, a c-Myc-activated long non-coding RNA, suppresses glycolysis by promoting Fbxw7-mediated c-Myc degradation. *EMBO Rep.* **2016**, *17*, 1204–1220. [CrossRef]
81. Liu, X.; Ma, J.; Xu, F.; Li, L. TINCR suppresses proliferation and invasion through regulating miR-544a/FBXW7 axis in lung cancer. *Biomed. Pharmacother.* **2018**, *99*, 9–17. [CrossRef] [PubMed]
82. Janssens, S.; Tinel, A.; Lippens, S.; Tschopp, J. PIDD mediates NF-kappaB activation in response to DNA damage. *Cell* **2005**, *123*, 1079–1092. [CrossRef] [PubMed]
83. De Gregoriis, G.; Ramos, J.A.; Fernandes, P.V.; Vignal, G.M.; Brianese, R.C.; Carraro, D.M.; Monteiro, A.N.; Struchiner, C.J.; Suarez-Kurtz, G.; Vianna-Jorge, R.; et al. DNA repair genes PAXIP1 and TP53BP1 expression is associated with breast cancer prognosis. *Cancer Biol. Ther.* **2017**, *18*, 439–449. [CrossRef]
84. Inano, S.; Sato, K.; Katsuki, Y.; Kobayashi, W.; Tanaka, H.; Nakajima, K.; Nakada, S.; Miyoshi, H.; Knies, K.; Takaori-Kondo, A.; et al. RFWD3-Mediated Ubiquitination Promotes Timely Removal of Both RPA and RAD51 from DNA Damage Sites to Facilitate Homologous Recombination. *Mol. Cell.* **2017**, *66*, 622–634.e8. [CrossRef]
85. Ricciuti, B.; Recondo, G.; Spurr, L.F.; Li, Y.Y.; Lamberti, G.; Venkatraman, D.; Umeton, R.; Cherniack, A.D.; Nishino, M.; Sholl, L.M.; et al. Impact of DNA Damage Response and Repair (DDR) Gene Mutations on Efficacy of PD-(L)1 Immune Checkpoint Inhibition in Non-Small Cell Lung Cancer. *Clin. Cancer Res.* **2020**, *26*, 4135–4142. [CrossRef]
86. Abbotts, R.; Thompson, N.; Madhusudan, S. DNA repair in cancer: Emerging targets for personalized therapy. *Cancer Manag. Res.* **2014**, *6*, 77–92. [CrossRef]
87. Carlsen, L.; El-Deiry, W.S. Anti-cancer immune responses to DNA damage response inhibitors: Molecular mechanisms and progress toward clinical translation. *Front. Oncol.* **2022**, *12*, 998388. [CrossRef] [PubMed]
88. Ruth, K.S.; Day, F.R.; Hussain, J.; Martinez-Marchal, A.; Aiken, C.E.; Azad, A.; Thompson, D.J.; Knoblochova, L.; Abe, H.; Tarry-Adkins, J.L.; et al. Genetic insights into biological mechanisms governing human ovarian ageing. *Nature* **2021**, *596*, 393–397. [CrossRef]
89. Shaw, A.; Gullerova, M. Home and Away: The Role of Non-Coding RNA in Intracellular and Intercellular DNA Damage Response. *Genes* **2021**, *12*, 1475. [CrossRef] [PubMed]
90. Cerna, K.; Oppelt, J.; Chochola, V.; Musilova, K.; Seda, V.; Pavlasova, G.; Radova, L.; Arigoni, M.; Calogero, R.A.; Benes, V.; et al. MicroRNA miR-34a downregulates FOXP1 during DNA damage response to limit BCR signalling in chronic lymphocytic leukaemia B cells. *Leukemia* **2019**, *33*, 403–414. [CrossRef]
91. Michelini, F.; Pitchiaya, S.; Vitelli, V.; Sharma, S.; Gioia, U.; Pessina, F.; Cabrini, M.; Wang, Y.; Capozzo, I.; Iannelli, F.; et al. Damage-induced lncRNAs control the DNA damage response through interaction with DDRNAs at individual double-strand breaks. *Nat. Cell Biol.* **2017**, *19*, 1400–1411. [CrossRef] [PubMed]
92. Lukas, J.; Altmeyer, M. A lncRNA to repair DNA. *EMBO Rep.* **2015**, *16*, 1413–1414. [CrossRef] [PubMed]
93. Qiu, M.; Zhang, N.; Yao, S.; Zhou, H.; Chen, X.; Jia, Y.; Zhang, H.; Li, X.; Jiang, Y. DNMT3A-mediated high expression of circ_0057504 promotes benzo[a]pyrene-induced DNA damage via the NONO-SFPQ complex in human bronchial epithelial cells. *Environ. Int.* **2022**, *170*, 107627. [CrossRef] [PubMed]
94. Li, P.P.; Li, R.G.; Huang, Y.Q.; Lu, J.P.; Zhang, W.J.; Wang, Z.Y. LncRNA OTUD6B-AS1 promotes paclitaxel resistance in triple negative breast cancer by regulation of miR-26a-5p/MTDH pathway-mediated autophagy and genomic instability. *Aging* **2021**, *13*, 24171–24191. [CrossRef] [PubMed]

genes

MDPI

Article

Sequencing of the Pituitary Transcriptome after GnRH Treatment Uncovers the Involvement of lncRNA-m23b/miR-23b-3p/CAMK2D in FSH Synthesis and Secretion

Tian Wang [1,†], Guokun Zhao [1,†], Song Yu [1], Yi Zheng [1], Haixiang Guo [1], Haoqi Wang [1], Peisen Zhao [1], Wenyin Xie [1], Wenzhi Ren [2,*] and Bao Yuan [1,*]

1 Department of Laboratory Animals, College of Animal Sciences, Jilin University, Changchun 130062, China
2 Jilin Provincial Key Laboratory of Animal Model, Jilin University, Changchun 130062, China
* Correspondence: renwz@jlu.edu.cn (W.R.); yuan_bao@jlu.edu.cn (B.Y.); Tel.: +86-431-8783-6562 (W.R.); +86-431-8783-6536 (B.Y.)
† These authors contributed equally to this work.

Abstract: The pituitary gland is a key participant in the hypothalamic–pituitary–gonadal axis, as it secretes a variety of hormones and plays an important role in mammalian reproduction. Gonadotrophin-releasing hormone(GnRH) signaling molecules can bind to GnRH receptors on the surfaces of adenohypophysis gonadotropin cells and regulate the expression of follicle-stimulating hormone(FSH) and luteinizing hormone(LH) through various pathways. An increasing number of studies have shown that noncoding RNAs mediate the regulation of GnRH signaling molecules in the adenohypophysis. However, the expression changes and underlying mechanisms of genes and noncoding RNAs in the adenohypophysis under the action of GnRH remain unclear. In the present study, we performed RNA sequencing (RNA-seq) of the rat adenohypophysis before and after GnRH treatment to identify differentially expressed mRNAs, lncRNAs, and miRNAs. We found 385 mRNAs, 704 lncRNAs, and 20 miRNAs that were significantly differentially expressed in the rat adenohypophysis. Then, we used a software to predict the regulatory roles of lncRNAs as molecular sponges that compete with mRNAs to bind miRNAs, and construct a GnRH-mediated ceRNA regulatory network. Finally, we enriched the differentially expressed mRNAs, lncRNA target genes, and ceRNA regulatory networks to analyze their potential roles. Based on the sequencing results, we verified that GnRH could affect FSH synthesis and secretion by promoting the competitive binding of lncRNA-m23b to miR-23b-3p to regulate the expression of Calcium/Calmodulin Dependent Protein Kinase II Delta(CAMK2D). Our findings provide strong data to support exploration of the physiological processes in the rat adenohypophysis under the action of GnRH. Furthermore, our profile of lncRNA expression in the rat adenohypophysis provides a theoretical basis for research on the roles of lncRNAs in the adenohypophysis.

Keywords: GnRH; rat adenohypophysis cells; lncRNA-m23b; miR-23b-3p; CAMK2D; FSH; animal reproduction

Citation: Wang, T.; Zhao, G.; Yu, S.; Zheng, Y.; Guo, H.; Wang, H.; Zhao, P.; Xie, W.; Ren, W.; Yuan, B. Sequencing of the Pituitary Transcriptome after GnRH Treatment Uncovers the Involvement of lncRNA-m23b/miR-23b-3p/CAMK2D in FSH Synthesis and Secretion. *Genes* **2023**, *14*, 846. https://doi.org/10.3390/genes14040846

Academic Editor: Miloš Macholán

Received: 3 March 2023
Revised: 24 March 2023
Accepted: 28 March 2023
Published: 31 March 2023

1. Introduction

The pituitary gland is an important part of the mammalian endocrine and reproductive systems. It is located in the sella turcica at the base of the skull and includes the anterior, middle, and posterior lobes [1]. The posterior pituitary, also known as the neurohypophysis, mainly stores and secretes vasopressin (VP) and oxytocin (OT), which are synthesized by neurons in the hypothalamus [2]. The anterior and middle pituitary glands together form the adenohypophysis, which secretes thyroid-stimulating hormone (TSH), growth hormone (GH), follicle-stimulating hormone (FSH), luteinizing hormone (LH), prolactin (PRL), and other hormones [3]. The proportions of different cell types in the adenohypophysis are affected by various hormones in the body [4]. However, GH-secreting cells usually account for approximately 50% of cells and are the most abundant cells in the pituitary gland; the remaining four cell types have similar proportions of approximately 10–20% each [5]. The gonadotropins that secrete LH and FSH are the last cells to differentiate among anterior adenohypophysis cells, which reflects the importance of gonadotropins to pituitary function [6]. Gonadotropins include FSH and LH, which consist of a specific β subunit and a common α subunit. *Cga* encodes the α subunit [7], and *Lhb* and *Fshb* encode the β subunit [8]. In addition, studies have shown that gonadotropin-releasing hormone (GnRH) secreted by the hypothalamus can act on the pituitary to regulate gonadotropin synthesis [9,10].

GnRH is a neurohormone that is synthesized and secreted by hypothalamic neurons (GnRH neurons). It acts on pituitary gonadotropins through the vascular portal system and promotes the release of FSH and LH [11]. GnRH can bind to the GnRH receptor (GnRHR), located on the surfaces of adenohypophysis gonadotropins, to induce the *Cga* promoter, *Fshb*, and *Lhb* to promote gene transcription [12–14]. GnRH signaling molecules have been found to regulate the expression of *Fshb* and *Lhb* through the cAMP, CAMK II, and MAPK signaling pathways, which are important for the accurate regulation of GnRH [15–17]. Genes such as CAMK2D, MAPK3, and CAMK2A are also involved in the GnRH signaling pathway [18]. Ca^{2+}/CAMK II plays a key role in the transmission of GnRH signals from the plasma membrane to the rat *Fshb* and *Lhb* subunit genes [19]. However, critical validation of how GnRH activates Ca^{2+}/CAMK II and thus affects FSH synthesis and secretion is lacking.

GnRH and its analogs have an important application value in animal production. For example, immunocastration by inoculation with GnRH can replace surgical castration to promote animal development [20]. Treatments with a certain concentration of GnRH can effectively increase the concentration of progesterone in dairy cows, which is conducive to the establishment and maintenance of pregnancy [21]. However, the effects of GnRH administration vary among individuals, and the molecular mechanism of action of GnRH has not been completely clarified. The rat is a common experimental animal model for research on the effect of GnRH on the pituitary gland. Various studies have used rat experimental animal models to explore the expression changes of gonadotropins in the adenohypophysis after a GnRH treatment [22–24].

Previous studies have reported that GnRH can regulate the expression of noncoding RNAs (ncRNAs) in mammals. For example, in rats with liver fibrosis, GnRH treatment can significantly upregulate the expression of miR-200b and liver fibrosis markers in the liver [25]. GnRH inhibits miR-7 expression in the mouse hypothalamic–pituitary–ovarian axis [26]. Related studies have reported that GnRH treatment can alter microRNA (miRNA) expression profiles in mouse LβT2 gonadotropic cells and porcine anterior pituitary cells [9,27]. Our previous study also revealed that after GnRH treatment, the levels of circRNAs in the rat adenohypophysis are changed, and 14 circRNAs are significantly differentially expressed [22]. NcRNAs in the pituitary can play important roles. CircAkap17b competitively binds miR-7 family members to regulate FSH synthesis and secretion in rat adenohypophysis cells by promoting the expression of *Fshb* [28]. Long ncRNAs (LncRNAs) are a type of RNA that can participate in various biological processes by affecting gene transcription, protein activity, and epigenetic modification [29]. Our previous study showed that lncRNAs differ between immature and mature rat adenohypophyses [30]. In addition, we found that lncRNA-m433s1 can target miR-433 binding to regulate FSH secretion [31] and lncRNA-m18as1 can bind competitively to miR-18a-5p to regulate follicle-stimulating hormone secretion through the Smad2/3 pathway [32]. However, whether GnRH treatment can change the expression of lncRNAs in the rat adenohypophysis and the downstream regulatory mechanisms has not yet been reported. Additionally, the functional resolution of lncRNAs lacks relevant validation, and in addition, there is a lack of data supporting the regulatory network regarding the integrated regulatory relationships of lncRNAs, miRNAs, and mRNAs before and after GnRH treatment. In addition, it remains unclear whether noncoding RNAs are involved in the activation of Ca^{2+}/CAMK II by GnRH to affect FSH synthesis and secretion.

In this study, we performed RNA sequencing of the rat adenohypophysis before and after GnRH treatment and detected the differential expression of mRNA, miRNA, and lncRNA in the adenohypophysis. Meanwhile, using bioinformatics software, we predicted the ceRNA regulatory network and experimentally elucidated the molecular mechanism of lncRNA regulation of FSH secretion. Our results provide data supporting the molecular mechanisms in the regulation of pituitary function by GnRH and complement the endocrine mechanisms of GnRH regulation of pituitary FSH secretion at the epigenetic level.

2. Materials and Methods

2.1. Ethics Statement

The experiments were carried out in strict accordance with the Guidelines for the Care and Use of Laboratory Animals of Jilin University. The animal experiments in this study were carried out in the Animal Experiment Center of Jilin University, and the experimental protocol of this study was approved by the Institutional Animal Care and Use Committee of Jilin University with the license number SY201809010.

2.2. Animals and GnRH Treatment

The animals used in the experiments were 8-week-old male Sprague Dawley (SD) rats, provided by the laboratory Animal Center of Jilin University. The grouping strategy used for this study has been described previously [22]. Briefly, 12 SD rats were equally divided into four groups: two control groups and two experimental groups. The rats in the experimental group were injected intra-abdominally twice with 0.2 mL of 1 μg/mL GnRH solution, with an interval of 2 h between injections. The rats in the control group were injected with the same dose of normal saline. All SD rats were euthanized with carbon dioxide ten minutes after the second injection, and the isolated adenohypophyses were placed in tubes containing TRIzol and stored at −80 °C until RNA extraction.

2.3. RNA Extraction

The rat adenohypophyses were removed from the −80 °C freezer, and total RNA was extracted by the TRIzol method. Then, a NanoDrop 2000 spectrophotometer (Thermo Fisher Scientific, Wilmington, DE, USA) was used to measure the concentration and purity of the total RNA in the rat adenohypophysis tissue. Agarose gel electrophoresis was performed to ensure that the total RNA from the rat adenohypophysis tissue was not contaminated. Finally, we used an RNA Nano 6000 Assay Kit and an Agilent Bioanalyzer 2100 system (Agilent Technologies, CA, USA) to detect whether the RNA was intact.

2.4. Library Construction

We set the GnRH-treated rats as the experimental group (L1, L2, S1, S2) and the saline-treated rats as the control group (L3, L4, S3, S4). Three adenohypophyses were used to construct two types of anterior adenohypophysis pools for each sample.

To construct cDNA libraries of lncRNAs and mRNAs, the ribosomal RNA (rRNA) in samples was removed using a Ribo-Zero kit (Epicenter, Madison, WI, USA), and then random fragmentation of the rRNA-depleted RNA was performed with fragmentation buffer. After the rRNA in the sample was removed, first-strand cDNA was synthesized using reverse transcriptase and random primers using the treated RNA as a template. Then, RNase H and DNA polymerase were added to the system to synthesize the second-strand cDNA. The DNA fragments were ligated to NEBNext Adapters after 3′ end adenylation. AMPure XP Beads (Beckman Coulter, Beverly, MA, USA) were used to purify the library fragments to select insertion fragments of 150~200 bp. Then, 3 μL of USER Enzyme (Beckman Coulter, Beverly, MA, USA) was used to process the cDNA to complete the ligation at 37 °C for 15 min. Finally, PCR was used to obtain the cDNA library.

A miRNA small RNA library was constructed with an NEBNext® Small RNA Library Prep Set for Illumina (NEB, USA). Briefly, the 3′SR Adapter was first ligated; then, to prevent the formation of adapter dimers, we used SR-RT primers to bind the remaining 3′SR Adapter. Finally, the 5′ adapter was connected, and the first-strand cDNA was synthesized through reverse transcription. PCR was used to amplify the first-strand cDNA, and fragment screening was completed via polyacrylamide gel electrophoresis (PAGE).

BioMarker Technologies (Beijing, China) used the Illumina HiSeq X Ten platform to sequence our constructed libraries.

2.5. Sequencing, Clustering, Quality Control, and Inter-Sample Correlation Comparison

For lncRNAs and mRNAs, a TruSeq PE Cluster Kit v3-cBot-HS (Illumina, San Diego, CA, USA) was used to cluster the index-coding samples on the cBot Cluster Generation System. After clustering, the library preparation was sequenced on the Illumina HiSeq platform, and paired-end reads were generated.

MiRNAs were clustered on index-encoded samples using the TruSeq PE Cluster Kit v4-cBot-HS (Illumina) on the cBot Cluster Generation System. After cluster generation, library preparations were sequenced on the Illumina platform, and single-end reads were generated.

Quality control of the sequencing results was conducted. We first processed the raw data (raw reads) with a Perl script. In this process, clean reads are obtained by deleting low-quality data from the raw reads. At the same time, we calculated the Q30, Q20, and GC content of the clean reads based on the sequencing results [33]. The downstream analyses of the sequencing results were all performed on high-quality clean reads. For miRNA, it was necessary to screen for clean reads by deleting sequences shorter than 15 nt or longer than 35 nt.

Pearson's correlation coefficient (r) was calculated to evaluate the correlations of biological replicates. The closer the correlation coefficient is to 1, the higher the similarity of expression patterns between samples.

2.6. Identification of lncRNAs, mRNAs, and miRNAs

We identified lncRNAs, mRNAs, and miRNAs through several methods using the rat reference genome (Rnor_6.0) to complete the alignment and analysis of the sequencing results.

For lncRNAs and mRNAs, we used StringTie [34] for transcriptome assembly, followed by the gff compare program for annotation of the assembled transcripts. By combining several computational tools, such as CPC2, CNCI, Pfam, and CPAT [35–38], we sorted the ncRNA candidates from the unknown transcripts and selected transcripts with lengths greater than 200 nt and with more than two exons as lncRNAs.

For miRNA, Bowtie [39] software was used to align the clean reads with reads in four databases: Silva, GtRNAdb, Repbase, and Rfam. The unannotated reads containing miRNAs were obtained. Then, Bowtie software was used to compare the unannotated reads and the abovementioned rat reference genome to obtain information on the positions of the reads in the rat genome. After the alignment was complete, we aligned the reads with miRBase to identify known miRNAs and finally used Randfold software to predict new miRNAs from the sequences of unknown miRNAs.

2.7. Differential Expression Analysis

The expression levels of mRNAs and lncRNAs in the sequencing results were calculated based on the fragments per kilobase of transcript per million mapped reads (FPKM) index, and then the DESeq [40] R package was used to analyze the differential expression of genes between the experimental group and the control group. To prevent false-positive results, we used the Benjamini and Hochberg method to adjust the resulting p values. Our criteria for screening differentially expressed mRNAs/lncRNAs were a $|\log2(\text{fold change, FC})| \geq 1$ and a p value ≤ 0.01.

For miRNAs, we used the transcripts per million (TPM) algorithm [41] for normalization, followed by the DESeq2 [42] R package to identify differentially expressed miRNAs between the experimental and control groups. Again, the resulting p values were adjusted using the Benjamini and Hochberg method. Our criteria for identifying differentially expressed miRNAs were a $|\log2(FC)| \geq 0.5$ and a p value ≤ 0.05.

2.8. Target Gene Prediction

For lncRNA target gene prediction, we defined adjacent genes within 100 kb of lncRNAs as lncRNA cis-target genes, and we defined genes complementary to lncRNA sequences as lncRNA trans-target genes.

To investigate whether lncRNAs can affect gene expression by binding to specific miRNAs, we used the miRanda (animal), RNAhybrid, and TargetScan databases to predict the miRNAs targeting lncRNAs and mRNAs. We used the lncRNA, mRNA, and miRNA base sequence files from our RNA-seq results as input. Finally, we selected the differentially expressed lncRNAs, mRNAs, and miRNAs to construct a ceRNA regulatory network.

2.9. Enrichment Analysis

The differentially expressed genes, lncRNA cis-target genes, lncRNA trans-target genes, and miRNA target genes in our samples were used for enrichment analysis. ClusterProfiler was used to determine the enriched terms in the biological process, molecular function, and cellular component categories. The enrichment analysis employed a hypergeometric test approach to find GO entries that were significantly enriched for the differentially expressed compared to the whole genomic background. GO cluster analysis enabled us to investigate the clusters of biological functions of the differentially expressed genes. The Kyoto Encyclopedia of Genes and Genomes (KEGG) database was used to further interpret the functions of genes by performing pathway annotation.

2.10. Cell Culture

Rat adenohypophysis cells were isolated and digested from adenohypophysis tissue that was obtained from 8-week-old male SD rats. D/F12 (DMEM/F12, HyClone, Logan, UT, USA), containing 10% fetal bovine serum (FBS, Gibco, New York, NY, USA), as well as 1% penicillin and streptomycin, was used to culture rat adenohypophysis cells. Experiments were performed in a sterile environment, and adenohypophysis cells were cultured in a 5% CO_2 incubator at 37 °C. The detailed methods for rat primary cell culture were consistent with those described in previous studies [43].

2.11. Cell Transfection

Cells were inoculated into 12-well plates at a cell count of 3×10^5 cells per well, and the cell density after inoculation was 70%. Cells were cultured in DMEM/F12 containing 10% fetal bovine serum (FBS) and 1% penicillin and streptomycin (Gibco) to reach 50% cell density. First, we added buffer, reagents, mimics, and inhibitors to the centrifuge tubes. Then, we briefly centrifuged the samples using a vortex shaker to ensure homogeneous mixing. Next, we transfected the mixture into a 12-well plate filled with adenohypophysis cell suspension. Finally, we placed the 12-well plates into a 5% CO_2 incubator to provide sufficient time for the reaction. The siRNA, mimic, and inhibitor for the transfection process were obtained from Ribo (Guangzhou, China), and the transfection process was carried out strictly according to the protocol recommended by Ribo.

2.12. Real-Time Quantitative PCR Assay

We obtained cDNA by reverse transcription using a FastQuant RT Kit (with gDNase, Tiangen, Beijing, China). Next, we detected the mRNA expression of related genes by RT-qPCR using MonAmpTM ChemoHS qPCR Mix (Mona Bio, Wuhan, China). All primers were designed by the NCBI (Table S19). The reference gene for mRNA was GHPDH. The reference gene for miRNA was U6.

2.13. Dual Luciferase Reporter Gene Assay

The dual luciferase reporter used the pmiR-RB-REPORT™ dual luciferase reporter vector. The reporter fluorophore of the vector was the sea kidney luciferase gene (hRluc). The internal reference was the firefly luciferase gene (hluc) and clones the target gene downstream of the hRluc gene. Since the predicted obtained miRNAs may target the target gene, miRNA mimics were co-transfected with the constructed vectors in cells, and the interaction between miRNAs and target genes was verified by the downregulation of the relative fluorescence values of reporter genes. The plasmids used in the transfection process were obtained from Ribo (Guangzhou, China), and the transfection process was performed strictly according to the protocol recommended by Ribo.

2.14. Western Blot Analysis

After we lysed the cells, the concentration of protein samples was first determined. Next, after separating the proteins by PAGE gel electrophoresis, the proteins were transferred to PVDF membranes (Millipore, Carlsbad, CA, USA), and the PVDF membranes were blocked at room temperature for 15 min. Then, the PVDF membranes were incubated with primary antibodies. The next day, the membrane was washed three times using TBST and incubated with secondary antibodies for 1 h. After washing again, the membrane was developed on a Tanon 5200 Multifunctional System (Tanon, Shanghai, China). The antibodies used in this experiment were anti-GAPDH (1:1000, Cell Signaling Technology, Shanghai, China, 2118S), anti-CAMK2D (1:1000, Abcam, Cambridge, MA, USA, ab181052), and anti-rabbit IgG (H+L)-HRP (1:3000; Cell Signaling Technology, Shanghai, China, 7074S). All details are provided in Support File S1.

2.15. ELISA

After collecting the supernatant, we used the Rat FSH ELISA kit (EnzymeLink Biotechnology Co., Ltd., Shanghai, China) to detect changes in the levels of FSH. All steps were performed strictly according to the manufacturer's protocol. The standard curve was drawn according to the concentrations of the standards provided by the kit, and the concentration of each sample was calculated.

2.16. Statistical Analysis

The data are presented as the mean from three independent experiments ± standard deviation of the mean. Significant differences between groups were determined using SPSS 22.0 software and one-way ANOVA, and $p < 0.05$ was considered to indicate statistical significance.

3. Results

3.1. Expression Analysis of mRNAs, lncRNAs, and miRNAs in the Rat Adenohypophysis

We performed quality control on the original sequencing reads to obtain high-quality clean reads. In this study, a total of 4 libraries with a total of 71.51 Gb of clean data were analyzed for lncRNAs (Table S1). For miRNAs, a total of 78.83 M clean reads were obtained from the sequencing results (Table S2). Afterward, we aligned the clean reads of each sample with the rat reference genome (Rattus_norvegicus.Rnor6.0) to analyze the sequence feature information of each sample. We compared the lncRNA sequencing results from rats with the rat reference genome and found that at least 94.77% of the reads mapped to the rat genome (Table S3). For miRNAs, statistical analysis revealed that at least 67.44% of the unannotated reads were miRNAs (Table S4). We used the Pearson correlation coefficient r to evaluate the biological replicate correlations of the samples, and the r2 statistics of the lncRNA, mRNA, and miRNA samples are shown in Supporting Tables S5–S7.

We performed RNA-seq to examine the expression of lncRNAs, mRNAs, and miRNAs in the adenohypophyses of rats before and after GnRH treatment. We found that a total of 23,742 lncRNAs were expressed in the rat adenohypophysis, including 1849 antisense lncRNAs (8%), 6559 intronic lncRNAs (28%), 12,952 long intergenic ncRNAs (lincRNAs)

(55%), and 2382 sense lncRNAs (10%) (Figure 1A). The lncRNA lengths were distributed between 400 and 3000+ nt, but most lncRNAs were distributed in the 3000+ nt range, while the others were mainly distributed in the 400–1000 nt range (Figure 1B). The lncRNAs were mainly distributed on chromosomes 1–8 (Figure 1C). We also found that these lncRNAs were expressed in all chromosomes of the rat genome but were mainly distributed on chromosomes 1–10 (Figure 1D). Finally, we mapped the sequenced lncRNAs into the genome and drew a Circos map (Figure 1E).

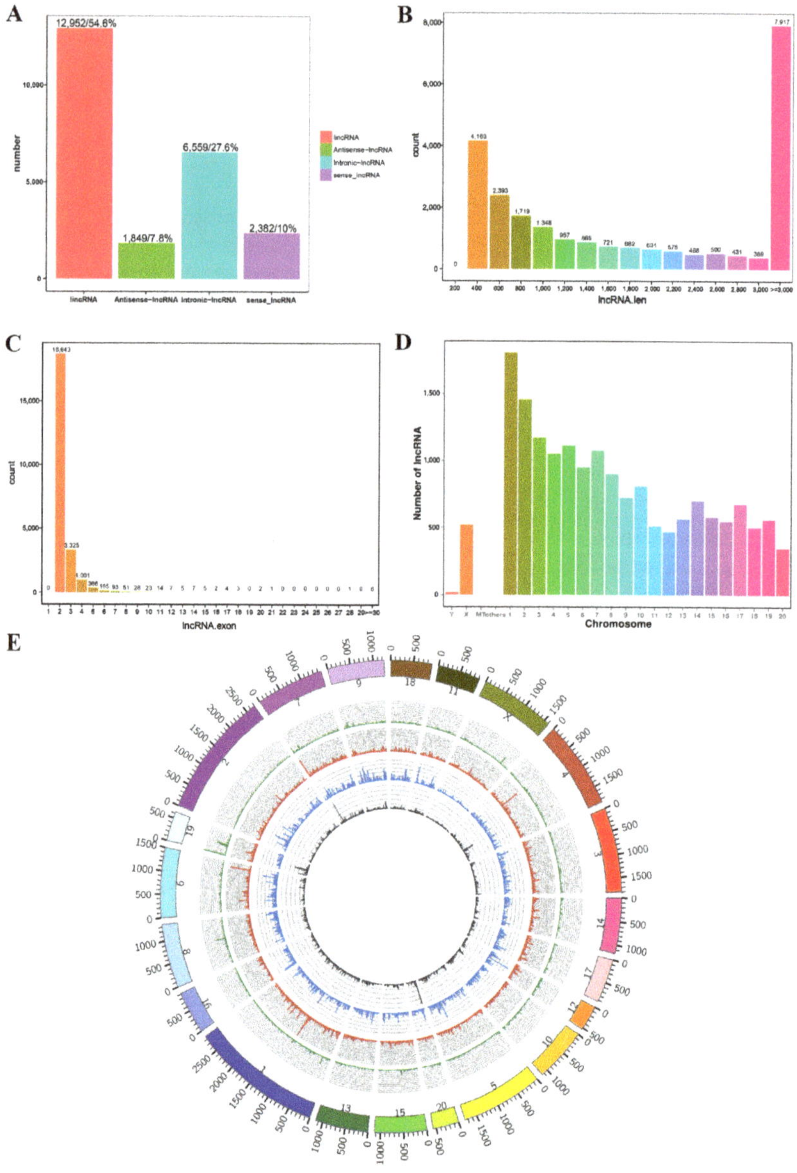

Figure 1. Results of lncRNA sequencing in the adenohypophysis of rats after GnRH treatment. (**A**) LncRNA types and numbers. (**B**) Statistical analysis of LncRNA lengths. (**C**) Statistical analysis of LncRNA exon number. (**D**) Locational distribution of lncRNAs on rat chromosomes. (**E**) Circos plot containing lncRNA data.

For mRNAs, we found that a total of 27,367 genes were expressed in the rat adenohypophysis. Then, based on the alignment results we discovered a total of 9638 new genes, of which 6084 were functionally annotated (Table S8). The mRNA lengths were distributed between 200 and 3000+ nt, but most lncRNAs were distributed in the 400–800 nt range, while the others were mainly distributed in the 3000+ nt range (Figure 2A). These mRNAs were distributed on all chromosomes in the rat genome, but were most highly expressed on chromosome 1 (Figure 2B). Most mRNAs contained exons clustered between 1 and 2 (Figure 2C). Upon comparing the sequenced lncRNAs and mRNAs, we found that the number of alternatively spliced isoforms of mRNAs was higher than the overall number of lncRNAs (Figure 2D). The boxplots clearly show that the expression levels of mRNA were slightly lower than those of lncRNA transcripts (Figure 2E).

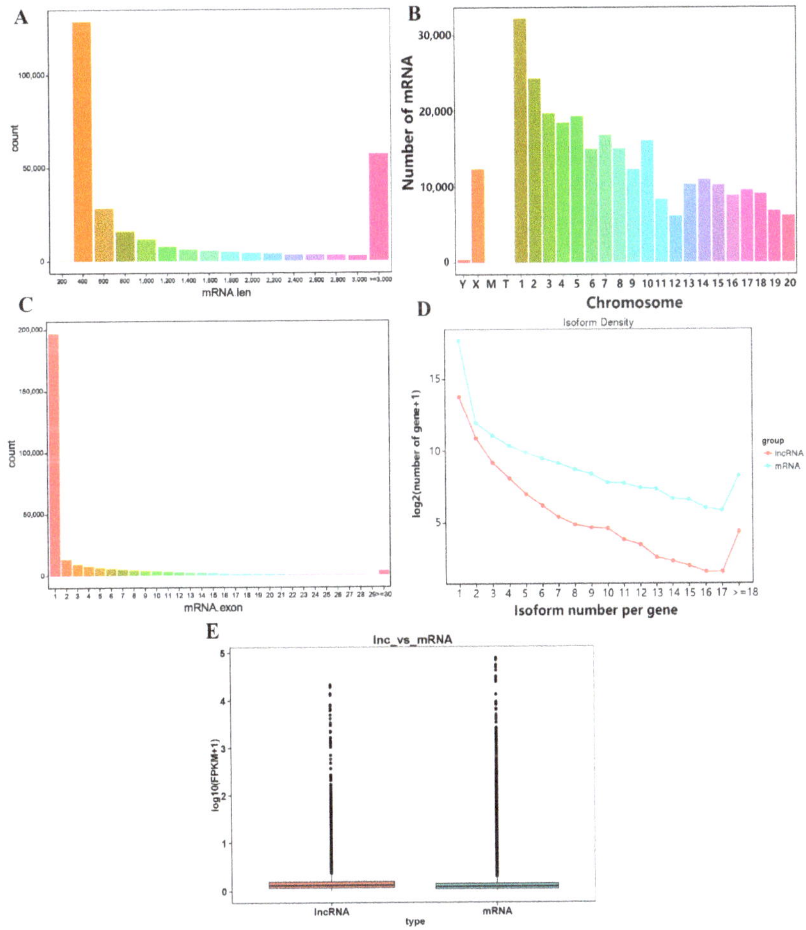

Figure 2. Results of mRNA sequencing in the adenohypophysis of rats after GnRH treatment. (**A**) Statistical analysis of mRNA length. (**B**) Location distribution of mRNA on rat chromosomes. (**C**) Statistical analysis of mRNA exon numbers. (**D**) mRNA/lncRNA alternatively spliced isoform levels. (**E**) Comparison of mRNA/lncRNA expression levels.

A total of 742 miRNAs were also detected in the RNA-seq results, including 395 known mRNAs and 347 newly predicted miRNAs (Table S9). The identified known miRNAs and new miRNAs as well as the overall miRNA length were mainly concentrated in the range of 20 nt to 24 nt, with 22 nt predominant (Figure 3A,B). Among the detected miRNAs, a small number of miRNAs were expressed at high levels, and the expression of most of the miRNAs was still relatively low (Figure 3C). In addition, the correlation coefficients between all four groups of samples were above 0.93, which implies similar expression trends and high replication correlations between the samples (Figure 3D).

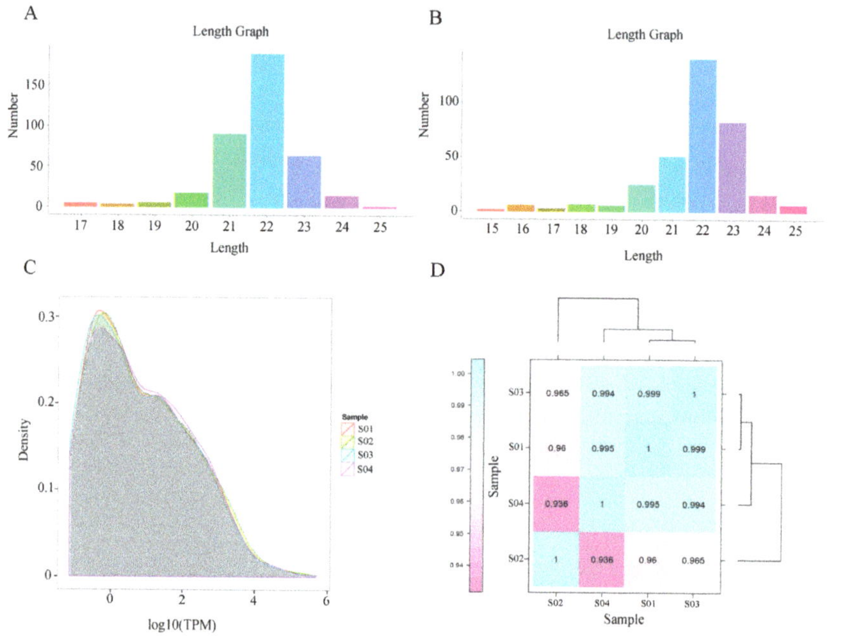

Figure 3. Results of miRNA sequencing in the adenohypophysis of rats after GnRH treatment. (**A**,**B**) Statistical analysis of miRNA length. (**C**) Overall distribution of miRNA expression. The curves of different colors in the figure represent different samples, the horizontal coordinates of the points on the curves indicate the logarithmic values of the TPM of the corresponding samples, and the vertical coordinates of the points indicate the probability density. (**D**) Sample correlation relationship chart. Different colors in the graph represent different correlation coefficient values. The horizontal and vertical coordinates represent different samples.

3.2. Analysis of Differential Gene Expression before and after GnRH Treatment

After statistical analysis of the expression of all lncRNAs, mRNAs, and miRNAs in the samples, excluding RNAs that were expressed individually in only a single sample, DESeq was used to detect differential expressions of RNAs after GnRH treatment. For detection of significantly differentially expressed lncRNAs, a p value ≤ 0.01 and $|\log2(FC)| \geq 1$ were used as the screening criteria (Figure 4A). Ultimately, we obtained 704 significantly differentially expressed lncRNAs, of which 366 were significantly upregulated after GnRH treatment and 338 were significantly downregulated after GnRH treatment. (Figure 4B and Table S10).

We used a p value ≤ 0.01 and a $|\log2(FC)| \geq 1$ as screening criteria for significantly differentially expressed mRNAs (Figure 4C). Ultimately, 385 differentially expressed mRNAs were obtained, of which 180 mRNAs were significantly upregulated after GnRH treatment and 205 mRNAs were significantly downregulated after GnRH treatment (Figure 4D and Table S11).

Likewise, we used DESeq2 to detect differentially expressed miRNAs before and after GnRH treatment. Ultimately, 20 differentially expressed miRNAs were obtained, of which 13 miRNAs were significantly upregulated after GnRH treatment and 7 miRNAs were significantly downregulated after GnRH treatment (Figure 4E and Table S12).

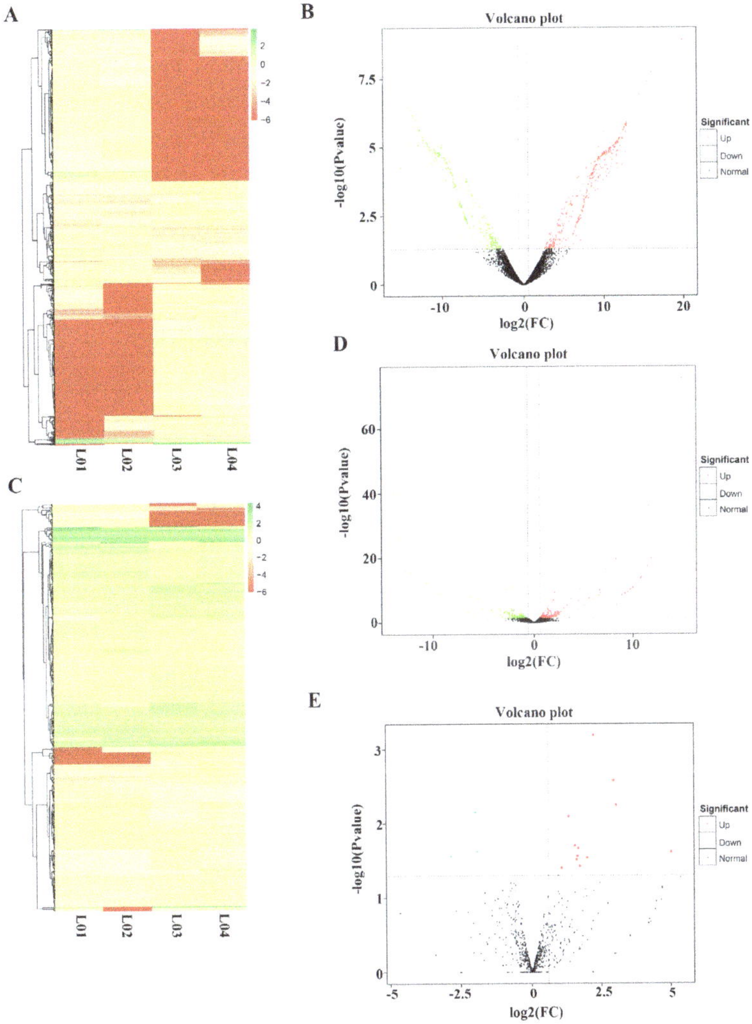

Figure 4. Analysis of differential gene expression before and after GnRH treatment. (**A**) Differential expression analysis of four groups of lncRNAs. (**B**) LncRNA expression volcano plot. (**C**) Differential expression analysis of four groups of mRNAs. (**D**) mRNA expression volcano plot. (**E**) miRNA expression volcano plot.

3.3. Interactions among lncRNAs, mRNAs, and miRNAs

Since mRNAs and lncRNAs contain multiple miRNA response elements (MREs), software can be used to predict miRNAs that may target lncRNAs and mRNAs. We performed targeting relationship prediction for differentially expressed lncRNAs, mRNAs, and miRNAs after GnRH treatment. The prediction results showed that 366 upregulated lncRNAs, 7 downregulated miRNAs, and 190 upregulated mRNAs constituted a GnRH-promoted ceRNA regulatory network (Table S13); in addition, 338 downregulated lncRNAs, 13 upregulated miRNAs, and 203 downregulated mRNAs constituted a GnRH-repressive ceRNA regulatory network (Table S14). Finally, Cytoscape software was used to select the ten lncRNAs with the highest expression levels to generate a circular lncRNA–miRNA–mRNA regulatory network map (Figure 5A,B).

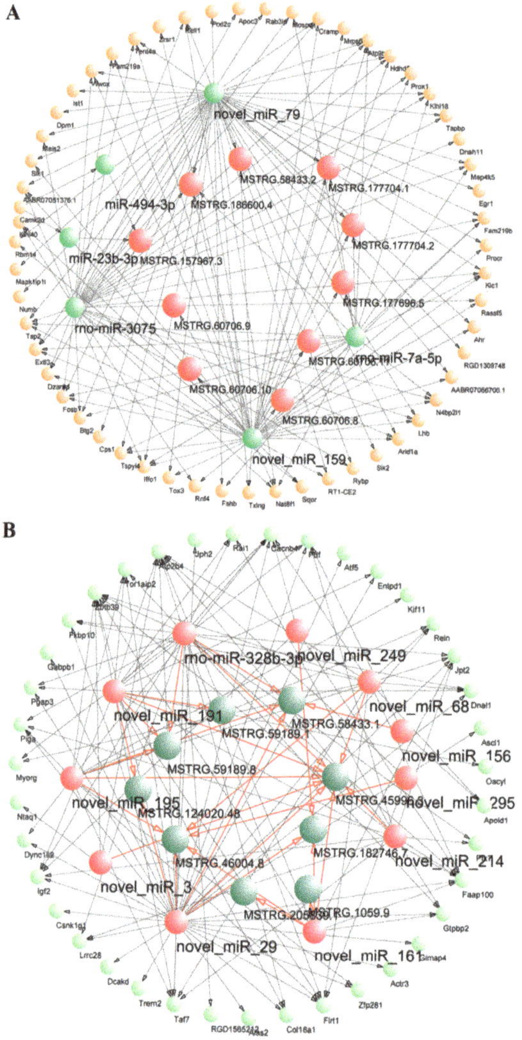

Figure 5. Interactions among lncRNAs, mRNAs, and miRNAs. (**A**) GnRH-promoting ceRNA regulatory network in the rat adenohypophysis. (**B**) GnRH-inhibitory ceRNA regulatory network in the rat adenohypophysis.

The GO analysis results for the GnRH-promoted ceRNA regulatory network showed that the differentially expressed genes were mainly enriched with 51 GO terms, including cellular processes, intracellular anatomical structures, and protein bindings (Table S15). KEGG showed that the ceRNA target genes were mainly enriched in the MAPK signaling pathway, the calcium signaling pathway, the cAMP signaling pathway, and other pathways (Table S16). The GO analysis results for the GnRH-mediated inhibition of the ceRNA regulatory network showed that the differentially expressed genes were mainly enriched with 73 GO terms, including intracellular organelles, bindings, and metabolic processes (Table S17). KEGG showed that the ceRNA target genes were mainly enriched in the GnRH signaling pathway, antigen processing, and presentation (Table S18).

3.4. GnRH Promotes the Target Binding of lncRNA-m23b to miR-23b-3p

Based on the above results, we selected nine lncRNAs as candidates for this study. RT-qPCR results showed that the expression of MSTRG.157967.3, MSTRG.134648.6, MSTRG.199232.3, MSTRG.57325.10, and MSTRG.199839.1 was significantly increased after GnRH treatment (Figure 6A). Based on the ceRNA regulatory network and validation results, lncRNA-MSTRG.157967.3 was finally identified as our candidate for study and renamed lncRNA-m23b. lncRNA-m23b is an intergenic lncRNA on chromosome 4, consisting of two introns (Figure 6B). We predicted its coding ability using the Coding Potential Calculator (CPC) 2.0 (http://cpc2.gao-lab.org, accessed on 3 January 2022), and the prediction showed that lncRNA-m23b has no coding ability (Figure 6C). We further investigated the expression of lncRNAs in different tissues of male rats. The results showed that lncRNA expression was the highest in the adenohypophysis gland, followed by the pineal gland, spleen, and kidneys. The expression was lower in other tissues (Figure 6D). We next identified the expression of lncRNA-m23b in the adenohypophysis of rats at different ages, and the results showed that the expression of lncRNA-m23b gradually increased with age (Figure 6E).

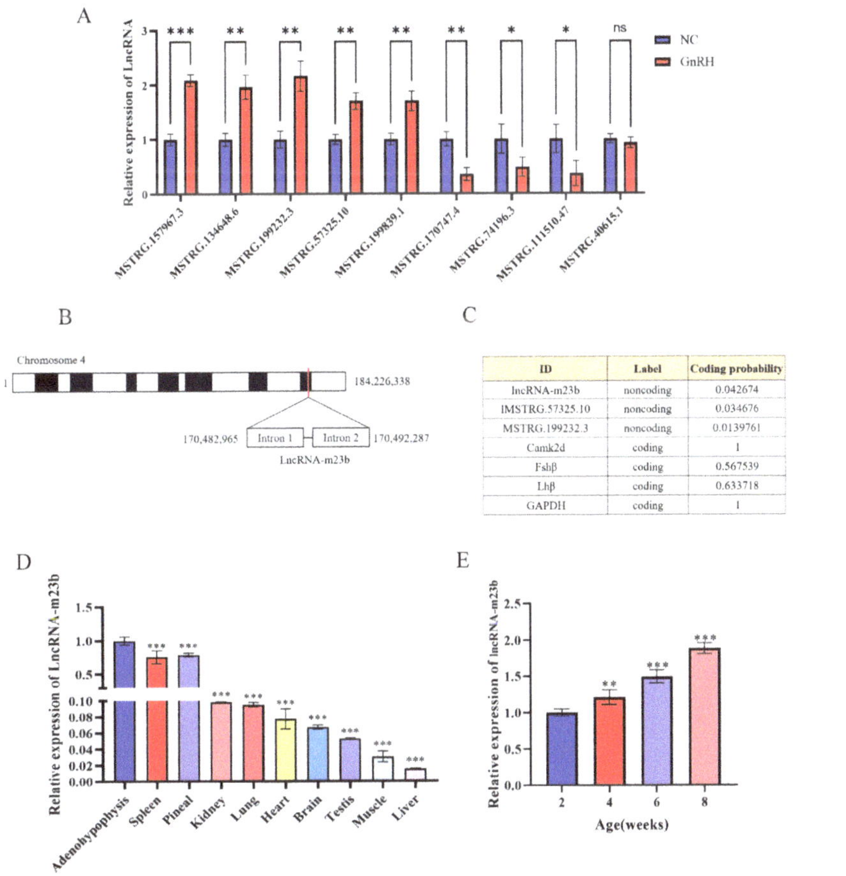

Figure 6. Identification of LncRNA-m23b. (**A**) Detection of differentially expressed lncRNAs by RT-qPCR. (**B**) LncRNA-m23b chromosomal information. (**C**) Prediction of coding potential by CPC 2.0. (**D**) Expression of lncRNA-m23b in different tissues of male rats, lncRNA-m23b expression in the adenohypophysis was used as a control. (**E**) Expression of lncRNA-m23b in the adenohypophysis of rats at different developmental stages. ns, $p > 0.05$; *, $p < 0.05$; **, $p < 0.01$; ***, $p < 0.001$.

To investigate the function of lncRNAs and how they function in cells, we used TargetScan, miRanda, and RNAhybrid to predict miRNAs that may have a targeting relationship with lncRNA-m23b. To determine the relationship between lncRNA-m23b and miRNAs, we performed dual luciferase reporter gene experiments, which showed that co-transfection of pmiR-lncRNA-m23b-WT with rno-miR-23b-3p resulted in a significant decrease of luciferase activity (Figure 7A). Then, based on the base complementation region information predicted by the TargetScan (Figure 7B), we mutated the targeted complementary sequences to construct the reporter gene mutant plasmid pmiR-lncRNA-m23b-MUT. After co-transfection, the relative luciferase activity of the mutant reporter vector was significantly restored compared to that of the wild-type reporter vector (Figure 7C). We next performed siRNA transfection of lncRNA-m23b in rat adenoidal cells, and the results showed that siRNA-lncRNA-m23b-2 had the best transfection effect after siRNA transfection (Figure 7D), so we chose siRNA-lncRNA-m23b-2 for subsequent transfection experiments. lncRNA-m23b knockdown was followed by a significant increase in the miR-23b-3p expression (Figure 7E).

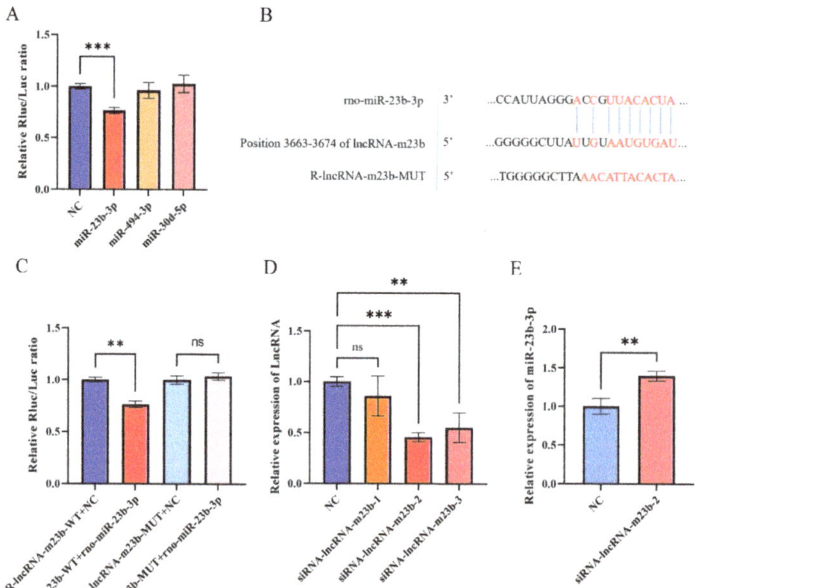

Figure 7. (**A**) Identification of miRNAs that might directly target lncRNA-m23b. (**B**) TargetScan's anticipated miR-23b-3p base complementary pairing sequence with lncRNA-m23b is highlighted in red. (**C**) Changes in relative luciferase activity after the transfection of plasmids with NC/mimics. (**D**) Screening of siRNAs. (**E**) RT-qPCR after knockdown of lncRNA-m23b. ns, $p > 0.05$; **, $p < 0.01$; ***, $p < 0.001$.

3.5. LncRNA-m23b Is Involved in the Regulation of the FSH Synthesis and Secretion by Targeting the Camk2d Expression through Competitive Binding with miR-23b-3p

We used TargetScan to predict the mRNAs that might target miR-23b-3p and found that the binding sites between miR-23b-3p and *Camk2d* were similar to those between miR-23b-3p and lncRNA-m23b (Figure 8A). Next, we co-transfected the constructed pmiR-Camk2d-3′UTR-WT reporter plasmid with the miR-23b-3p mimic and the negative control (NC), and found that transfection of the miR-23b-3p mimic resulted in a significant decrease of luciferase activity (Figure 8B). We mutated the targeted complementary sequence to construct a reporter gene mutant plasmid, pmiR-Camk2d-3′UTR-MUT, and the relative luciferase activity was significantly restored after co-transfection (Figure 8B). Next, we next performed miR-23b-3p mimic/inhibitor transfection in rat adenoidal cells. RT-qPCR results showed that overexpression of miR-23b-3p significantly downregulated *Camk2d* mRNA expression and inhibition of miR-23b-3p upregulated *Camk2d* mRNA expression (Figure 8C), while Western blotting results showed that miR-23b-3p overexpression resulted in a significant decrease in CAMK2D protein expression level, and inhibition of miR-23b-3p expression resulted in a significant increase in CAMK2D protein expression level (Figure 8D).

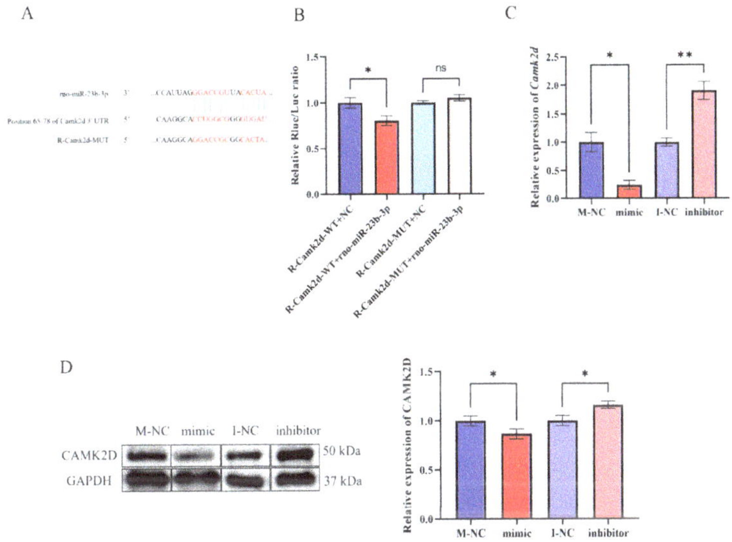

Figure 8. MiR-23b-3p targets Camk2d-3′UTR and regulates its gene expression. (**A**) TargetScan's anticipated miR-23b-3p base complementary pairing sequence with lncRNA-m23b is highlighted in red. (**B**) Changes in relative luciferase activity after the transfection of plasmids with NC/mimic. (**C**,**D**) *Camk2d* mRNA and CAMK2D protein changes after miR-23b-3p mimic/inhibitor transfection. ns, $p > 0.05$; *, $p < 0.05$; **, $p < 0.01$.

In addition, we found that *Camk2d* mRNA expression was significantly decreased after the lncRNA-m23b knockdown (Figure 9A). Western blotting results showed that CAMK2D protein expression was significantly decreased after the lncRNA-m23b knockdown (Figure 9B). We next performed rescue experiments to co-transfect siRNA-lncRNA-m23b-2 with a miR-23b-3p inhibitor. RT-qPCR results showed that inhibition of miR-23b-3p alleviated the inhibitory effect of lncRNA-m23b knockdown on *Camk2d* mRNA (Figure 9C). Western blotting results showed that inhibition of miR-23b-3p attenuated the inhibitory effect of lncRNA-m23b knockdown on the CAMK2D protein (Figure 9D). These results demonstrate that lncRNA-m23b regulates *Camk2d* expression through competitive binding of miR-23b-3p. In addition, we found that knockdown of lncRNA-m23b resulted in a significant decrease in FSH expression (Figure 9A). The ELISA results showed that FSH secretion was significantly decreased after lncRNA-m23b knockdown (Figure 9E). The overexpression of the miR-23b-3p mimic significantly downregulated *Fshb* mRNA expression, and inhibition of miR-23b-3p upregulated *Fshb* mRNA expression (Figure 9F). ELISA results showed that miR-23b-3p overexpression resulted in a significant decrease of FSH secretion, and inhibition of miR-23b-3p expression resulted in a significant increase of FSH secretion (Figure 9G).

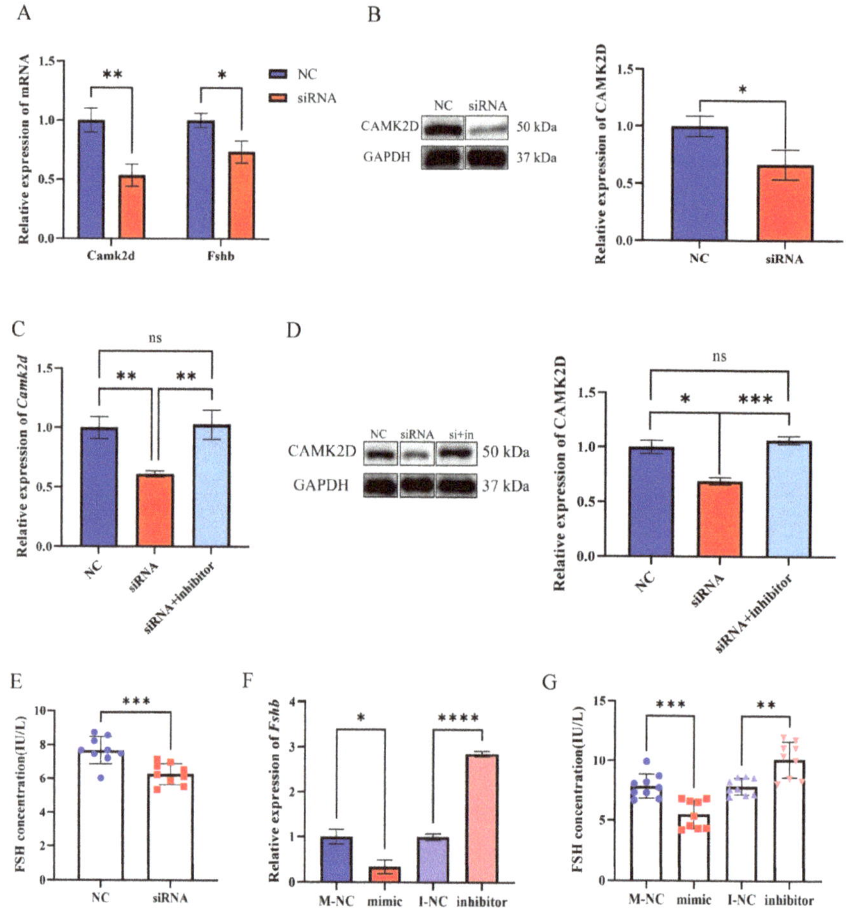

Figure 9. LncRNA-m23b is involved in the regulation of the FSH synthesis and secretion by targeting *Camk2d* expression through competitive binding with miR-23b-3p. (**A**) mRNA changes after lncRNA-m23b knockdown. (**B**) The CAMK2D protein changes after lncRNA-m23b knockdown. (**C**,**D**) *Camk2d* mRNA and CAMK2D protein changes after co-transfection of siRNA-lncRNA-m23b-2 with miR-23b-3p inhibitor. (**E**) FSH secretion changes after lncRNA-m23b knockdown. (**F**,**G**) *Fshb* mRNA and FSH secretion changes after miR-23b-3p mimic/inhibitor transfection. ns, $p > 0.05$; *, $p < 0.05$; **, $p < 0.01$; ***, $p < 0.001$; ****, $p < 0.0001$.

4. Discussion

GnRH is involved in the regulation of the entire hypothalamic–pituitary–gonadal axis and plays an important role in reproductive regulation [44]. Our previous research revealed significant differences in methylation levels in the adenohypophyses of rats after GnRH treatment [24]. In addition, 14 circRNAs were found to be significantly differentially expressed in the rat adenohypophysis [22]. However, the overall regulatory effects of GnRH on the expression of other ncRNAs and mRNAs in the rat adenohypophysis have not been reported. Therefore, in this study, RNA-seq was used to detect the expression changes of lncRNAs, mRNAs, and miRNAs in the adenohypophyses of rats before and after GnRH treatment.

Regarding the differentially expressed mRNAs, RNA-seq showed that genes including *Egr1*, *Map4k5*, *Fosb*, *Myh11*, and *Pkm* were significantly upregulated in the adenohypophyses of rats after GnRH treatment. The regulation of some differentially expressed mRNAs by GnRH has been verified in other studies. For example, different frequencies of GnRH pulses can upregulate the expression of *Egr1* through MAPK 8/9 and MAPK1/3, and early growth factor (EGR1) is crucial for *Lhb* transcription [45]. During GnRH-mediated regulation of gonadotropin expression, MAPK affects gonadotropin transcription and plays an important role in intracellular signal transduction [46]. Other upregulated genes also play important roles in reproductive regulation after GnRH treatment [47–49]. Previous research has shown that GnRH treatment with different pulse frequencies can significantly increase the levels of phospho-CAMK2, which can mediate intracellular calcium signaling pathways [16,50]. In addition, GnRH can regulate the expression of *Fshb* by activating the MAPK signaling pathway [16].

Regarding the differentially expressed miRNAs, the sequencing results showed that rno-miR-3075 and rno-miR-7a-5p were significantly downregulated after GnRH treatment. Moreover, there is a targeting relationship between miR-3075 and *Fosb* in rats [51]. Another study has demonstrated that GnRH treatment can significantly upregulate the expression of *Fosb* [52]. In addition, rno-miR-7a-5p has been shown to regulate the expression of *Fshb* in the rat adenohypophysis [53]. The above findings support the accuracy of our RNA-seq results. However, other studies have found that various miRNAs, including rno-miR-186-5p, rno-miR-433, and rno-miR-488, can regulate the synthesis of *Fshb* [23,31,54], and our sequencing results did not indicate significant differential expression of these miRNAs, possibly because the differences in the expression of these miRNAs did not meet our threshold for significance or because GnRH did not have a regulatory effect on these miRNAs.

LncRNAs are ncRNAs, and increasing evidence indicates that lncRNAs are involved in the mammalian reproductive regulation process [55–57]. In general, lncRNAs can act as molecular sponges to competitively bind miRNAs to regulate gene expression; this is referred to as the intracellular ceRNA regulatory mechanism [58]. An increasing number of studies have shown that ceRNA regulatory networks play important roles in the reproductive process. For example, the lncRNA PATR can competitively bind to miR-101-3p to regulate the expression of ZEB1, thereby promoting epithelial–mesenchymal transition (EMT) [59]. In addition, studies have found that in the rat adenohypophysis, lncRNA-m433s1 can act as a molecular sponge for miR-433 and participate in reproductive processes by regulating the expression of *Fshb* [31]. The results of our enrichment analysis indicated that the lncRNAs and mRNAs that were differentially expressed after GnRH treatment may play important roles in reproductive regulation. Therefore, we screened out the differentially expressed mRNAs, lncRNAs, and miRNAs and constructed a GnRH-promoted ceRNA regulatory network and a GnRH-inhibited ceRNA regulatory network for rat glands based on their regulatory relationships. We performed GO and KEGG enrichment analyses on the genes in the two ceRNA regulatory networks. KEGG enrichment analysis revealed that the target genes in the two ceRNA regulatory networks were enriched in GnRH targeting-related pathways. Together, the above enrichment results indicate that the differentially expressed lncRNAs may be regulated by GnRH and may participate in various biological processes in the rat adenohypophysis through different pathways.

Generally, lncRNAs are able to regulate gene expression by competitively binding to miRNAs in the cytoplasm [60]. Our results showed that GnRH treatment was able to induce an increase in lncRNA-m23b expression levels after GnRH treatment, while our study established the function of lncRNA-m23b as a miRNA sponge. First, the dual-luciferase reporter assay showed that lncRNA-m23b has a targeting relationship with miR-23b-3p, and the 3′UTR of *Camk2d* mRNA also has a targeting binding site for miR-23b-3p. Second, knockdown of lncRNA-m23b leads to an upregulation of miR-23b-3p expression and a repression of *Camk2d* mRNA expression. Moreover, overexpression of miR-23b-3p leads to repression of *Camk2d* mRNA expression, and conversely, inhibition of miR-23b-3p leads to elevated expression of *Camk2d* mRNA. To demonstrate that the inhibition of *Camk2d* mRNA expression was caused by the increase in miR-23b-3p, we performed co-transfection experiments of siRNA-lncRNA-m23b with a miR-23b-3p inhibitor in cells. Expression of

Camk2d was significantly recovered by inhibiting the increase in miR-23b-3p levels that would normally occur in the absence of lncRNA-m23b. Therefore, based on the ceRNA hypothesis, we identified a miR-23b-3p-mediated ceRNA mechanism between lncRNA-m23b and *Camk2d*.

Calcium/calmodulin-dependent kinase II (Ca^{2+}/CAMK II) is an important mediator of calcium signaling in various cell types and is also able to regulate gene expression and hormone secretion [61,62]. It has been reported that GnRH can affect *Fshb* gene expression and thus enhance FSH secretion by activating Ca^{2+}/CAMK II [19]. In the present study, we found that *Camk2d* is a target gene of miR-23b-3p and that lncRNA-m23b regulates *Camk2d* by competitively binding miR-23b-3p. We also examined the secretion of *Fshb* mRNA as well as FSH hormone after transfection. The results showed that the expression trend of *Fshb* was consistent with the expression trend of *Camk2d*. Therefore, we hypothesized that GnRH treatment could cause an increase in the expression level of lncRNA-m23b, which could compete with miR-23b-3p to increase the expression of *Camk2d* mRNA and ultimately promote the synthesis and secretion of FSH. Due to the many pathways involved in the regulation of FSH synthesis and secretion by Ca^{2+}/CAMK II, we did not investigate them in depth in this paper. In future studies, we will further reveal the mechanism of GnRH regulation of FSH synthesis and secretion.

Due to the complexity of the female physiological cycle and the vast differences in the regulation between gonadotropin-releasing hormone and gonadotropins in females at different physiological stages, the most fundamental issue of our research is the synthesis of gonadotropins. This adds to the difficulties of our study. At the same time, it is possible that the results we obtained are only appropriate for a particular physiological period. However, there are fewer factors involved in male than in female animals. Therefore, we chose only male rats for our study to obtain a more generalized result and thus enrich the molecular mechanism of GnRH regulation of FSH secretion. As our research into the mechanisms of FSH synthesis progresses, we will consider exploring more detailed molecular mechanisms of FSH synthesis in animals at different physiological stages in the future.

In conclusion, this is the first study to analyze the overall expression changes of mRNAs/lncRNAs/miRNAs in the adenohypophyses of rats after GnRH treatment. We identified mRNAs, lncRNAs, and miRNAs in the rat adenohypophysis before and after GnRH treatment and constructed a ceRNA regulatory network composed of the differentially expressed mRNAs, lncRNAs, and miRNAs in the rat adenohypophysis. This network may play an important regulatory role in GnRH-mediated regulation of the rat adenohypophysis and reproduction. In addition, we performed enrichment analysis on the differentially expressed mRNAs, lncRNA target genes, and ceRNA regulatory networks to analyze their potential roles in the rat adenohypophysis. We demonstrated for the first time that GnRH can affect FSH synthesis and secretion by regulating the overexpression of lncRNA-m23b and promoting the competitive binding of lncRNA-m23b to miR-23b-3p to regulate *Camk2d* (Figure 10). We hope that our findings will provide a reference for researchers who are interested in adeno-pituitary ncRNAs and aid in further exploration of the regulatory role of GnRH in the pituitary.

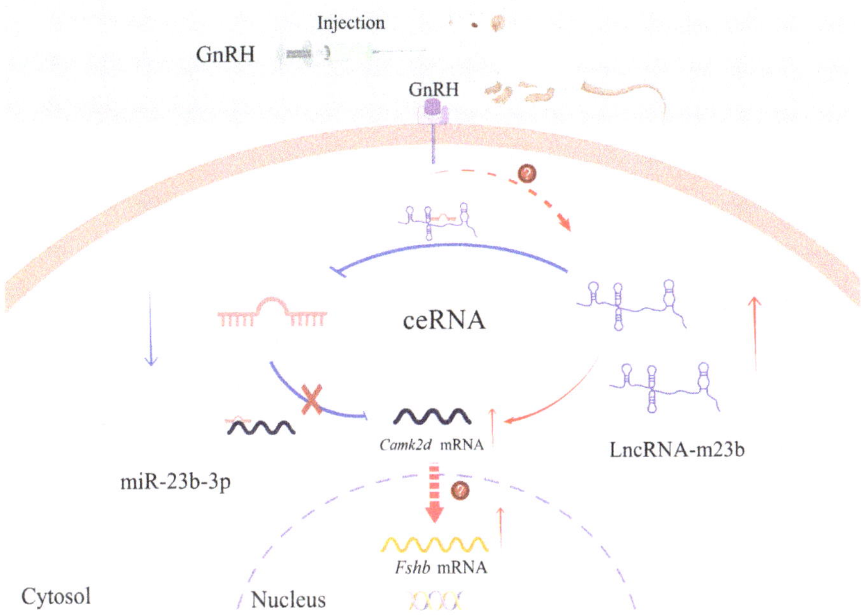

Figure 10. Schematic of the proposed lncRNA-m23b mechanism in rat anterior adenohypophysis cells. The blue arrows on the diagram represent inhibition, and the red arrows represent facilitation. This figure was drawn using Figdraw (https://www.figdraw.com/static/index.html#/, accessed on 2 March 2023). The unique authorization code is UOWUU765ca.

5. Conclusions

In conclusion, this study analyzed the overall lncRNA/mRNA/miRNA expression changes in the rat adenohypophysis after GnRH treatment. It was also confirmed that GnRH can affect the synthesis and secretion of FSH by overexpressing lncRNA-m23b and promoting the competitive binding of lncRNA-m23b to miR-23b-3p, which in turn regulates the overexpression of *Camk2d*. Our study provides a theoretical basis for resolving the regulatory role of GnRH in the adenohypophysis gland.

Supplementary Materials: The following supporting information can be downloaded at: https://www.mdpi.com/article/10.3390/genes14040846/s1, Support File S1: Western blot analysis. Support table: Table S1: Statistical table of LncRNA sequencing data evaluation, Table S2: Statistical table of miRNA sequencing data evaluation, Table S3: Statistical table of LncRNA Clean Data and rat genome comparison results, Table S4: Statistical table of miRNA Clean Data and rat genome comparison results, Table S5: LncRNA sample correlation coefficient, Table S6: mRNA sample correlation coefficient, Table S7: miRNA sample correlation coefficient, Table S8: Annotation of new genes, Table S9: miRNA expression, Table S10: lncRNAs with differential expression, Table S11: mRNAs with differential expression, Table S12: miRNAs with differential expression, Table S13: GnRH-promoting ceRNA regulatory network, Table S14: GnRH-inhibitory ceRNA regulatory network, Table S15: GO annotation of the GnRH-promoting ceRNA regulatory network, Table S16: KEGG annotation of GnRH-promoting ceRNA regulatory network, Table S17: GO annotation of GnRH-inhibitory ceRNA regulatory network, Table S18: KEGG annotation of GnRH-inhibitory ceRNA regulatory network, Table S19: All primers.

Author Contributions: Conceptualization, W.R. and B.Y.; methodology, Y.Z.; software, T.W.; validation, H.G., S.Y. and G.Z.; formal analysis, W.X. and P.Z.; investigation, B.Y.; resources, W.R.; data curation, H.W.; writing—original draft preparation, T.W. and G.Z.; writing—review and editing, B.Y.; visualization, T.W.; supervision, G.Z.; project administration, W.R.; funding acquisition, B.Y. All authors have read and agreed to the published version of the manuscript.

Funding: This research was funded by the National Natural Science Foundation of China (32172729).

Institutional Review Board Statement: We performed animal experiments in the Laboratory Animal Center of Jilin University. All experimental protocols have been approved by the Institutional Animal Care and Use Committee of Jilin University (license number: SY201809010). The animals were first anesthetized with carbon dioxide and then euthanized with carbon dioxide.

Informed Consent Statement: Not applicable.

Data Availability Statement: All sequencing data from this study can be found in the supporting documentation.

Conflicts of Interest: The authors declare no conflict of interest.

References

1. Amar, A.P.; Weiss, M.H. Pituitary anatomy and physiology. *Neurosurg. Clin. N. Am.* **2003**, *14*, 11–23. [CrossRef] [PubMed]
2. Brownstein, M.J.; Russell, J.T.; Gainer, H. Synthesis, transport, and release of posterior pituitary hormones. *Science* **1980**, *207*, 373–378. [CrossRef] [PubMed]
3. Kaplan, S.A. The pituitary gland: A brief history. *Pituitary* **2007**, *10*, 323–325. [CrossRef]
4. Edwards, W.; Raetzman, L.T. Complex integration of intrinsic and peripheral signaling is required for pituitary gland development. *Biol. Reprod.* **2018**, *99*, 504–513. [CrossRef] [PubMed]
5. Horvath, E.; Kovacs, K. Fine structural cytology of the adenohypophysis in rat and man. *J. Electron. Microsc. Tech.* **1988**, *8*, 401–432. [CrossRef]
6. Cohen, L.E.; Radovick, S. Molecular basis of combined pituitary hormone deficiencies. *Endocr. Rev.* **2002**, *23*, 431–442. [CrossRef] [PubMed]
7. Weiss, J.; Duca, K.A.; Crowley, W.F., Jr. Gonadotropin-releasing hormone-induced stimulation and desensitization of free alpha-subunit secretion mirrors luteinizing hormone and follicle-stimulating hormone in perifused rat pituitary cells. *Endocrinology* **1990**, *127*, 2364–2371. [CrossRef]
8. Ciccone, N.A.; Kaiser, U.B. The biology of gonadotroph regulation. *Curr. Opin. Endocrinol. Diabetes Obes.* **2009**, *16*, 321–327. [CrossRef]
9. Ye, R.S.; Xi, Q.Y.; Qi, Q.; Cheng, X.; Chen, T.; Li, H.; Kallon, S.; Shu, G.; Wang, S.B.; Jiang, Q.Y.; et al. Differentially expressed miRNAs after GnRH treatment and their potential roles in FSH regulation in porcine anterior pituitary cell. *PLoS ONE* **2013**, *8*, e57156. [CrossRef]
10. Kanasaki, H.; Purwana, I.N.; Miyazaki, K. Possible role of PACAP and its PAC1 receptor in the differential regulation of pituitary LHbeta- and *Fshb*eta-subunit gene expression by pulsatile GnRH stimulation. *Biol. Reprod.* **2013**, *88*, 35. [CrossRef]
11. Oleari, R.; Lettieri, A.; Paganoni, A.; Zanieri, L.; Cariboni, A. Semaphorin Signaling in GnRH Neurons: From Development to Disease. *Neuroendocrinology* **2019**, *109*, 193–199. [CrossRef] [PubMed]
12. Melamed, P.; Haj, M.; Yosefzon, Y.; Rudnizky, S.; Wijeweera, A.; Pnueli, L.; Kaplan, A. Multifaceted Targeting of the Chromatin Mediates Gonadotropin-Releasing Hormone Effects on Gene Expression in the Gonadotrope. *Front. Endocrinol.* **2018**, *9*, 58. [CrossRef] [PubMed]
13. Shalev, D.; Melamed, P. The role of the hypothalamus and pituitary epigenomes in central activation of the reproductive axis at puberty. *Mol. Cell. Endocrinol.* **2020**, *518*, 111031. [CrossRef]
14. Stamatiades, G.A.; Kaiser, U.B. Gonadotropin regulation by pulsatile GnRH: Signaling and gene expression. *Mol. Cell. Endocrinol.* **2018**, *463*, 131–141. [CrossRef]
15. Dobkin-Bekman, M.; Naidich, M.; Pawson, A.J.; Millar, R.P.; Seger, R.; Naor, Z. Activation of mitogen-activated protein kinase (MAPK) by GnRH is cell-context dependent. *Mol. Cell. Endocrinol.* **2006**, *252*, 184–190. [CrossRef]
16. Burger, L.L.; Haisenleder, D.J.; Aylor, K.W.; Marshall, J.C. Regulation of intracellular signaling cascades by GNRH pulse frequency in the rat pituitary: Roles for CaMK II, ERK, and JNK activation. *Biol. Reprod.* **2008**, *79*, 947–953. [CrossRef] [PubMed]
17. Yeh, D.M.; Coss, D. PACAP induces *Fshb*eta gene expression via EPAC. *Mol. Cell. Endocrinol.* **2019**, *492*, 110438. [CrossRef]
18. Tahir, M.S.; Porto-Neto, L.R.; Gondro, C.; Shittu, O.B.; Wockner, K.; Tan, A.W.L.; Smith, H.R.; Gouveia, G.C.; Kour, J.; Fortes, M.R.S. Meta-Analysis of Heifer Traits Identified Reproductive Pathways in Bos indicus Cattle. *Genes* **2021**, *12*, 768. [CrossRef]
19. Haisenleder, D.J.; Burger, L.L.; Aylor, K.W.; Dalkin, A.C.; Marshall, J.C. Gonadotropin-releasing hormone stimulation of gonadotropin subunit transcription: Evidence for the involvement of calcium/calmodulin-dependent kinase II (Ca/CAMK II) activation in rat pituitaries. *Endocrinology* **2003**, *144*, 2768–2774. [CrossRef]
20. Xu, M.; Xu, C.; Liu, F.; Shen, X.; Meng, J.; Chen, H.; Yang, J.; Zhou, P.; Gao, R.; Gan, S. Effects of active immunization with newly modified GnRH peptides on spermatogenesis and production performance of Holstein bulls. *Biol. Reprod.* **2018**, *99*, 461–472. [CrossRef]
21. Fields, S.D.; Perry, B.L.; Perry, G.A. Effects of GnRH treatment on initiation of pulses of LH, LH release, and subsequent concentrations of progesterone. *Domest. Anim. Endocrinol.* **2009**, *37*, 189–195. [CrossRef] [PubMed]
22. Guo, H.X.; Yuan, B.; Su, M.T.; Zheng, Y.; Zhang, J.Y.; Han, D.X.; Wang, H.Q.; Huang, Y.J.; Jiang, H.; Zhang, J.B. Identification of Circular RNAs in the Anterior Pituitary in Rats Treated with GnRH. *Animals* **2021**, *11*, 2557. [CrossRef] [PubMed]

23. Wang, H.Q.; Wang, W.H.; Chen, C.Z.; Guo, H.X.; Jiang, H.; Yuan, B.; Zhang, J.B. Regulation of FSH Synthesis by Differentially Expressed miR-488 in Anterior Adenohypophyseal Cells. *Animals* **2021**, *11*, 3262. [CrossRef] [PubMed]
24. Wang, H.Q.; Zhang, J.B.; Zheng, Y.; Zhang, W.D.; Guo, H.X.; Cong, S.; Ding, Y.; Yuan, B. Comprehensive analysis of differences in N6-methyladenosine RNA methylomes in the rat adenohypophysis after GnRH treatment. *FASEB J.* **2022**, *36*, e22204. [CrossRef]
25. Kyritsi, K.; Meng, F.; Zhou, T.; Wu, N.; Venter, J.; Francis, H.; Kennedy, L.; Onori, P.; Franchitto, A.; Bernuzzi, F.; et al. Knockdown of Hepatic Gonadotropin-Releasing Hormone by Vivo-Morpholino Decreases Liver Fibrosis in Multidrug Resistance Gene 2 Knockout Mice by Down-Regulation of miR-200b. *Am. J. Pathol.* **2017**, *187*, 1551–1565. [CrossRef]
26. He, J.; Xu, S.; Ji, Z.; Sun, Y.; Cai, B.; Zhang, S.; Wang, P. The role of miR-7 as a potential switch in the mouse hypothalamus-pituitary-ovary axis through regulation of gonadotropins. *Mol. Cell. Endocrinol.* **2020**, *518*, 110969. [CrossRef]
27. Yuen, T.; Ruf, F.; Chu, T.; Sealfon, S.C. Microtranscriptome regulation by gonadotropin-releasing hormone. *Mol. Cell. Endocrinol.* **2009**, *302*, 12–17. [CrossRef]
28. Wang, C.J.; Gao, F.; Huang, Y.J.; Han, D.X.; Zheng, Y.; Wang, W.H.; Jiang, H.; Gao, Y.; Yuan, B.; Zhang, J.B. circAkap17b acts as a miR-7 family molecular sponge to regulate FSH secretion in rat pituitary cells. *J. Mol. Endocrinol.* **2020**, *65*, 135–148. [CrossRef]
29. Huang, Y. The novel regulatory role of lncRNA-miRNA-mRNA axis in cardiovascular diseases. *J. Cell. Mol. Med.* **2018**, *22*, 5768–5775. [CrossRef]
30. Han, D.X.; Sun, X.L.; Fu, Y.; Wang, C.J.; Liu, J.B.; Jiang, H.; Gao, Y.; Chen, C.Z.; Yuan, B.; Zhang, J.B. Identification of long non-coding RNAs in the immature and mature rat anterior pituitary. *Sci. Rep.* **2017**, *7*, 17780. [CrossRef]
31. Han, D.X.; Sun, X.L.; Wang, C.J.; Yu, Z.W.; Zheng, Y.; Huang, Y.J.; Wang, W.H.; Jiang, H.; Gao, Y.; Yuan, B.; et al. Differentially expressed lncRNA-m433s1 regulates FSH secretion by functioning as a miRNA sponge in male rat anterior pituitary cellsdagger. *Biol. Reprod.* **2019**, *101*, 416–425. [CrossRef] [PubMed]
32. Zhang, W.; Ren, W.; Han, D.; Zhao, G.; Wang, H.; Guo, H.; Zheng, Y.; Ji, Z.; Gao, W.; Yuan, B. LncRNA-m18as1 competitively binds with miR-18a-5p to regulate follicle-stimulating hormone secretion through the Smad2/3 pathway in rat primary pituitary cells. *J. Zhejiang Univ. Sci. B* **2022**, *23*, 502–514. [CrossRef] [PubMed]
33. Ewing, B.; Green, P. Base-calling of automated sequencer traces using phred. II. Error probabilities. *Genome Res.* **1998**, *8*, 186–194. [CrossRef]
34. Kim, D.; Langmead, B.; Salzberg, S.L. HISAT: A fast spliced aligner with low memory requirements. *Nat. Methods* **2015**, *12*, 357–360. [CrossRef] [PubMed]
35. Sun, L.; Luo, H.; Bu, D.; Zhao, G.; Yu, K.; Zhang, C.; Liu, Y.; Chen, R.; Zhao, Y. Utilizing sequence intrinsic composition to classify protein-coding and long non-coding transcripts. *Nucleic Acids Res.* **2013**, *41*, e166. [CrossRef]
36. Kong, L.; Zhang, Y.; Ye, Z.Q.; Liu, X.Q.; Zhao, S.Q.; Wei, L.; Gao, G. CPC: Assess the protein-coding potential of transcripts using sequence features and support vector machine. *Nucleic Acids Res.* **2007**, *35*, W345–W349. [CrossRef] [PubMed]
37. Wang, L.; Park, H.J.; Dasari, S.; Wang, S.; Kocher, J.P.; Li, W. CPAT: Coding-Potential Assessment Tool using an alignment-free logistic regression model. *Nucleic Acids Res.* **2013**, *41*, e74. [CrossRef]
38. Finn, R.D.; Bateman, A.; Clements, J.; Coggill, P.; Eberhardt, R.Y.; Eddy, S.R.; Heger, A.; Hetherington, K.; Holm, L.; Mistry, J.; et al. Pfam: The protein families database. *Nucleic Acids Res.* **2014**, *42*, D222–D230. [CrossRef]
39. Langmead, B.; Trapnell, C.; Pop, M.; Salzberg, S.L. Ultrafast and memory-efficient alignment of short DNA sequences to the human genome. *Genome Biol.* **2009**, *10*, R25. [CrossRef]
40. Robinson, M.D.; McCarthy, D.J.; Smyth, G.K. EdgeR: A Bioconductor package for differential expression analysis of digital gene expression data. *Bioinformatics* **2010**, *26*, 139–140. [CrossRef]
41. Fahlgren, N.; Howell, M.D.; Kasschau, K.D.; Chapman, E.J.; Sullivan, C.M.; Cumbie, J.S.; Givan, S.A.; Law, T.F.; Grant, S.R.; Dangl, J.L.; et al. High-throughput sequencing of Arabidopsis microRNAs: Evidence for frequent birth and death of MIRNA genes. *PLoS ONE* **2007**, *2*, e219. [CrossRef]
42. Love, M.I.; Huber, W.; Anders, S. Moderated estimation of fold change and dispersion for RNA-seq data with DESeq2. *Genome Biol.* **2014**, *15*, 550. [CrossRef] [PubMed]
43. Han, D.X.; Sun, X.L.; Xu, M.Q.; Chen, C.Z.; Jiang, H.; Gao, Y.; Yuan, B.; Zhang, J.B. Roles of differential expression of microRNA-21-3p and microRNA-433 in FSH regulation in rat anterior pituitary cells. *Oncotarget* **2017**, *8*, 36553–36565. [CrossRef] [PubMed]
44. Kaprara, A.; Huhtaniemi, I.T. The hypothalamus-pituitary-gonad axis: Tales of mice and men. *Metabolism* **2018**, *86*, 3–17. [CrossRef]
45. Burger, L.L.; Haisenleder, D.J.; Aylor, K.W.; Marshall, J.C. Regulation of Lhb and Egr1 gene expression by GNRH pulses in rat pituitaries is both c-Jun N-terminal kinase (JNK)- and extracellular signal-regulated kinase (ERK)-dependent. *Biol. Reprod.* **2009**, *81*, 1206–1215. [CrossRef]
46. Harris, D.; Bonfil, D.; Chuderland, D.; Kraus, S.; Seger, R.; Naor, Z. Activation of MAPK cascades by GnRH: ERK and Jun N-terminal kinase are involved in basal and GnRH-stimulated activity of the glycoprotein hormone LHbeta-subunit promoter. *Endocrinology* **2002**, *143*, 1018–1025. [CrossRef]
47. Okumu, L.A.; Braden, T.D.; Vail, K.; Simon, L.; Goyal, H.O. Low androgen induced penile maldevelopment involves altered gene expression of biomarkers of smooth muscle differentiation and a key enzyme regulating cavernous smooth muscle cell tone. *J. Urol.* **2014**, *192*, 267–273. [CrossRef] [PubMed]

48. Alen, F.; Gomez-Redondo, I.; Rivera, P.; Suarez, J.; Ramos-Ibeas, P.; Pericuesta, E.; Fernandez-Gonzalez, R.; Perez-Cerezales, S.; Horiuchi, K.; Orio, L.; et al. Sex-Dimorphic Behavioral Alterations and Altered Neurogenesis in U12 Intron Splicing-Defective Zrsr1 Mutant Mice. *Int. J. Mol. Sci.* **2019**, *20*, 3543. [CrossRef] [PubMed]
49. Farshori, P.Q.; Shah, B.H.; Arora, K.K.; Martinez-Fuentes, A.; Catt, K.J. Activation and nuclear translocation of PKCdelta, Pyk2 and ERK1/2 by gonadotropin releasing hormone in HEK293 cells. *J. Steroid Biochem. Mol. Biol.* **2003**, *85*, 337–347. [CrossRef]
50. Melamed, P.; Savulescu, D.; Lim, S.; Wijeweera, A.; Luo, Z.; Luo, M.; Pnueli, L. Gonadotrophin-releasing hormone signalling downstream of calmodulin. *J. Neuroendocrinol.* **2012**, *24*, 1463–1475. [CrossRef]
51. Qian, X.; Lin, G.; Wang, J.; Zhang, S.; Ma, J.; Yu, B.; Wu, R.; Liu, M. CircRNA_01477 influences axonal growth via regulating miR-3075/FosB/Stat3 axis. *Exp. Neurol.* **2022**, *347*, 113905. [CrossRef]
52. Chen, J.; An, B.S.; Cheng, L.; Hammond, G.L.; Leung, P.C. Gonadotropin-releasing hormone-mediated phosphorylation of estrogen receptor-alpha contributes to fosB expression in mouse gonadotrophs. *Endocrinology* **2009**, *150*, 4583–4593. [CrossRef] [PubMed]
53. Wang, C.J.; Guo, H.X.; Han, D.X.; Yu, Z.W.; Zheng, Y.; Jiang, H.; Gao, Y.; Yuan, B.; Zhang, J.B. Pituitary tissue-specific miR-7a-5p regulates FSH expression in rat anterior adenohypophyseal cells. *PeerJ* **2019**, *7*, e6458. [CrossRef] [PubMed]
54. Han, D.X.; Xiao, Y.; Wang, C.J.; Jiang, H.; Gao, Y.; Yuan, B.; Zhang, J.B. Regulation of FSH expression by differentially expressed miR-186-5p in rat anterior adenohypophyseal cells. *PLoS ONE* **2018**, *13*, e0194300. [CrossRef]
55. Li, K.; Xu, J.; Luo, Y.; Zou, D.; Han, R.; Zhong, S.; Zhao, Q.; Mang, X.; Li, M.; Si, Y.; et al. Panoramic transcriptome analysis and functional screening of long noncoding RNAs in mouse spermatogenesis. *Genome Res.* **2021**, *31*, 13–26. [CrossRef]
56. Wang, J.; Wang, L.; Feng, G.; Wang, Y.; Li, Y.; Li, X.; Liu, C.; Jiao, G.; Huang, C.; Shi, J.; et al. Asymmetric Expression of LincGET Biases Cell Fate in Two-Cell Mouse Embryos. *Cell* **2018**, *175*, 1887–1901.e18. [CrossRef] [PubMed]
57. Wang, X.; Zhang, X.; Dang, Y.; Li, D.; Lu, G.; Chan, W.Y.; Leung, P.C.K.; Zhao, S.; Qin, Y.; Chen, Z.J. Long noncoding RNA HCP5 participates in premature ovarian insufficiency by transcriptionally regulating MSH5 and DNA damage repair via YB1. *Nucleic Acids Res.* **2020**, *48*, 4480–4491. [CrossRef]
58. Wang, R.; Zhang, S.; Chen, X.; Li, N.; Li, J.; Jia, R.; Pan, Y.; Liang, H. CircNT5E Acts as a Sponge of miR-422a to Promote Glioblastoma Tumorigenesis. *Cancer Res.* **2018**, *78*, 4812–4825. [CrossRef]
59. Liang, H.; Yu, T.; Han, Y.; Jiang, H.; Wang, C.; You, T.; Zhao, X.; Shan, H.; Yang, R.; Yang, L.; et al. LncRNA PTAR promotes EMT and invasion-metastasis in serous ovarian cancer by competitively binding miR-101-3p to regulate ZEB1 expression. *Mol. Cancer* **2018**, *17*, 119. [CrossRef]
60. Salmena, L.; Poliseno, L.; Tay, Y.; Kats, L.; Pandolfi, P.P. A ceRNA hypothesis: The Rosetta Stone of a hidden RNA language? *Cell* **2011**, *146*, 353–358. [CrossRef]
61. Hanson, P.I.; Schulman, H. Neuronal Ca^{2+}/calmodulin-dependent protein kinases. *Annu. Rev. Biochem.* **1992**, *61*, 559–601. [CrossRef] [PubMed]
62. Cui, Z.J.; Hidaka, H.; Dannies, P.S. KN-62, a calcium/calmodulin-dependent protein kinase II inhibitor, inhibits high potassium-stimulated prolactin secretion and intracellular calcium increases in anterior pituitary cells. *Biochim. Biophys. Acta (BBA)-Mol. Cell Res.* **1996**, *1310*, 343–347. [CrossRef]

genes

MDPI

Article

WNT Co-Receptor LRP6 Is Critical for Zygotic Genome Activation and Embryonic Developmental Potential by Interacting with Oviductal Paracrine Ligand WNT2

Fusheng Yao †, Jia Hao †, Zhaochen Wang, Meiqiang Chu, Jingyu Zhang, Guangyin Xi, Zhenni Zhang, Lei An and Jianhui Tian *

State Key Laboratory of Animal Biotech Breeding, National Engineering Laboratory for Animal Breeding, Key Laboratory of Animal Genetics, Breeding and Reproduction of the Ministry of Agriculture and Rural Affairs, College of Animal Science and Technology, China Agricultural University, No. 2 Yuanmingyuan West Road, Beijing 100193, China
* Correspondence: tianjh@cau.edu.cn
† These authors contributed equally to this work.

Abstract: Mammalian preimplantation development depends on the interaction between embryonic autocrine and maternal paracrine signaling. Despite the robust independence of preimplantation embryos, oviductal factors are thought to be critical to pregnancy success. However, how oviductal factors regulate embryonic development and the underlying mechanism remain unknown. In the present study, focusing on WNT signaling, which has been reported to be essential for developmental reprogramming after fertilization, we analyzed the receptor-ligand repertoire of preimplantation embryonic WNT signaling, and identified that the WNT co-receptor LRP6 is necessary for early cleavage and has a prolonged effect on preimplantation development. LRP6 inhibition significantly impeded zygotic genome activation and disrupted relevant epigenetic reprogramming. Focusing on the potential oviductal WNT ligands, we found WNT2 as the candidate interacting with embryonic LRP6. More importantly, we found that WNT2 supplementation in culture medium significantly promoted zygotic genome activation (ZGA) and improved blastocyst formation and quality following in vitro fertilization (IVF). In addition, WNT2 supplementation significantly improved implantation rate and pregnancy outcomes following embryo transfer. Collectively, our findings not only provide novel insight into how maternal factors regulate preimplantation development through maternal-embryonic communication, but they also propose a promising strategy for improving current IVF systems.

Keywords: WNT signaling; LRP6; WNT2; preimplantation embryos; zygotic genome activation; in vitro fertilization

Citation: Yao, F.; Hao, J.; Wang, Z.; Chu, M.; Zhang, J.; Xi, G.; Zhang, Z.; An, L.; Tian, J. WNT Co-Receptor LRP6 Is Critical for Zygotic Genome Activation and Embryonic Developmental Potential by Interacting with Oviductal Paracrine Ligand WNT2. *Genes* **2023**, *14*, 891. https://doi.org/10.3390/genes14040891

Academic Editor: Bao Yuan

Received: 2 January 2023
Revised: 2 February 2023
Accepted: 4 April 2023
Published: 10 April 2023

1. Introduction

Mammalian preimplantation development is orchestrated through a series of consecutive cellular and molecular events that are finetuned via the synergistic interaction of embryonic autocrine and maternal paracrine signaling. The interaction involves a wide range of growth factor ligands secreted from the embryo or the oviduct, as well as their interacting receptors expressed in preimplantation embryos [1,2].

Embryo-maternal communication through autocrine or paracrine factors initiates from the earliest stages of embryonic development. The robust independence of preimplantation embryo development in vitro suggests that the developmental program is autonomous [2]. However, increasing evidence from our own [3,4] and other groups [5,6] has challenged this conclusion: although the essential role of embryonic autocrine factors in supporting preimplantation development has been well accepted, oviductal paracrine factors are also critical for developmental potential of preimplantation embryos. Despite the importance of the synergistic cooperation of autocrine and paracrine factors in preimplantation development, however, how they cooperate and modulate downstream intracellular pathways during this critical developmental window remains an open question for the developmental biology.

Among the signaling pathways that are essential for cellular survival and growth, canonical WNT signaling plays a critical role in acquiring development potential [7,8]. Activation of canonical WNT signaling involves both Frizzleds (Fzds) and its co-receptor low-density lipoprotein receptor-related protein5/6 (LRP5/6) [9,10]. Autocrine or paracrine WNT ligands can bind to Fzds or LRP5/6, resulting in β-catenin stabilization and accumulation. β-catenin is involved in transcription upregulation by binding to transcriptional factors or to a component of SWI/SNF, a chromatin-remodeling complex that regulates chromatin accessibility [11,12]. When WNT signaling is inactivated due to the absence of extracellular ligands, β-catenin is phosphorylated by glycogen synthase kinase 3 (GSK3) and is ubiquitylated, thus leading to rapid proteasomal degradation. In contrast, when WNT signaling is activated via ligand-induced receptor phosphorylation, β-catenin ubiquitination is prevented, and then, stabilized β-catenin can translocate into the nucleus and act as a coactivator of transcription factors [13]. Although the developmental roles of many ligands and receptors of WNT signaling have been identified in early embryos [14,15], the receptor–ligand repertoire and coupled intracellular pathways remain largely unknown. In particular, the role of oviductal WNT ligands in modulating embryonic WNT signaling has never been determined.

In the present study, we aimed to profile the ligand-receptor repertoire of WNT signaling during preimplantation development and found that embryonic WNT co-receptor LRP6 and its interacting oviductal ligand WNT2 may be the candidates critical for modulating WNT signaling during early development. To test this hypothesis, we functionally detected the developmental role of LRP6 and WNT2, as well as their effect on landmark molecular and cellular events during preimplantation development. Our study will not only present novel insight into understanding embryo–maternal interactions during preimplantation development but will also provide a new strategy for improving current in vitro fertilization (IVF) systems.

2. Results

2.1. *WNT Co-Receptor LRP6 Is Critical to Early Cleavage of Preimplantation Development*

To decipher the receptor–ligand repertoire of embryonic WNT signaling, we first analyzed the dynamic expression patterns of WNT signaling receptors in in vivo conceived embryos and their temporally corresponding oviduct/uterus, using our previously published RNA-seq data [16]. Several WNT signaling receptors, such as *Fzd5*, *Fzd6*, etc., showed higher expression levels in oviduct/uterus, indicating that these receptors may function in the oviduct/uterus (Figure 1A). By contrast, *Fzd2*, *Fzd9* and *Lrp6* were primarily expressed in embryos. Of note, *Fzd9* and *Lrp6* were specifically expressed in embryos, and *Lrp6* showed a consistently higher expression throughout preimplantation development. We also validated LRP6 expression on the protein level using immunofluorescent analysis (Figure 1B).

Next, we focused on the stage-specific developmental role of LRP6 in preimplantation embryos by using salinomycin, a potent selective LRP6 inhibitor that can block WNT-induced LRP6 phosphorylation and cause degradation of LRP6 protein [17]. Salinomycin supplementation showed a dose-dependent effect on compromising the preimplantation developmental rate; 20 and 40 μM salinomycin significantly decreased the ratio of embryos that developed to the blastocyst stage and increased the proportion of embryos arrested at the two-cell stage (Figure 1C,D). LRP6 inhibition from the four-cell stage onward did not affect embryo development until blastocyst formation, and LRP6 inhibition from either the eight-cell or morular stage onward has no impact on preimplantation development (Figure 1E–G). These results suggest that LRP6 may be preferentially critical to early cleavage of preimplantation development, thus affecting subsequent blastocyst formation.

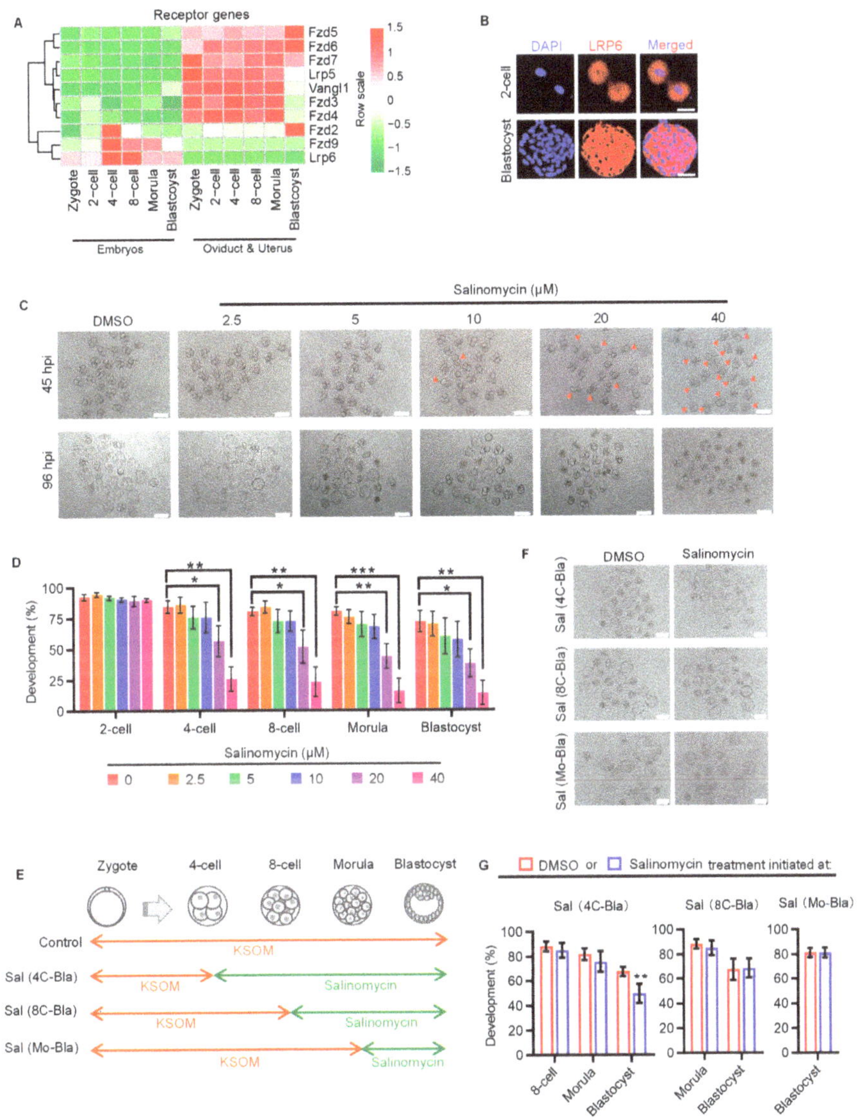

Figure 1. LRP6 plays an important role in early cleavage stages. (**A**) Heatmap showing the dynamic expression patterns of receptors of WNT signaling in IVO embryos and corresponding oviductal and uterus tissues. (**B**) Representative images of LRP6 staining in 2-cell and blastocyst. Scale bar = 50 μm. (**C**) Representative images of 4-cell and blastocyst following culture from zygotes in DMSO or salinomycin. Red arrows indicated arrested 2-cell. Salinomycin is a potential LRP6 inhibitor. hpi, hours post insemination. Scale bar = 200 μm. (**D**) Developmental progression of IVF preimplantation treated with or without salinomycin. * $p < 0.05$, ** $p < 0.01$, *** $p < 0.001$. (**E**) Experimental design of embryo culture treated with or without salinomycin initiated at 4-cell, 8-cell and morula. Sal, salinomycin; 4C, 4-cell; 8C, 8-cell; Mo, morula; Bla, blastocyst. Related to (**F**,**G**). (**F**) Representative images of blastocysts following culture of 4-cell, 8-cell and morula stage embryos in 40 μM salinomycin. Scale bar = 200 μm. (**G**) Percentage of embryos to reach the various preimplantation embryo stages following culture of 4-cell, 8-cell and morula stage embryos in 40 μM salinomycin. ** $p < 0.01$.

2.2. LRP6 Inhibition Impedes Zygotic Genome Activation and Disrupts Relevant Epigenetic Reprogramming

Having confirmed the critical role of LRP6 in early cleavage, we next asked if LRP6 is essential for zygotic genome activation (ZGA), a hallmark event that occurs around the early cleavage stage and is of prime importance for subsequent development. Mouse ZGA includes minor and major waves, which occur before and after second-round DNA replication, respectively [18]. To test if minor or major ZGA impairment was responsible for LRP6 inhibition-induced two-cell arrest, we designed a set of experiments based on the stage-specific salinomycin exposure (Figure 2A). Compared with genetic knockout or knockdown, chemical-induced inhibition can ensure a transient but not consistent or prolonged blockage of LRP6, thus determining its functional window. Using 5-ethynyl uridine (EU) incorporation assay, which specifically labels nascent RNA from de novo transcription [19,20], we found that LRP6 inhibition from the zygote to late two-cell stage significantly impeded both minor and major ZGA (Experiments 1 and 2) (Figure 2B,C). We next showed that LRP6 inhibition-induced ZGA impairment was reversible when embryos were transferred to salinomycin-free medium before 5-EU incorporation (Experiments 3 and 4) (Figure 2D,E). This finding is in line with the developmental consequence of salinomycin exposure at the corresponding stage. LRP6 inhibition that covered minor ZGA resulted in a 50% reduction in blastocyst rate, while the LRP6 inhibition that covered both minor and major ZGA resulted in severe developmental arrest at the two-cell stage, as well as almost complete failure of subsequent development, similar with the consistent exposure throughout the preimplantation stage (Figure 2F,G). Given that WNT signaling plays an important role in modulating various histone epigenetic modifications, such as H3K4me3, H3K9me3, etc. [21], which are also the prerequisites for transcriptional regulation of ZGA [22,23], we next asked if salinomycin-induced two-cell arrest was associated with disrupted histone modifications. Immunofluorescent analyses showed that salinomycin-treated two-cell embryos exhibited aberrantly high levels of H3K4me3 and H3K9me3 (Figure 2H,I), suggesting that LRP6 may play an important role in prompting ZGA via modulating epigenetic reprogramming.

2.3. Oviductal Paracrine WNT2 Is Critical for Activating LRP6 and Prompting ZGA

Having confirmed the role of LRP6 in regulating ZGA and supporting preimplantation development, we next attempted to determine the effect of oviductal paracrine factors in activating WNT-LRP6 signaling and prompting embryonic development. To this end, we first compared the activity of WNT signaling between in vivo (IVO) preimplantation embryos and their counterparts generated under standardized in vitro fertilization (IVF) conditions, which were used as the model that lacks oviductal paracrine ligands because currently used commercial culture medium does not include any WNT ligands [24]. By gene set variation analysis (GSVA), we found that IVO embryos exhibited high-level WNT signaling at the early cleavage and blastocyst stages. By contrast, IVF embryos showed lower activity of WNT signaling at these stages (Figure 3A). Gene set enrichment analysis (GSEA) also supports this result. WNT signaling tended to be enriched in IVO embryos (Figure 3B). Then, we attempted to determine the paracrine ligands from the oviduct/uterus that could interact with LRP6 to activate WNT signaling. By integrating UNIPROT-derived secreted proteins and the STRING database, we screened out seven candidate ligands that could interact with LRP6, including WNT3, DKK2, WNT2, etc. (Figure 3C). All seven ligands were specifically expressed in the oviduct/uterus (Figure 3D). Next, we selected WNT2 as the candidate because its expression was more enriched at the early cleavage and blastocyst stages, which temporally coincided with WNT signaling dynamics during preimplantation development. To explore the developmental role of oviductal paracrine WNT, we supplemented WNT2 in culture medium throughout the preimplantation stage, based on mouse embryo assay (MEA) by using in vivo fertilized and in vitro cultured embryos. Although having no obvious impact on embryo preimplantation developmental rate, WNT2 supplementation showed a dose-dependent effect on improving increasing inner cell mass (ICM) cells and ICM:TE (trophectoderm) ratio of blastocysts (Figure 3E–I). Of note, we also found that WNT2 supplementation significantly promoted transcriptional upregulation of both minor and major ZGA, but not in blastocyst activation (Figure 3J–L), a process that is critical for blastocyst–uterine communication and blastocyst implantation [25]. These results indicate that oviductal paracrine WNT2 has an important role in promoting embryonic ZGA, probably via the interaction with its membrane receptor LRP6.

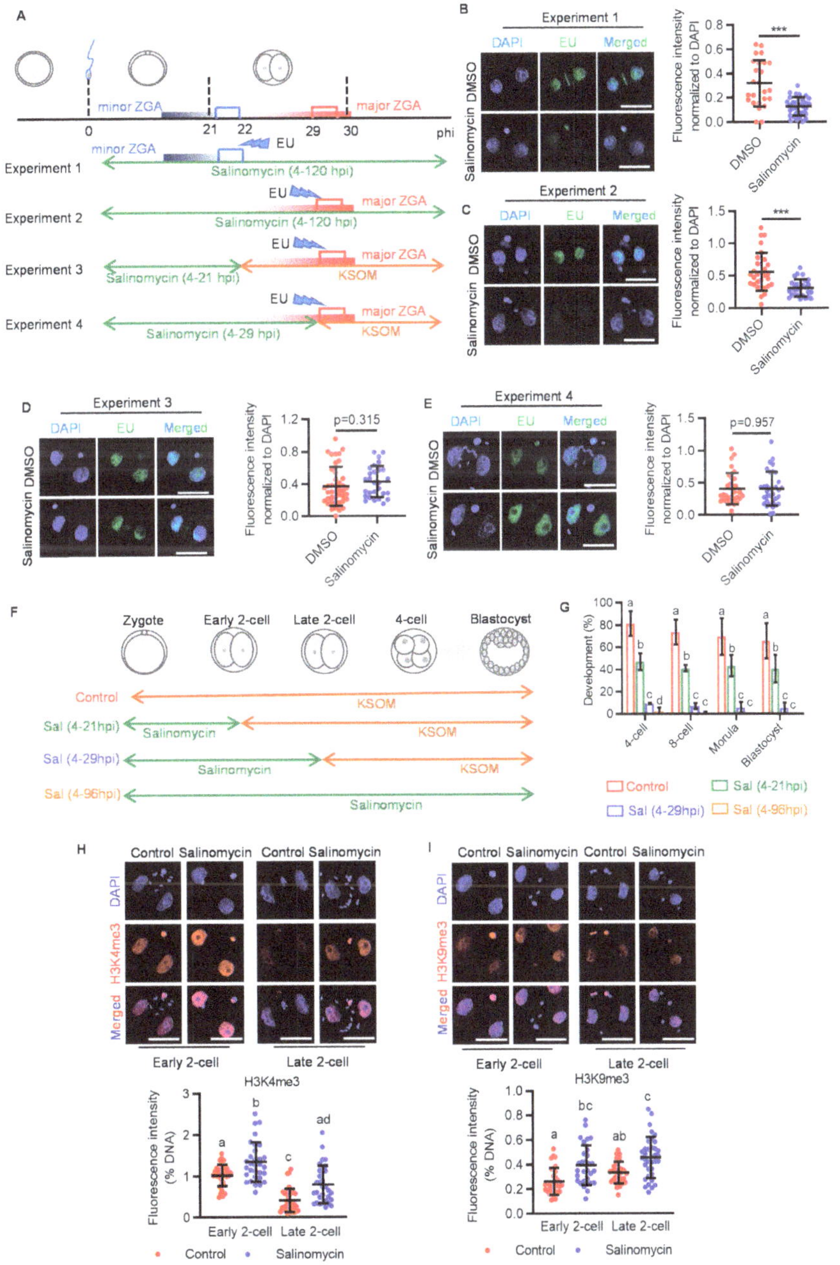

Figure 2. Inhibition of LRP6 impairs ZGA. (**A**) Experimental design of EU incorporation in salinomycin-treated embryos at various periods. (**B–E**) Representative images of EU staining at

2-cell embryos (**upper panel**) and quantification of the EU signal intensity (**lower panel**). Each plot indicates relative staining intensity in each embryo. Scale bar = 50 µm. Error bar = means ± SD. *** $p < 0.001$. (**F**) Experimental design of salinomycin addition at different time points. (**G**) Development rate of embryos to reach the various preimplantation embryo stages. Scale bar = 50 µm. Error bar = means ± SD. (**H**,**I**) Representative images of H3K4me3 (H, **upper panel**) and H3K9me3 ((**I**) **upper panel**) staining in early or late 2-cell treated with or without salinomycin. Scatter plot showing the quantification of relative H3K4me3 (H, **lower panel**) and H3K9me3 ((**I**) **lower panel**) signal intensity. Scale bar = 50 µm. Error bar = means ± SD. Different letters (a, b, c, d) indicate a significant difference ($p < 0.05$).

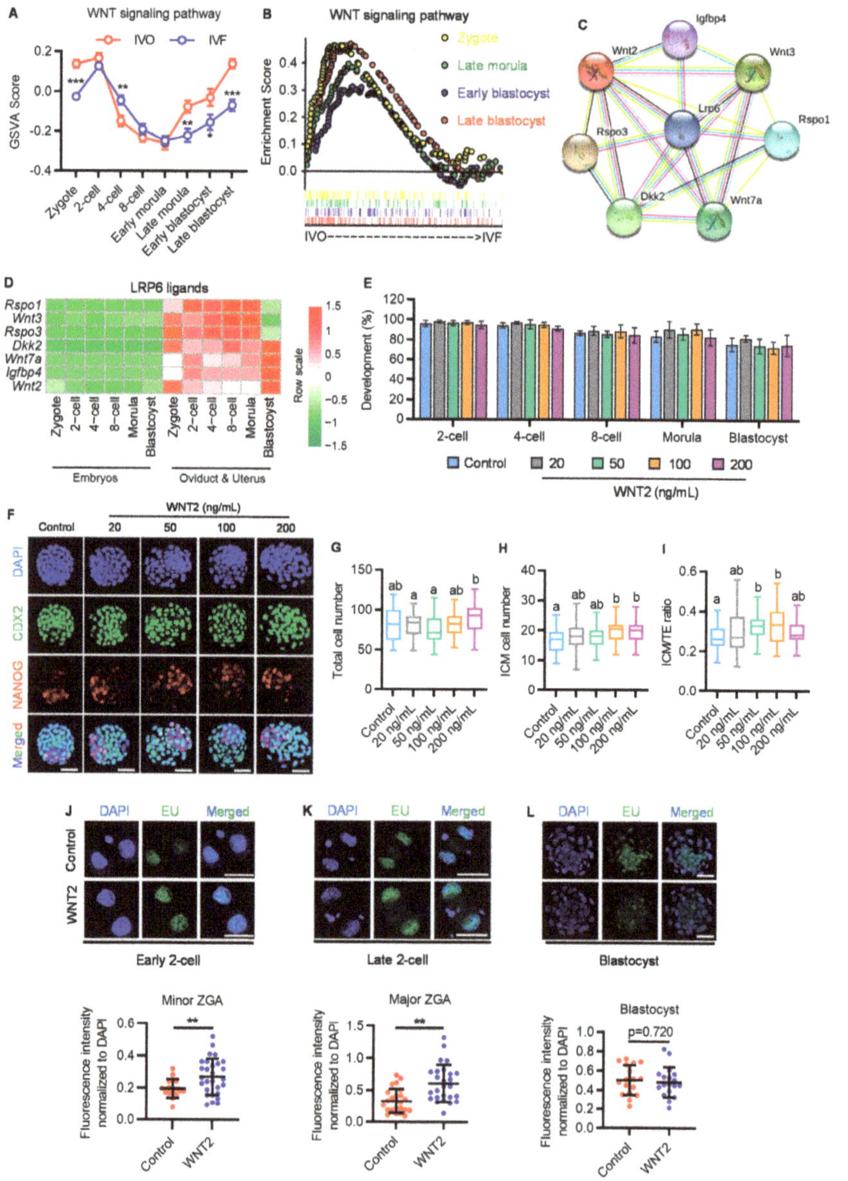

Figure 3. Exogenous WNT2 promotes embryo development via interacting with LRP6. (**A**) GSVA plot showing the dynamic activity of WNT signaling in IVO and IVF preimplantation embryos.

* $p < 0.05$, ** $p < 0.01$, *** $p < 0.001$. (**B**) GSEA plot showing the enrichment of WNT signaling pathway in IVO and IVF preimplantation embryos. (**C**) The interaction network of LRP6 with its ligands based on STRING database. (**D**) Heatmap showing the dynamic expression patterns of LRP6 ligands in IVO preimplantation embryos and corresponding oviductal and uterus tissues. (**E**) Development rate of embryos at different stages treated with or without WNT2. (**F**) Representative images of CDX2 (green), NANOG (red) and DAPI (blue) staining in control and WNT2-exposed IVF blastocysts. Scale bar = 50 μm. (**G–I**) Quantification of total cell number (**G**), ICM cell number (**H**) and ICM:TE ratio (**I**) of blastocysts is indicated; each plot represents one blastocyst. Related to Figure 3F. Different letters (a, b) indicate a significant difference ($p < 0.05$). (**J–L**) Representative images of EU staining in WNT2 treated or without treated early 2-cell ((**J**) **upper panel**), late 2-cell. ((**K**) **upper panel**) and blastocyst ((**L**) **upper panel**). Scatter plot showing the relative signal intensity of EU staining. Scale bar = 50 μm. Error bar = means ± SD. ** $p < 0.01$.

2.4. Supplementation of Oviductal Paracrine WNT2 Improves Embryo Quality and Pregnancy Outcomes

Given that oviductal paracrine WNT2 can promote both minor and major ZGA, we next explored the effect of exogenous WNT2 supplementation on the developmental potential of IVF embryos. Partially distinct from the results of MEA, we found that 50 ng/mL WNT2 significantly promoted blastocyst formation (Figure 4A,B), and the beneficial effect was completely abrogated by blocking LRP6 (Figure 4C). Moreover, immunofluorescence analyses showed that WNT2 supplementation has no effect on total cell number, but it increased the ICM cell number and ICM:TE ratio (Figure 4D–G).

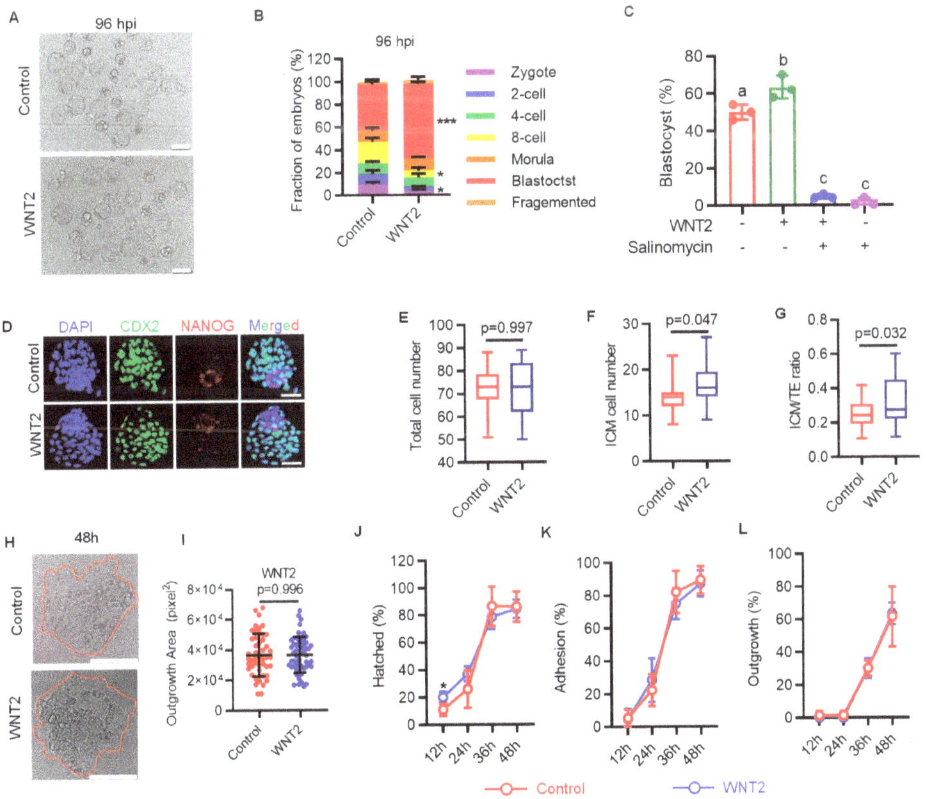

Figure 4. Exogenous WNT2 increases the implantation potentials of embryo. (**A**) Representative images of IVF blastocysts treated with or without WNT2 from zygote stage. Scale bar = 200 μm.

(**B**) Fraction of embryos at blastocyst stage after treated with or without WNT2. Error bar = means + SD. * $p < 0.05$, *** $p < 0.001$. (**C**) Effect of combing WNT2 and salinomycin on embryo development. Data are shown as means ± SD. Different letters (a, b, c) indicate a significant difference ($p < 0.05$). (**D**) Representative images of CDX2 (green), NANOG (red) and DAPI (blue) staining in control and WNT2-exposed IVF blastocysts. Scale bar = 50 μm. (**E–G**) Quantification of total cell number (**E**), ICM cell number (**F**) and ICM:TE ratio (**G**) of blastocysts, each plot represents one blastocyst. Error bar = means ± SD. Related to Figure 4C. (**H,I**) Representative images of IVF blastocysts after 48 h in vitro culture (**H**) and quantification of outgrowth area (**I**). Red circle represents outgrowth. Scale bar = 200 μm. Error bar = means ± SD. (**J–L**) The ratio of IVF blastocysts hatched (**J**), adhesion (**K**) and outgrowth (**L**) after 12, 24, 36 and 48 h in vitro culture. Error bar = means ± SD. * $p < 0.05$.

Furthermore, we evaluated the subsequent embryo implantation potential using in vitro embryo outgrowth, a model for evaluating embryo implantation potential [26]. Embryos exposed to WNT2 throughout the preimplantation stage showed significantly accelerated blastocyst hatching at 12 h. However, the percentage of blastocysts that attached to and outgrew on the fibronectin, as well as the final outgrowth area, were similar to that in the control group (Figure 4H–L). Finally, we confirmed the beneficial effect of WNT2 supplementation on subsequent pregnancy outcomes following embryo transfer. WNT2 supplementation throughout in vitro culture significantly increased both the implantation and live birth rates. Importantly, WNT2 supplementation did not lead to any detectable fetal or placental defects (Table 1). These results suggest that WNT2 supplementation has a positive role in preimplantation development and subsequent pregnancy outcomes but no detectable adverse effect on fetal growth and health.

Table 1. Pregnancy outcomes of IVF blastocysts treated with WNT2 at E19.5. The mice were sacrificed at E19.5, and the fetuses were collected to measure all indicators.

	Control	**WNT2**
Recipients (*n*)	31	18
Implantations (%)	66.13 ± 23.37	82.87 ± 15.33 *
Fetuses (% of transferred)	30.11 ± 13.50	39.81 ± 17.25 *
Fetuses (% of implanted)	50.56 ± 23.60	48.42 ± 19.74
Fetal weight (g)	1.54 ± 0.30	1.56 ± 0.27
Occipito-frontal diameter (mm)	6.79 ± 1.26	6.88 ± 1.38
Crown rump length (mm)	24.44 ± 2.22	24.80 ± 2.21
Tail length (mm)	12.61 ± 1.79	12.90 ± 1.25
Placental weight (g)	0.16 ± 0.03	0.15 ± 0.04
Placental diameter (mm)	8.89 ± 1.17	8.63 ± 1.10

Data are shown as means ± SD. * $p < 0.05$.

3. Discussion

Mammalian preimplantation embryos undergo a series of consecutive events, such as ZGA, cellular division, lineage specification, etc., all of which are essential for subsequent development and pregnancy success. These processes were tightly regulated by multiple signal transduction pathways, and each pathway participates in one or several related biological events [27]. Among these pathways, canonical WNT signaling has been reported to play an important role in influencing the developmental potential of preimplantation embryos [7,15]. However, mechanisms responsible for the modulation of embryonic WNT signaling activity, in particular, how oviductal paracrine ligands regulate embryonic WNT signaling during preimplantation development, remain elusive.

Our results provided the evidence supporting the essential role of LRP6, as the critical receptor of WNT signaling, in determining preimplantation developmental consequence (Figure 1). Of note, using the model of stage-specific LRP6 inhibition, we identified that activation of LRP6 is necessary for minor and major ZGA (Figure 2A–E). LRP6 inhibition covering both minor and major ZGA resulted in a large proportion of two-cell arrest, while LRP6 inhibition that only covered minor ZGA led to a considerable decrease in blastocyst formation (Figure 2F,G). This finding was in line with previously reported developmental consequences following transient inhibition of minor ZGA by DRB, a reversible inhibitor of Pol II-mediated transcription [18]. These facts indicate that LRP6-mediated WNT signaling activation plays an important role in ZGA and is the prerequisite for subsequent development.

Changes in histone modifications are hallmark reprogramming events during the process of ZGA, and disruption in histone reprogramming could directly impair ZGA [18,28,29]. In our study, LRP6 inhibition resulted in disruption of both H3K4me3 and H3K9me3 reprogramming at the two-cell stage (Figure 2H–I). As an active histone mark, H3K4me3 generally appears as a sharp status at the promoter regions to facilitate transcription factor incorporation by recruiting CHD1, a histone remodeler, to "loosen" the chromatin structure [30,31]. In mature oocytes, the genome is highly enriched with non-canonical flat H3K4me3 domains, which can inhibit deposition of H3K27ac and combing of transcriptional factors [22]. The removal of non-canonical flat H3K4me3 was thought to be necessary for the re-establishment of canonical sharp H3K4me3 peaks at the promoter regions, which are critical for triggering ZGA [15,32]. Our results suggest that LRP6 inhibition impeded the removal of oocyte-deposited non-canonical H3K4me3 (Figure 2H). Similarly, another histone modification, H3K9me3, also showed higher levels due to LRP6 inhibition (Figure 2I). As a repressive histone mark, H3K9me3 is tightly involved in the formation of heterochromatin and maintenance of gene silencing in mature oocytes [33]. Thus, post-fertilization removal of H3K9me3 is crucial for initiating ZGA and epigenetic reprogramming that confers developmental potential [34,35]. These results, taken together, imply that LRP6-mediated WNT signaling may participate in the removal of oocyte-deposited histone marks that are necessary for ZGA initiation soon after fertilization.

Highlighting the important role of paracrine factors from the maternal oviduct, we also propose a strategy for improving current IVF systems by supplementing exogenous WNT2 in culture medium. We found that exogenous supplementation of WNT2 not only promoted blastocyst formation (Figure 4A,B), but also improved embryonic lineage commitment, revealed by an increased ratio of ICM:TE (Figure 4D–G). More importantly, we also showed that WNT2 supplementation could increase the pregnancy success rates and final pregnancy outcomes after embryo transfer (Table 1). Our concept is in line with the results of previous studies: epigenomic and gene expression patterns of IVF embryos can be partially corrected by supplementing culture medium with oviductal fluid [36,37]. However, considering the potential risk of disease transmission, the strategy of oviductal fluid addition may be practical to in vitro embryo production in domestic and laboratory animals. By contrast, the chemically defined culture medium using the functional growth factors or cytokines that are present in oviductal fluid should be a more reasonable strategy, especially in the context of clinical use of human-assisted reproductive technologies.

Collectively, based on our results, we propose the model to illustrate the important role of LRP6 and its interacting oviductal ligand WNT2 in supporting preimplantation development. Our data proved that WNT co-receptor LRP6 is necessary for early cleavage and has a prolonged effect on preimplantation development. Activation of LRP6 through oviductal or embryonic WNTs promotes ZGA by affecting H3K4me3 and H3K9me3 modifications (Figure 5, left panel). In contrast, the absence of paracrine WNT ligands in in vitro culture medium may contribute to impaired embryo developmental potential, and aberrant H3K4me3 and H3K9me3 modifications and impeded ZGA may participate in this impairment (Figure 5, right panel).

Overall, focusing on the maternal paracrine factors and their embryonic interacting receptors, our study not only provides novel insight into the critical role and mechanism of the oviductal WNT ligand in finetuning preimplantation developmental reprogramming, but it also supports the concept highlighting the usage of oviductal cytokines or growth factors in increasing pregnancy success rates of IVF embryos.

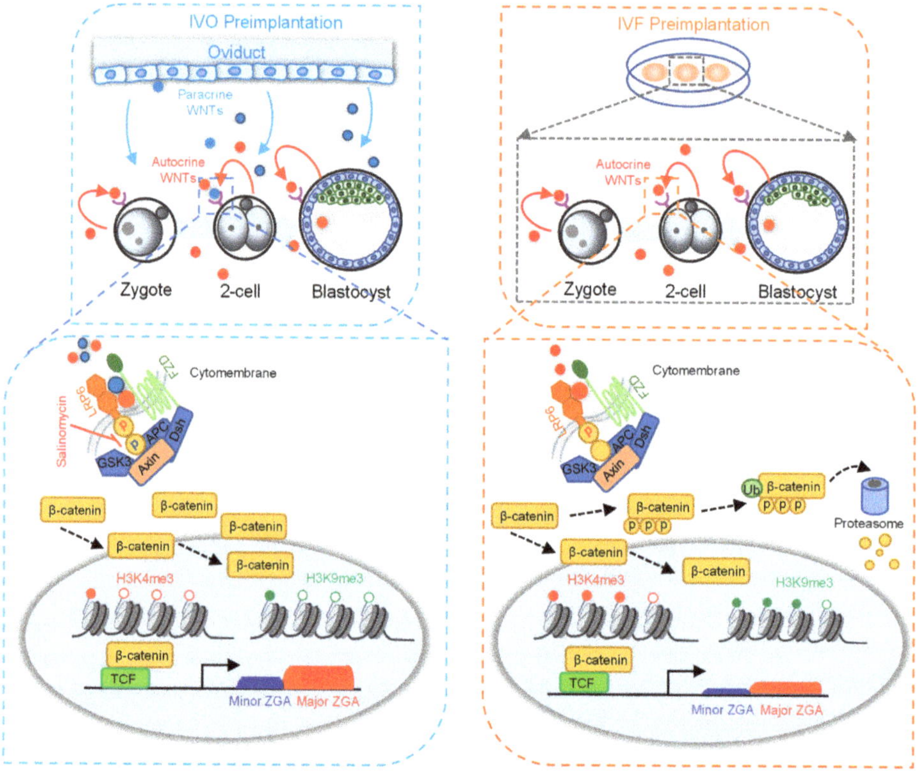

Figure 5. A model illustrating the important roles of LRP6 and paracrine ligands in preimplantation development. The absence of paracrine ligands from the oviduct/uterus in in vitro culture medium impaired embryo developmental potential with aberrant high levels of H3K4me3 and H3K9me3 modification at the ZGA period. Activation of LRP6 through oviductal and embryonic WNTs promotes in vivo ZGA by affecting H3K4me3 and H3K9me3 reprogramming (**left panel**). The inactivation of LRP6 in the absence of paracrine WNTs from the oviduct/uterus in vitro culture medium impaired ZGA accompanied with aberrant high levels of H3K4me3 and H3K9me3 modification (**right panel**).

4. Materials and Methods

4.1. Animals

Eight-week-old ICR mice were kept in a temperature-controlled room with 12 h alternating light/dark. All mice experiments were approved by the Institutional Animal Care and Use Committee of China Agricultural University. The ethical clearance approval number is AW11107102-1-2.

4.2. Oocyte Source, In Vitro Fertilization (IVF) and Embryo Culture

Approximately 600 female mice were super-ovulated by peritoneal injection with 5 IU pregnant mare serum gonadotropin (PMSG, Ningbo, China) and 5 IU human chorionic gonadotrophin (HCG, Ningbo, China) after 48 h. For IVC, the super-ovulated female mice were cocaged individually with male mice after HCG injection. The next morning, mice with vaginal plug were identified as successfully mated and were sacrificed to collect embryo–cumulus complexes. To isolate fertilized zygotes, the embryo–cumulus complexes were treated with 300 mg/mL of hyaluronidase to disperse the cumulus cells, after being washed in M2 medium, the zygotes with pronuclear were transferred to the potassium simplex optimized medium containing amino acids (KSOM + AA; Millipore, Darmstadt, Germany) for culture.

For IVF, the sperm were released from cauda epididymis and capacitated for 1 h in modified Krebs–Ringer bicarbonate medium (TYH). Female mice were sacrificed, and cumulus–oocyte complexes (COCs) were pulled out from oviduct and transferred to modified human tubal fluid (mHTF) for 30 min. Then, sperm were added to mHTF medium for fertilization. After 4 h of incubation, embryos were washed in mHTF and then transferred into KSOM for culture. Early 2-cell, late 2-cell and blastocysts were collected at 22, 29 and 96 h post insemination (hpi).

Unless otherwise stated, embryos were cultured in KSOM medium containing WNT2 (H00007472-P01, Abnova, China) or salinomycin (S8129, SELLECK, Houston, TX, USA) through preimplantation period.

4.3. *Analysis of Transcription Activity*

Embryos were incubated with KSOM containing 500 μM 5-ethynyl uridine (EU; RiboBio, Guangzhou, China) for 1 h at 37 °C and fixed in 4% paraformaldehyde (PFA) for 30 min. Then, embryos were permeabilized in 0.5% Triton X-100 (Sigma-Aldrich, St. Louis, MO, USA) for 20 min and incubated with Cell-Light^{TM} Apollo 488 Imaging Kit (Thermo Fisher Scientific, Waltham, MA, USA) for 30 min. After being washed in 0.5% Triton X-100, then treated for another 10 min. in 0.1% PBS-PVA. The embryos were incubated with 4′,6-diamidino-2-phenylindole (DAPI) for 15 min and mounted on glass slides. Images were captured by using fluorescence microscope (BX51TRF; Olympus, Tokyo, Japan) at a wavelength of 488 nm. The fluorescence intensity was quantified by using Image J software (https://imagej.net/ij/index.html (accessed on 1 January 2023), Rawak Software Inc., Stuttgart, Germany).

4.4. *Embryo Transfer*

On the day of IVF, approximately 60 pseudo-pregnant female mice were mated with vasectomized male mice. The next morning, mice with vaginal plug were recipients and consider as day 0.5. For embryo transfer, six well-developed blastocysts were transferred to each uterine horn of recipient. At embryonic day (E)19.5 (14 days after embryo transfer), recipients were sacrificed to evaluate the development ability of each embryo.

4.5. *Blastocyst Outgrowth*

First, 24-well plates were coated with 500 μL PBS containing 10 μg/mL fibronectin and incubated at 37 °C for 4 h. Then, wells were washed two times with PBS, and Dulbecco's Modified Eagle's medium (DMEM; 11965092, Waltham, MA, USA) containing 5% fetal bovine serum (FBS; Gibco) was added to wells for further equilibration at 37 °C for 4 h. At 96 hpi, well-developed blastocysts were transferred to wells and incubated for 48 h. The hatching rate, adhesion rate and outgrowth rate were calculated at 12, 24, 36 and 48 h following transfer to the outgrowth plate. Images of the outgrowing embryos were captured with an inverted microscope (IX71, Olympus, Tokyo, Japan) at 48 h. The area covered by the embryo was recorded as the outgrowth area.

4.6. *Immunofluorescence Staining*

Embryos were washed three times in 0.1%PBS-PVA, and the zona pellucida was removed by acidic Tyrode's solution (T1788, Sigma, St. Louis, MO, USA), and then, embryos were transferred to 4% PFA for 1 h. After being washed with 0.5% Triton X-100, embryos were blocked in 1% BSA for 1 h and incubated with primary antibodies overnight at 4 °C. Followed wash three times in 0.5% Triton X-100, the embryos were incubated with secondary antibodies for 1 h. Finally, the embryos were incubated with DAPI for 15 min to and fluorescence signals were observed under BX51 microscope (Olympus, Tokyo, Japan).

The following primary antibodies used in this research are listed as follows: anti-LRP6 (1:500 dilution, T58345S, Abmart, Shanghai, China), anti-NANOG (1:500 dilution, ab80892, Abcam, Cambridge, UK), anti-CDX2 (1:500 dilution, MU392A-UC, BioGenex

Laboratories, San Francisco, CA, USA), anti-H3K4me3 (1:500 dilution, ab8580, Abcam) and anti-H3K9me3 (1:1000 dilution, ab8898, Abcam).

4.7. GSVA and GSEA Analysis

WNT signaling correlated gene set was downloaded from KEGG databases (KEGG: Kyoto Encyclopedia of Genes and Genomes). GSVA analysis was performed by R package of GSVA, and GraphPad was used to visualize the result (https://www.graphpad.com/features, accessed on 1 January 2023). GSEA was performed by using javaGSEA software (https://www.gsea-msigdb.org/gsea/index.jsp, accessed on 1 January 2023), and R package ggplot2 was used to visualize these results.

4.8. Statistical Analysis

Two-tailed t test was used to analyze significant differences between the two groups. For multiple comparisons, one-way ANOVA was used to compare differences among groups by using IBM SPSS 18; significance was set as * for $p < 0.05$, ** for $p < 0.01$, *** for $p < 0.001$.

Author Contributions: Conceptualization, J.T., L.A., G.X. and F.Y.; methodology, J.T., L.A. and G.X.; software, F.Y., J.H. and J.Z.; validation, F.Y., J.H., Z.W. and M.C.; formal analysis, F.Y., J.H. and J.Z.; investigation, Z.Z.; resources, J.T. and L.A.; data curation, F.Y. and M.C.; writing—original draft preparation, F.Y., L.A. and J.H.; writing—review and editing, L.A., F.Y. and J.H.; visualization, F.Y. and J.Z.; supervision, J.T. and L.A.; project administration, J.T., L.A. and G.X.; funding acquisition, J.T. and L.A. All authors have read and agreed to the published version of the manuscript.

Funding: This work was supported by grants from the National Key R&D Program (2021YFD1300901, 2017YFD0501901 and 2017YFD0501905), the National Natural Science Foundation of China (no. 31930103 and 3167246) and the Beijing Innovation Consortium of Livestock Research System (grant number BAIC05-2022).

Institutional Review Board Statement: All mice experiments were approved by the Institutional Animal Care and Use Committee of China Agricultural University. The ethical clearance approval number is AW11107102-1-2.

Informed Consent Statement: Not applicable.

Data Availability Statement: The data presented in this study are available on request from the corresponding author.

Acknowledgments: The authors thank all the staff involved in the mouse care and feeding.

Conflicts of Interest: The authors declare no conflict of interest.

References

1. Hardy, K.; Spanos, S. Growth factor expression and function in the human and mouse preimplantation embryo. *J. Endocrinol.* **2002**, *172*, 221–236. [CrossRef]
2. Kaye, P.L. Preimplantation growth factor physiology. *Rev. Reprod.* **1997**, *2*, 121–127. [CrossRef] [PubMed]
3. Tan, K.; An, L.; Miao, K.; Ren, L.; Hou, Z.; Tao, L.; Zhang, Z.; Wang, X.; Xia, W.; Liu, J.; et al. Impaired imprinted X chromosome inactivation is responsible for the skewed sex ratio following in vitro fertilization. *Proc. Natl. Acad. Sci. USA* **2016**, *113*, 3197–3202. [CrossRef] [PubMed]
4. Fu, W.; Yue, Y.; Miao, K.; Xi, G.; Zhang, C.; Wang, W.; An, L.; Tian, J. Repression of FGF signaling is responsible for Dnmt3b inhibition and impaired de novo DNA methylation during early development of in vitro fertilized embryos. *Int. J. Biol. Sci.* **2020**, *16*, 3085–3099. [CrossRef] [PubMed]
5. Liu, X.; Hao, Y.; Li, Z.; Zhou, J.; Zhu, H.; Bu, G.; Liu, Z.; Hou, X.; Zhang, X.; Miao, Y.L. Maternal Cytokines CXCL12, VEGFA, and WNT5A Promote Porcine Oocyte Maturation via MAPK Activation and Canonical WNT Inhibition. *Front. Cell Dev. Biol.* **2020**, *8*, 578. [CrossRef]
6. Hiradate, Y.; Hara, K.; Tanemura, K. Effect of neurotensin on cultured mouse preimplantation embryos. *J. Reprod. Dev.* **2020**, *66*, 421–425. [CrossRef]
7. Gambini, A.; Stein, P.; Savy, V.; Grow, E.J.; Papas, B.N.; Zhang, Y.; Kenan, A.C.; Padilla-Banks, E.; Cairns, B.R.; Williams, C.J. Developmentally Programmed Tankyrase Activity Upregulates beta-Catenin and Licenses Progression of Embryonic Genome Activation. *Dev. Cell* **2020**, *53*, 545–560. [CrossRef]

8. Xie, H.; Tranguch, S.; Jia, X.; Zhang, H.; Das, S.K.; Dey, S.K.; Kuo, C.J.; Wang, H. Inactivation of nuclear Wnt-beta-catenin signaling limits blastocyst competency for implantation. *Development* **2008**, *135*, 717–727. [CrossRef]
9. Albrecht, L.V.; Tejeda-Muñoz, N.; De Robertis, E.M. Cell Biology of Canonical Wnt Signaling. *Annu. Rev. Cell Dev. Biol.* **2021**, *37*, 369–389. [CrossRef]
10. Gordon, M.D.; Nusse, R. Wnt signaling: Multiple pathways, multiple receptors, and multiple transcription factors. *J. Biol. Chem.* **2006**, *281*, 22429–22433. [CrossRef]
11. Barker, N.; Hurlstone, A.; Musisi, H.; Miles, A.; Bienz, M.; Clevers, H. The chromatin remodelling factor Brg-1 interacts with beta-catenin to promote target gene activation. *EMBO J.* **2001**, *20*, 4935–4943. [CrossRef] [PubMed]
12. Hecht, A.; Vleminckx, K.; Stemmler, M.P.; van Roy, F.; Kemler, R. The p300/CBP acetyltransferases function as transcriptional coactivators of beta-catenin in vertebrates. *EMBO J.* **2000**, *19*, 1839–1850. [CrossRef] [PubMed]
13. Macdonald, B.T.; Tamai, K.; He, X. Wnt/beta-catenin signaling: Components, mechanisms, and diseases. *Dev. Cell* **2009**, *17*, 9–26. [CrossRef] [PubMed]
14. Harwood, B.N.; Cross, S.K.; Radford, E.E.; Haac, B.E.; De Vries, W.N. Members of the WNT signaling pathways are widely expressed in mouse ovaries, oocytes, and cleavage stage embryos. *Dev. Dyn.* **2008**, *237*, 1099–1111. [CrossRef]
15. Wang, J.; Sinha, T.; Wynshaw-Boris, A. Wnt signaling in mammalian development: Lessons from mouse genetics. *Cold Spring Harb. Perspect. Biol.* **2012**, *4*, a007963. [CrossRef]
16. Chu, M.; Yao, F.; Xi, G.; Yang, J.; Zhang, Z.; Yang, Q.; Tian, J.; An, L. Vitamin C Rescues in vitro Embryonic Development by Correcting Impaired Active DNA Demethylation. *Front. Cell Dev. Biol.* **2021**, *9*, 784244. [CrossRef]
17. Lu, D.; Choi, M.Y.; Yu, J.; Castro, J.E.; Kipps, T.J.; Carson, D.A. Salinomycin inhibits Wnt signaling and selectively induces apoptosis in chronic lymphocytic leukemia cells. *Proc. Natl. Acad. Sci. USA* **2011**, *108*, 13253–13257. [CrossRef]
18. Abe, K.I.; Funaya, S.; Tsukioka, D.; Kawamura, M.; Suzuki, Y.; Suzuki, M.G.; Schultz, R.M.; Aoki, F. Minor zygotic gene activation is essential for mouse preimplantation development. *Proc. Natl. Acad. Sci. USA* **2018**, *115*, E6780–E6788. [CrossRef]
19. Chen, Z.; Zhang, Y. Loss of DUX causes minor defects in zygotic genome activation and is compatible with mouse development. *Nat. Genet.* **2019**, *51*, 947–951. [CrossRef]
20. Jao, C.Y.; Salic, A. Exploring RNA transcription and turnover in vivo by using click chemistry. *Proc. Natl. Acad. Sci. USA* **2008**, *105*, 15779–15784. [CrossRef]
21. Uribe-Etxebarria, V.; Garcia-Gallastegui, P.; Perez-Garrastachu, M.; Casado-Andres, M.; Irastorza, I.; Unda, F.; Ibarretxe, G.; Subiran, N. Wnt-3a Induces Epigenetic Remodeling in Human Dental Pulp Stem Cells. *Cells* **2020**, *9*, 652. [CrossRef] [PubMed]
22. Dahl, J.A.; Jung, I.; Aanes, H.; Greggains, G.D.; Manaf, A.; Lerdrup, M.; Li, G.; Kuan, S.; Li, B.; Lee, A.Y.; et al. Broad histone H3K4me3 domains in mouse oocytes modulate maternal-to-zygotic transition. *Nature* **2016**, *537*, 548–552. [CrossRef] [PubMed]
23. Xu, R.; Li, S.; Wu, Q.; Li, C.; Jiang, M.; Guo, L.; Chen, M.; Yang, L.; Dong, X.; Wang, H.; et al. Stage-specific H3K9me3 occupancy ensures retrotransposon silencing in human pre-implantation embryos. *Cell Stem Cell* **2022**, *29*, 1051–1066. [CrossRef] [PubMed]
24. Chronopoulou, E.; Harper, J.C. IVF culture media: Past, present and future. *Hum. Reprod. Update* **2015**, *21*, 39–55. [CrossRef] [PubMed]
25. Wang, H.; Dey, S.K. Roadmap to embryo implantation: Clues from mouse models. *Nat. Rev. Genet.* **2006**, *7*, 185–199. [CrossRef] [PubMed]
26. Wang, J.; Mayernik, L.; Armant, D.R. Integrin signaling regulates blastocyst adhesion to fibronectin at implantation: Intracellular calcium transients and vesicle trafficking in primary trophoblast cells. *Dev. Biol.* **2002**, *245*, 270–279. [CrossRef]
27. Zhang, Y.; Yang, Z.; Wu, J. Signaling pathways and preimplantation development of mammalian embryos. *FEBS J.* **2007**, *274*, 4349–4359. [CrossRef]
28. Xia, W.; Xie, W. Rebooting the Epigenomes during Mammalian Early Embryogenesis. *Stem Cell Rep.* **2020**, *15*, 1158–1175. [CrossRef]
29. Shin, S.W.; Tokoro, M.; Nishikawa, S.; Lee, H.H.; Hatanaka, Y.; Nishihara, T.; Amano, T.; Anzai, M.; Kato, H.; Mitani, T.; et al. Inhibition of the ubiquitin-proteasome system leads to delay of the onset of ZGA gene expression. *J. Reprod. Dev.* **2010**, *56*, 655–663. [CrossRef]
30. Sims, R.R.; Reinberg, D. Stem cells: Escaping fates with open states. *Nature* **2009**, *460*, 802–803. [CrossRef]
31. Liu, X.; Wang, C.; Liu, W.; Li, J.; Li, C.; Kou, X.; Chen, J.; Zhao, Y.; Gao, H.; Wang, H.; et al. Distinct features of H3K4me3 and H3K27me3 chromatin domains in pre-implantation embryos. *Nature* **2016**, *537*, 558–562. [CrossRef] [PubMed]
32. Zhang, B.; Zheng, H.; Huang, B.; Li, W.; Xiang, Y.; Peng, X.; Ming, J.; Wu, X.; Zhang, Y.; Xu, Q.; et al. Allelic reprogramming of the histone modification H3K4me3 in early mammalian development. *Nature* **2016**, *537*, 553–557. [CrossRef] [PubMed]
33. Lindeman, L.C.; Andersen, I.S.; Reiner, A.H.; Li, N.; Aanes, H.; Ostrup, O.; Winata, C.; Mathavan, S.; Muller, F.; Alestrom, P.; et al. Prepatterning of developmental gene expression by modified histones before zygotic genome activation. *Dev. Cell* **2011**, *21*, 993–1004. [CrossRef] [PubMed]
34. Sridharan, R.; Gonzales-Cope, M.; Chronis, C.; Bonora, G.; Mckee, R.; Huang, C.; Patel, S.; Lopez, D.; Mishra, N.; Pellegrini, M.; et al. Proteomic and genomic approaches reveal critical functions of H3K9 methylation and heterochromatin protein-1gamma in reprogramming to pluripotency. *Nat. Cell Biol.* **2013**, *15*, 872–882. [CrossRef] [PubMed]
35. Sankar, A.; Lerdrup, M.; Manaf, A.; Johansen, J.V.; Gonzalez, J.M.; Borup, R.; Blanshard, R.; Klungland, A.; Hansen, K.; Andersen, C.Y.; et al. KDM4A regulates the maternal-to-zygotic transition by protecting broad H3K4me3 domains from H3K9me3 invasion in oocytes. *Nat. Cell Biol.* **2020**, *22*, 380–388. [CrossRef]

36. Barrera, A.D.; Garcia, E.V.; Hamdi, M.; Sanchez-Calabuig, M.J.; Lopez-Cardona, A.P.; Balvis, N.F.; Rizos, D.; Gutierrez-Adan, A. Embryo culture in presence of oviductal fluid induces DNA methylation changes in bovine blastocysts. *Reproduction* **2017**, *154*, 1–12. [CrossRef]
37. Canovas, S.; Ivanova, E.; Romar, R.; Garcia-Martinez, S.; Soriano-Ubeda, C.; Garcia-Vazquez, F.A.; Saadeh, H.; Andrews, S.; Kelsey, G.; Coy, P. DNA methylation and gene expression changes derived from assisted reproductive technologies can be decreased by reproductive fluids. *Elife* **2017**, *6*, e23670. [CrossRef]

genes
MDPI

Article

Intrafollicular Retinoic Acid Signaling Is Important for Luteinizing Hormone-Induced Oocyte Meiotic Resumption

Fupeng Wang †, Yawen Tang †, Yijie Cai, Ran Yang, Zongyu Wang, Xiaodong Wang, Qianying Yang, Wenjing Wang, Jianhui Tian and Lei An *

State Key Laboratory of Animal Biotech Breeding, National Engineering Laboratory for Animal Breeding, Key Laboratory of Animal Genetics, Breeding and Reproduction of the Ministry of Agriculture and Rural Affairs, College of Animal Science and Technology, China Agricultural University, No. 2 Yuanmingyuan West Road, Beijing 100193, China
* Correspondence: anleim@cau.edu.cn
† These authors contributed equally to the work.

Abstract: It has been clear that retinoic acid (RA), the most active vitamin A (VA) derivative, plays a central role in governing oocyte meiosis initiation. However, it has not been functionally determined if RA participates in luteinizing hormone (LH)-induced resumption from long-lasting oocyte meiotic arrest, which is essential for haploid oocyte formation. In the present study, using well-established in vivo and in vitro models, we identified that intrafollicular RA signaling is important for normal oocyte meiotic resumption. A mechanistic study indicated that mural granulosa cells (MGCs) are the indispensable follicular compartment for RA-prompted meiotic resumption. Moreover, retinoic acid receptor (RAR) is essential for mediating RA signaling to regulate meiotic resumption. Furthermore, we found zinc finger protein 36 (ZFP36) is the transcriptional target of RAR. Both RA signaling and epidermal growth factor (EGF) signaling were activated in MGCs in response to LH surge, and two intrafollicular signalings cooperate to induce rapid *Zfp36* upregulation and *Nppc* mRNA decrease, which is critical to LH-induced meiotic resumption. These findings extend our understanding of the role of RA in oocyte meiosis: RA not only governs meiotic initiation but also regulates LH-induced meiotic resumption. We also emphasize the importance of LH-induced metabolic changes in MGCs in this process.

Keywords: LH; Retinoic acid; EGF; oocyte meiotic resumption; C-natriuretic peptide; zinc finger protein 36

Citation: Wang, F.; Tang, Y.; Cai, Y.; Yang, R.; Wang, Z.; Wang, X.; Yang, Q.; Wang, W.; Tian, J.; An, L. Intrafollicular Retinoic Acid Signaling Is Important for Luteinizing Hormone-Induced Oocyte Meiotic Resumption. *Genes* **2023**, *14*, 946. https://doi.org/10.3390/genes14040946

Academic Editor: Bao Yuan

Received: 28 February 2023
Revised: 15 April 2023
Accepted: 16 April 2023
Published: 20 April 2023

1. Introduction

Retinoic acid (RA), an active derivative of vitamin A (VA), is involved in a number of functions in vertebrates [1,2]. Among these, RA plays an important role in regulating cell cycle and differentiation, etc. [3], and is essential for regulating meiotic initiation in both male and female germ cells [4]. In female embryonic ovaries, RA triggers germ cells' initial transition from mitosis to meiosis by stimulating a critical meiotic marker, named stimulated by retinoic acid gene 8 (*Stra8*), which is expressed in embryonic ovaries just before meiotic initiation [5,6]. In addition, increasing in vivo and in vitro evidence indicates that RA also participates in the process of folliculogenesis, oocyte maturation and ovulation [2]. Of interest, a series of previous in vitro studies among species have suggested that RA may promote the resumption of oocyte meiosis [7–12], a critical event that is essential for haploid oocyte formation [13], despite the controversial results [14,15]. However, these studies are primarily based on in vitro supplementation assays, and the role of RA in regulating oocyte meiotic resumption has never been systematically confirmed via in vivo physiological analysis and in vitro functional experiments. In addition, most above-mentioned studies haven't focused on the mechanism by which RA prompts oocyte meiotic resumption, and the underlying mechanism remains unclear.

In mammals, oocytes are arrested at the diplotene stage of meiosis until the surge of luteinizing hormone (LH) restarts meiosis and initiates the ovulatory process. Thus, successful oocyte meiotic resumption is the prerequisite for female fertility. The progression of gonadotropin-controlled oocyte meiosis is tightly regulated by C-natriuretic peptide (CNP). Before the LH surge, CNP is primarily synthesized by mural granulosa cells (MGCs) and binds to NPR2 on cumulus cells (CCs) throughout the follicle to stimulate cGMP production, thus maintaining meiotic arrest [16,17]. Following the LH surge, CNP and its encoding messenger RNA (mRNA) natriuretic peptide precursor C (*Nppc*) decrease immediately and restart meiosis [18]. Our recently published study has shown that TTP, also known as zinc finger protein 36 (ZFP36), a ubiquitously expressed mRNA-destabilizing protein, is the critical mechanism that underlies the LH-induced rapid decrease in *Nppc* mRNA and oocyte meiotic resumption. *Zfp36* mRNA is transiently up-regulated in MGCs in response to the LH surge. Then, upregulated TTP binds to *Nppc* mRNA 3′ UTR and, in turn, degrades *Nppc* mRNA [19].

By reanalyzing a previously published transcriptome that presents the rapid effects of LH on gene expression in the MGCs of mouse preovulatory follicles [20], we found that LH triggered the activation of the retinol metabolism pathway. More importantly, we found several retinoic acid response element (RARE) motifs located in the transcriptional regulatory regions of *Zfp36*. Taken together, these facts imply that RA may participate in LH-induced oocyte meiotic resumption by prompting rapid upregulation of *Zfp36*.

In this study, we systematically confirmed the role of RA in prompting oocyte meiotic resumption using the in vivo VA-deficient mouse model and in vitro culture models, including in vitro cultured preovulatory follicles, in vitro matured cumulus-oocyte complexes (COCs) and denuded oocytes. In addition, focusing on the receptor that mediates RA signaling and its downstream target that regulates meiotic progression, we explored the underlying mechanism by which RA regulates oocyte meiotic resumption. Thus, our study updates the role of RA in regulating the progression of oocyte meiosis: RA is important not only for meiotic entry but also for meiotic resumption.

2. Material and Methods

2.1. Animals

Eight-week-old ICR female mice were kept in a temperature-controlled room with 12-h alternating light/dark. All experiments with mice were approved by the Institutional Animal Care and Use Committee of China Agricultural University. The ethical clearance approval number is AW11107102-1-2.

Weanling female mice used for in vivo experiments were fed a vitamin A deficient diet (VAD; Beijing HFK Bioscience, Beijing, China; Table S2) for up to 6 weeks.

2.2. Culture of Preovulatory Follicles

Female mice were i.p. injected with 5 IU Equine chorionic gonadotropin (eCG; Ningbo Sansheng Biotechnology, Zhejiang, China) to stimulate follicle development to the preovulatory stage, as previous studies reported [21–25]. After 48 h, in vivo oocyte meiotic resumption was induced by an i.p. injection of 5 IU human chorionic gonadotrophin (hCG; Ningbo Sansheng Biotechnology). For in vitro culture, preovulatory follicles (POFs) were microdissected from eCG-primed mouse ovaries with 25-gauge needles under a stereomicroscope in Leibowitz L15 medium (Thermo Fisher Scientific, Waltham, MA, USA) supplemented with 5% fetal bovine serum (FBS; Thermo Fisher Scientific), 100 U Penicillin-Streptomycin (Thermo Fisher Scientific) as previously described [21]. Then they were transferred through α-MEM (Thermo Fisher Scientific) supplemented with 10% FBS, 100 U penicillin G, and 100 μg streptomycin sulfate 3 times before treatment. Subsequently, the POFs (5–30 POFs/group) were cultured for the indicated time in fresh α-MEM supplemented with RA (10 μM), with or without BMS493 (10 μM; Selleck, Houston, TX, USA), under 95% O_2 and 5% CO_2. To assess the ratio of in vitro germinal vesicle breakdown (GVBD), POFs were punctured to release the cumulus-oocyte complexes (COCs)and placing them in supplemented α-MEM, oocytes were denuded of cumulus cells and evaluated for morphological evidence of GVBD under a stereomicroscope.

2.3. Culture of COCs

COCs were isolated by puncturing the ovaries from eCG-stimulated mice with 25-gauge needles in the supplemented Leibowitz L15 medium. After isolation, COCs with homogeneous cytoplasm and more than 3 layers of unexpanded cumulus cells were selected as good-quality oocytes, as mentioned previously [26,27]. Then COCs were washed in the final incubation medium and cultured for the indicated time. The culture medium was α-MEM supplemented with 3% (wt/vol) bovine serum albumin (BSA; Sigma-Aldrich, Burlington, USA), 100 U Penicillin-Streptomycin, with or without BMS493 (10 μM; Selleck), UVI3003 (10 μM; Selleck). Cultures were maintained in an atmosphere of 95% O_2 and 5% CO_2 at 37 °C. The ratio of in vitro germinal vesicle breakdown (GVBD) was assessed after culture and oocytes were denuded of cumulus cells and evaluated for morphological evidence of GVBD under a stereomicroscope which refers to the dissolution of the huge nucleus of an oocyte that is arrested in prophase of meiosis I. The oocyte nucleus presents an irregular envelope surrounding dispersed condensed chromatin and is marked by the extrusion of the first polar body [28,29].

2.4. Culture of Mouse MGCs

Ovaries were removed from the female mice, and the MGCs were extracted by puncturing ovaries. Collected MGCs were purified by filtration using a 70-μm cell strainer (BS-70-CS; Beijing Labgic Technology, Beijing, China) centrifuged at 200× *g* for 5 min. After brief centrifugation and washing with DMEM/F-12 medium (Thermo Fisher Scientific) supplemented with 1% FBS, 100 U Penicillin-Streptomycin, MGCs were re-suspended in DMEM/F-12 culture media supplemented with 10% FBS, 100 U Penicillin-Streptomycin in 6-wells plate at 37 °C and 5% CO_2 overnight to allow the cells to attach. Before treatment, cells were washed thoroughly with phosphate-buffered saline (PBS) to remove non-adherent cells. Subsequently, the cells were cultured in fresh medium supplemented with different final concentrations (0.7 nM, 7 nM, 70 nM, or 700 nM) of RA (Sigma-Aldrich), EGF (20 ng/mL; Sigma-Aldrich), 0.1 μM TTNPB (Selleck), 1 μM BMS493 (Selleck), 0.1 μM PD0325901 (Selleck) or various combinations. Standard conditions were maintained for the control group. After the indicated time, the cells were harvested for further experiments.

2.5. RNA Extraction and Quantitative RT-PCR Analysis (qRT-PCR)

The total RNA from in vitro cultured MGCs was isolated using TRIzol (Thermo Fisher Scientific) according to the manufacturer's protocol. The cDNA synthesis was run in 20 μL volumes containing 1000 ng of total RNA by using the HiScript III RT SuperMix for qPCR (Vazyme Biotech, Nanjing, China) according to the manufacturer's instructions. For quantitative PCR analysis, 10 ng of cDNA (total RNA equivalent) per well in duplicate was amplified using gene-specific primers in Table S1 and SsoFast EvaGreen Supermix (BioRad Laboratories, CA, USA) on a Bio-Rad CFX96 Real-Time PCR System. The amplification conditions were as follows: initial denaturation for 30 s at 95 °C and 40 cycles of 5 s at 95 °C, annealing and elongation for 5 s at 60 °C. For each gene, at least 3 technical replicates and three biological replicates were assayed. *Gapdh* mRNA was used as a loading control, and data were analyzed using the $2^{-\Delta\Delta Ct}$ method.

2.6. Western Blotting

Protein lysates were prepared from collected granulosa cells, and a cell homogenization solution (Beyotime Biotechnology, Beijing, China) containing 1% Halt Protease and Phosphatase Inhibitor Cocktail (Thermo Fisher Scientific) was used to isolate protein from granulosa cells. The supernatant was collected after 10 min centrifugation at 15,000× *g* at 4 °C, and protein was quantified using an Enhanced Bicinchoninic Acid Protein Assay Kit (Beyotime Biotechnology). The samples were denatured in the same volume of

2 × Laemmli sample loading buffer (Bio-Rad Laboratories) with 5% β-mercaptoethanol (Sigma-Aldrich) for 5 min at 100 °C and stored at −80 °C for future use. Approximately 10 μg of total protein sample was loaded in each well for 12% SDS-PAGE and then transferred onto microporous PVDF membranes (Millipore) at 4 °C. Membranes containing the transferred proteins were blocked using 5% milk powder (*w*/*v*) in Tris-buffered saline containing 0.1% Tween 20 (TBST; 20 mM Tris–HCl, 150 mM NaCl and 0.1% Tween 20, pH 7.6) for 2 h at room temperature (RT). Membranes were incubated overnight at 4° C with primary antibodies at pre-determined concentrations (1:200, TTP (A-8), Santa Cruz, Cat#SC-374305; 1:1000, GAPDH, CST, Cat#2118s). Blots were washed in TBST and then incubated in horseradish peroxidase (HRP)–conjugated secondary antibodies (1:5000, Golden Bridge, Beijing, China) for 1 h at RT. After three washes in TBST, blots were developed using enhanced chemiluminescence reagents (Millipore), exposed digitally with Tanon 5200 chemiluminescent imaging system (Tanon, China)

2.7. GSEA Analysis

The retinol metabolism, reactome signaling by EGFR, and the response to retinoic acid correlated gene set were downloaded from the Kyoto Encyclopedia of Genes and Genomes (KEGG) databases (KEGG: Kyoto Encyclopedia of Genes and Genomes). Gene Set Enrichment Analysis (GSEA) was performed with these 3 gene sets by using javaGSEA software.

2.8. Cleavage under Targets & Release Using Nuclease (CUT&RUN)

CUT&RUN was performed as previously described [30–32]. The qRT-PCR primer is shown in Table S1. Retinoic Acid Receptor α (SC 515796, Santa Cruz Biotechnology, CA, USA) antibody was used in this experiment. Granulosa cells were collected and resuspended in 600 μL cold nuclear extraction (NE) buffer (20 mM HEPES-KOH, pH 7.9, 10 mM KCl, 0.5 mM Spermidine, 0.1% Triton X-100, 20% glycerol, freshly added protease inhibitors) for 10 min. 20 μL Concanavalin A beads (Polysciences) were prepared per sample. Beads were washed and resuspended in 300 uL Binding Buffer (20 mM HEPES-KOH, pH 7.9, 10 mM KCl, 1 mM CaCl2, 1 mM MnCl2). After centrifugation, Nuclei were added to beads and incubated for 10 min at room temperature (RT). After nuclei binding, samples were placed on a magnetic stand for 5 min, and the beads were resuspended in 1 mL cold blocking buffer (20 mM HEPES, pH 7.5, 150 mM NaCl, 0.5 mM Spermidine, 0.1% BSA, 2 mM EDTA, freshly added protease inhibitors) for 5 min. Then samples were placed on a magnetic stand and resuspended in 250 uL cold Wash Buffer (20 mM HEPES, pH 7.5, 150 mM NaCl, 0.5 mM Spermidine, 0.1% BSA, freshly added protease inhibitors). The primary antibody was added to a final concentration of 1:100 at 4 °C for 2 h. IgG was used as a negative control. Protein A-micrococcal nuclease (pA-MN; 0.95 ug/uL pA-MN was a gift from Wei Xie's lab) was added to a final concentration of 1:750 at 4 °C for 1 h. Samples were resuspended in 150 uL cold wash buffer, and 3 uL 100 mM $CaCl_2$ was added to the sample for 30 min. The reactions were stopped using 2 X STOP buffer (200 mM NaCl, 20 mM EDTA, 4 mM EGTA, 50 ug/mL RNaseA, 40 ug/mL glycogen, 10 pg/mL yeast spike-in DNA) for 20 min at 37 °C, and 3 uL 10% SDS and 2.5 uL 20 mg/mL Proteinase K were added to the supernatants incubated at 70 °C for 10 min. DNA was purified using phenol/chloroform/isoamyl alcohol (PCI; Ameresco), and the sample could then be used for qRT-PCR analysis.

2.9. Reanalysis of Published RNA-Sequencing Data

The original down sequence (Raw Reads) was obtained from the GEO data repository (GSE167939). Raw RNA-seq reads after trimming by Trim-galore (version 0.6.5) were aligned to the mouse (GRCm38/mm10) genome using HISAT2 (version 2.2.1) with default parameters. Raw read counts for each gene were generated using the featureCounts program based on the mouse gene annotation (Mus_musculus.GRCm38.102.chr.gtf). The read counts for each gene were calculated, and the expression values of each gene were normalized by TPM. To identify differentially expressed genes in RNA-seq, the DESeq2 R package was used for differential gene expression analyses with cutoffs FoldChange > 2 and FDR values (P adjusted: Padj values) < 0.05.

2.10. Statistical Analysis

Experiments were independently replicated for a minimum of three times unless otherwise specified, and all data is reported as the mean ± standard error of the mean (SEM). Student's t-test was used to analyze the significant difference between the 2 groups. For multiple comparisons, data were analyzed using a 1-way ANOVA followed by Tukey's multiple comparison test using IBM SPSS 18. Statistically significant differences were defined as * for $p < 0.05$, ** for $p < 0.01$, and *** for $p < 0.001$.

3. Results

3.1. VA Deficiency Impairs Oocyte Meiotic Resumption

To explore the role of RA in oocyte meiotic resumption, we used an in vivo mouse model that was established by feeding a VA-deficient (VAD) diet for up to 6 weeks, which was frequently used to study the physiological functions of RA [2,33,34]. When oocyte meiotic resumption, scored as germinal vesicle breakdown (GVBD) percentage, was evaluated at 6 h post-hCG (Figure 1A), VAD mice showed impaired meiotic resumption compared with that in the wild-type controls (Figure 1B). Next, to determine the follicular compartment responsible for RA-prompted oocyte meiotic resumption, we isolated cumulus-oocyte complexes (COCs) and denuded oocytes (DOs) from preovulatory follicles of stimulated VAD and control mice (Figure 1C). After in vitro maturation (IVM), oocyte meiotic resumption was not affected by VA deficiency when culturing neither COCs nor DOs (Figure 1D). Thus, our in vivo and in vitro results suggest that VA deficiency impairs oocyte meiotic resumption in an MGC-dependent manner.

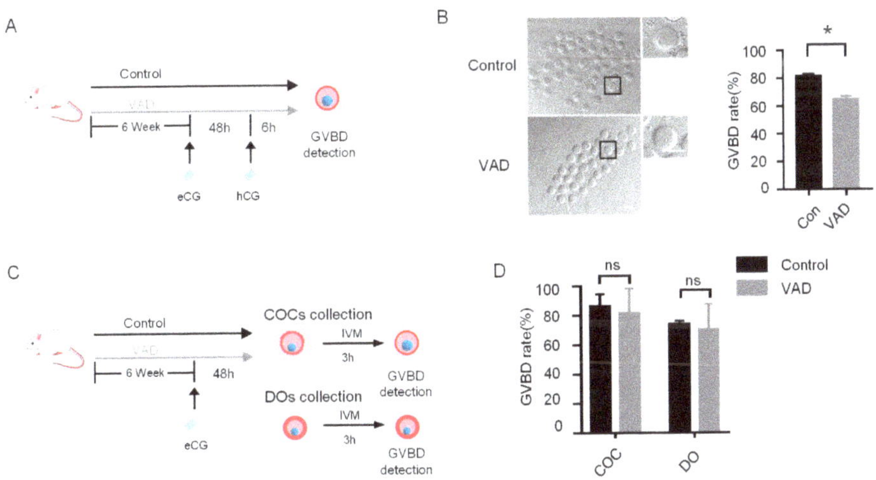

Figure 1. VA deficiency impairs oocyte meiotic resumption. (**A**) Flow chart of experimental design based on in vivo VAD mouse model. (**B**) Quantitative analysis of GVBD rates in control and VAD mouse. * $p < 0.05$. Left panel: representative image of GV and GVBD oocytes from control or VAD mouse. (**C**,**D**) Flow chart of experimental design in vitro maturation (IVM) of COCs and DOs from control and VAD mouse (**C**) and quantitative analysis of GVBD rates when using COCs or DOs. All data are presented as the mean ± SEMs of three independent experiments. ns, not significant ($p > 0.05$). * $p < 0.05$.

3.2. LH Activates RA Synthetic and Metabolic Pathway in Preovulatory MGCs

Having confirmed VA deficiency impairs oocyte meiotic resumption, we next attempted to ascertain whether RA, the most active VA derivative, is primarily responsible for these findings. We reanalyzed a previously published transcriptome that presents rapid effects of LH on gene expression in the MGCs of mouse preovulatory follicles [20]. Among differentially expressed genes responding to LH surge (Figure 2A) are functionally associated with retinoic acid signaling transduction (Figure 2B). GSEA analysis also indicated that the retinol metabolic pathway was activated in MGCs during LH-induced meiotic resumption (Figure 2C). In detail, genes associated with cellular retinol uptake and RA synthesis were generally upregulated during this process, whereas genes encoding RA-degrading enzymes were downregulated (Figure 2D), suggesting intrafollicular RA signaling may be responsible for VAD-induced impairment of oocyte meiotic resumption.

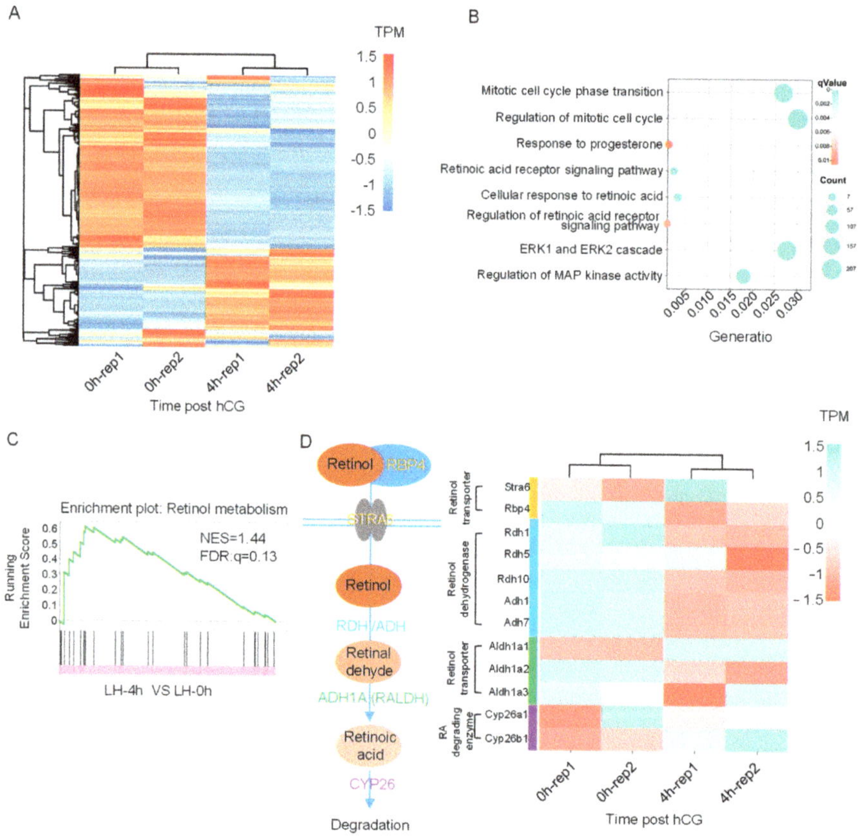

Figure 2. LH activates RA synthetic and metabolic pathways in preovulatory MGCs. Data are presented based on the reanalysis of previously published transcriptome data that present rapid effects of LH on gene expression in the MGCs of mouse preovulatory follicles. (**A**) Heat map of the hierarchical clustering results of DEGs in response to LH surge. (**B**) Bubble plot of the enriched GO terms of DEGs in response to LH. The *X*-axis in the bubble plot represents GeneRatio, while the *y*-axis indicates different biological processes. (**C**) GSEA analysis showing enrichment of Retinol metabolism, Normalized enrichment score (NES) and false discovery rate (FDR) q value are shown. (**D**) Schematic illustration of the retinoic acid metabolic pathway (Left panel) and the heat map of relevant genes in response to LH surge.

3.3. *RA Signaling Disruption Impairs Oocyte Meiotic Resumption*

Next, we attempted to provide direct evidence supporting the role of RA signaling in oocyte meiotic resumption. We isolated preovulatory follicles from eCG-primed mouse ovaries. Preovulatory follicles were cultured with LH for 4 h in the presence or absence of a RAR inhibitor, and we detected oocyte meiotic resumption after LH exposure (Figure 3A). We found that blockage of RA signaling by using 10 μM RAR pan-inhibitor BMS493 significantly impaired oocyte meiotic resumption scored as GVBD rate (Figure 3B). In contrast, RA supplementation during the culture significantly facilitated meiotic resumption, and the facilitating effect can be eliminated by RAR inhibition (Figure 3C,D and Figure S1A,B). In addition, in line with results from in vivo VAD model, RA signaling blockage in both in vitro cultured COCs and DOs (Figure 3E) showed that either RAR inhibition by BMS493 (Figure 3F,G, left panel) or RXR inhibition by UVI3003 (Figure 3F,G, right panel), didn't affect the progression of meiotic resumption. These observations further support the critical role of MGCs in mediating RA-prompted oocyte meiotic resumption.

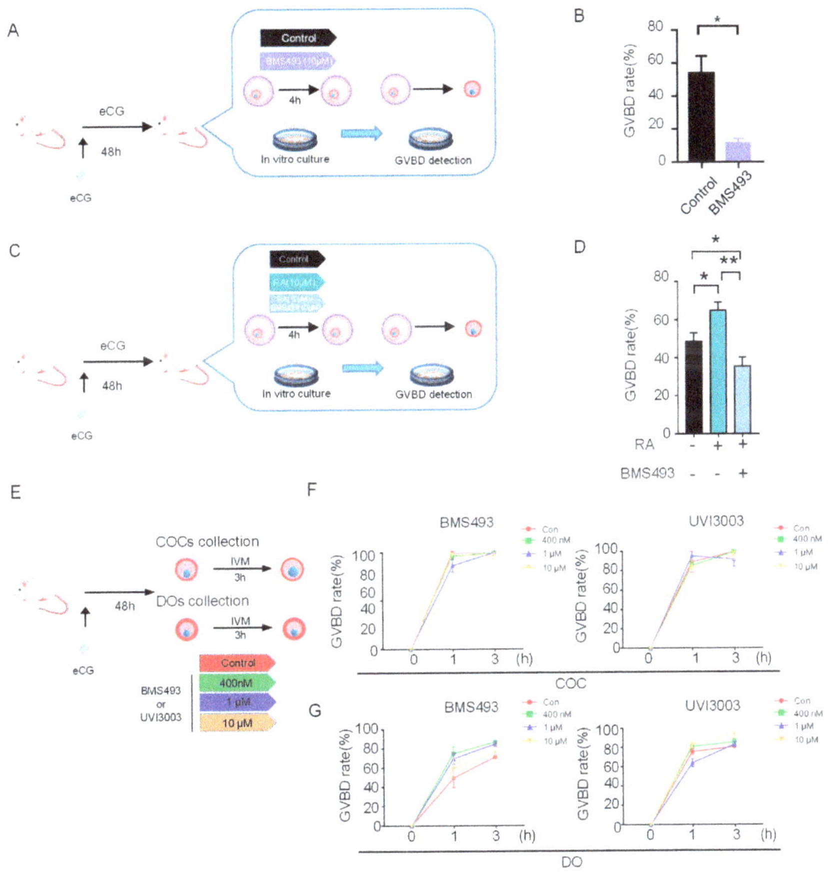

Figure 3. RA signaling disruption impairs oocyte meiotic resumption. (**A**) Flow chart of experimental design based on in vitro culture of eCG-primed preovulatory follicles treated with or without RAR pan-inhibitor BMS493. (**B**) Quantitative analysis of GVBD rates of oocytes collected from preovulatory follicles cultured under different IVM conditions. * $p < 0.05$ (**C**) Flow chart of experimental design based on in vitro culture of eCG-primed preovulatory follicles treated with RA alone or together with BMS493. (**D**) Quantitative analysis of GVBD rates of oocytes collected from preovulatory follicles cultured under different IVM conditions. * $p < 0.05$, ** $p < 0.01$. (**E**) Flow chart of experimental design based on IVM of COCs and DOs groups with or without different concentrations of BMS493 and UVI3003. (**F**,**G**) Quantitative analysis of GVBD rates of oocytes collected from COCs (**F**) and DOs (**G**) cultured with BMS493 or UVI3003. All data are presented as the mean ± SEM of three independent experiments. * $p < 0.05$.

3.4. *Zfp36* Is the Potential Transcriptional Target of RAR

Given MGC is critical to mediating RA-prompted oocyte meiotic resumption, we next examined the underlying mechanism in MGCs. Our recently published study showed that LH-induced rapid up-regulation of *Zfp36* in MGCs. *Zfp36* encodes an mRNA-destabilizing protein TTP that degrades "meiosis arrester" *Nppc* mRNA and, in turn, regulates oocyte meiotic resumption [19]. The spatiotemporal coincidence of LH-induced upregulation of RA signaling and *Zfp36* in MGCs (Figure 4A), as well as their functional role in prompting meiotic resumption, led us to ask if RA signaling may participate in Zfp36 upregulation. Supporting this hypothesis, the bioinformatic analysis showed TGAGTTCNAGGVCAGCCWG motif fitting the binding sequence for RARα and RARγ within the promoter region of Zfp36 (Figure 4B), in line with their expression patterns in preovulatory MGCs (Figure 4C). We noticed RAR motifs were located proximal to the binding sequences of EGR1 and ELK1, two identified transactional factors of *Zfp36* in a previous study [35] (Figure 4D), and reconstructed the protein-protein network of potential transcription factors also implied the interaction between EGR1, ELK1 and RAR (Figure 4E).

3.5. RA Prompts the Expression of *Zfp36* in MGC via RAR

To directly test if RA participates in LH-induced rapid upregulation of *Zfp36* in MGCs, we next isolated preovulatory MGCs and supplemented RA at different concentrations during in vitro culture. Results showed that a 7 nM RA supplementation medium could significantly upregulate *Zfp36* (Figure 5A) and led to a significant decrease in *Nppc* mRNA levels (Figure 5B), showing a negative correlation with those in the control group (Figure 5C). Furthermore, the stimulating effect of RA on *Zfp36* expression (Figure 5D and Figure S1C) can be eliminated by RAR inhibition, highlighting the essential role of RAR in mediating these effects. The role of RA in inducing Zfp36 upregulation and the mediating role of RAR was also confirmed at the protein levels (Figure 5E), as well as the corresponding inhibitory effect on *Nppc* mRNA levels (Figure 5F). To further confirmed the critical mediating role of RAR, we showed the RAR-specific agonist TTNPB could recapitulate the effect of RA supplementation on *Zfp36* upregulation (Figure 5G and Figure S1D) and *Nppc* decrease (Figure 5H). Finally, Cut&Run analysis provided direct evidence that RAR binding on the *Zfp36* promoter region was significantly increased in response to RA exposure (Figure 5I).

3.6. RA Signaling Cooperates with Parallel EGF Signaling to Upregulate *Zfp36* through an ERK-Dependent Pathway

Given our previous study has reported that EGF signaling is also the critical mediator of LH-induced upregulation of Zpf36 and oocyte meiotic resumption [19], and By reanalyzing a previously published transcriptome that presented rapid effects of LH on gene expression in the mural granulosa cells (MGCs) of preovulatory mouse follicles [20], our GSEA analysis showed that both EGFR signaling and RA signaling was activated in preovulatory MGCs in response to LH (Figure 6A). The spatiotemporal coincidence led us to examine the relationship between these two signalings in the regulation *Zfp36* upregulation. Using in vitro cultured preovulatory MGC model, we found RA supplementation further increased EGF-induced upregulation of *Zfp36* (Figure 6B), which was also confirmed on the protein level (Figure 6C). However, RA supplementation did not increase the expression levels of EGF signaling-dependent transcriptional factors *Elk1* and *Egr1* (Figure 6D), suggesting EGF and RA signalings may cooperate and are in parallel with each other to upregulate Zfp36. Moreover, given we have previously reported LH activates Zfp36 through the EGFR-ERK1/2–Dependent Pathway [19], we next wanted to know if the synergistic effect between RA and EGF on Zfp36 upregulation also depends on the ERK1/2 pathway. We found that when the ERK pathway was blocked by PD0325901, RA could not further enhance *Zfp36* upregulation in the presence of EGF, implying the dependence on the ERK pathway (Figure 6E).

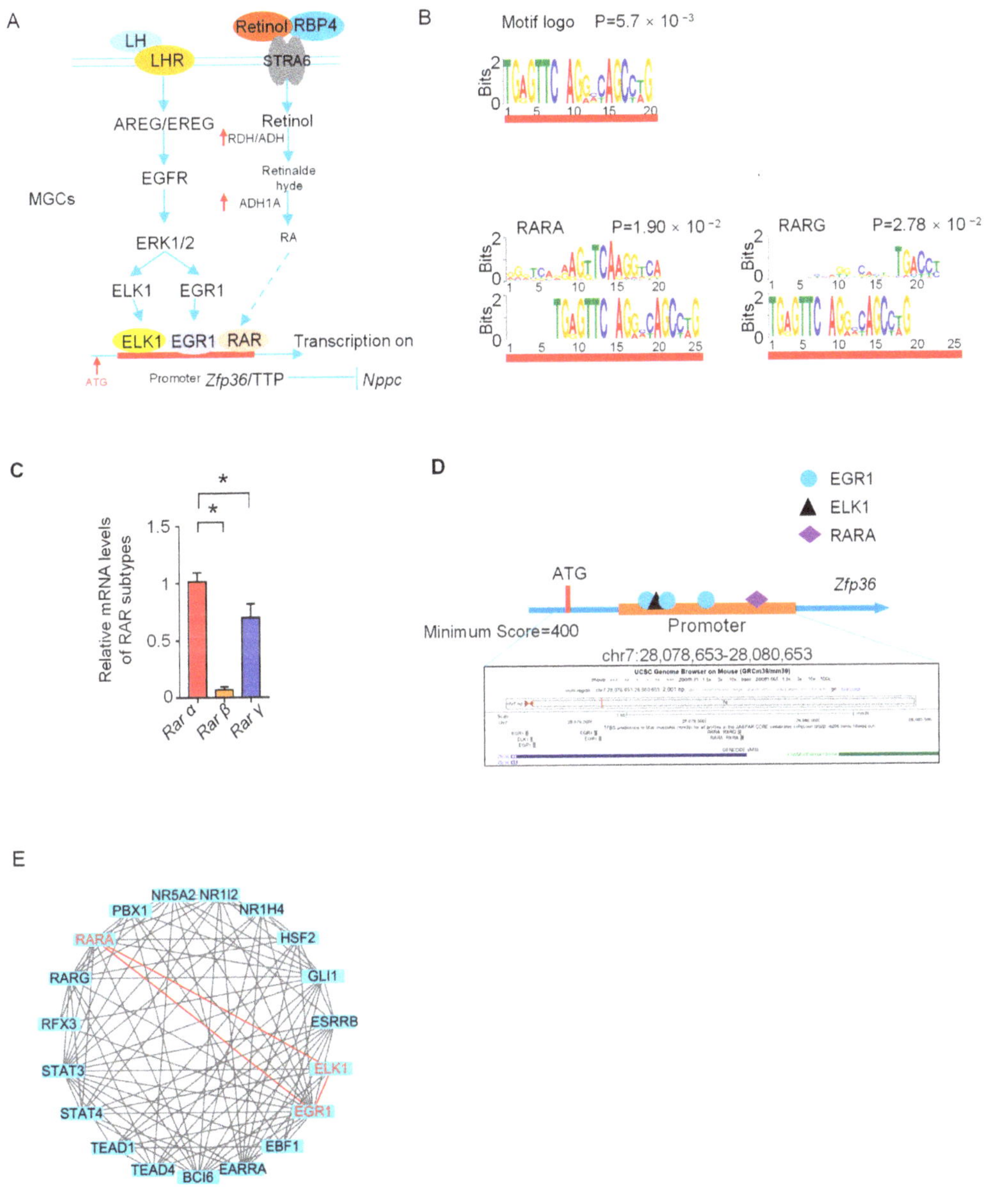

Figure 4. RAR is a potential transcription factor of *Zfp36*. (**A**) Schematic illustration of proposed parallel pathways that underlie the Zfp36 upregulation in MGCs in response to LH surge. (**B**) The most significant motifs based on comprehensive motif analysis on sequences located in the promoter regions of Zfp36 using MEME (v5.5.1). The identified motif was compared against a database of known mouse motifs named HOCOMOCO Mouse (v11 Core). The RAR-related DNA binding motifs were: RARA and RARG. (**C**) The relative mRNA levels of RAR subtypes in preovulatory MGCs. * $p < 0.05$. (**D**) Transcription factors EGR1 and ELK1 were predicted by the JASPAR2022 TFBS mm39 collection, which represented separate tracks in mouse genome assembly chr7:28,078,653-28,080,653. (**E**) STRING Protein-Protein Interaction (PPI) analysis of the transcription factors predicted by the MEME (v5.5.1). Red lines showed the interaction between the EGR1, ELK1 and RARA.

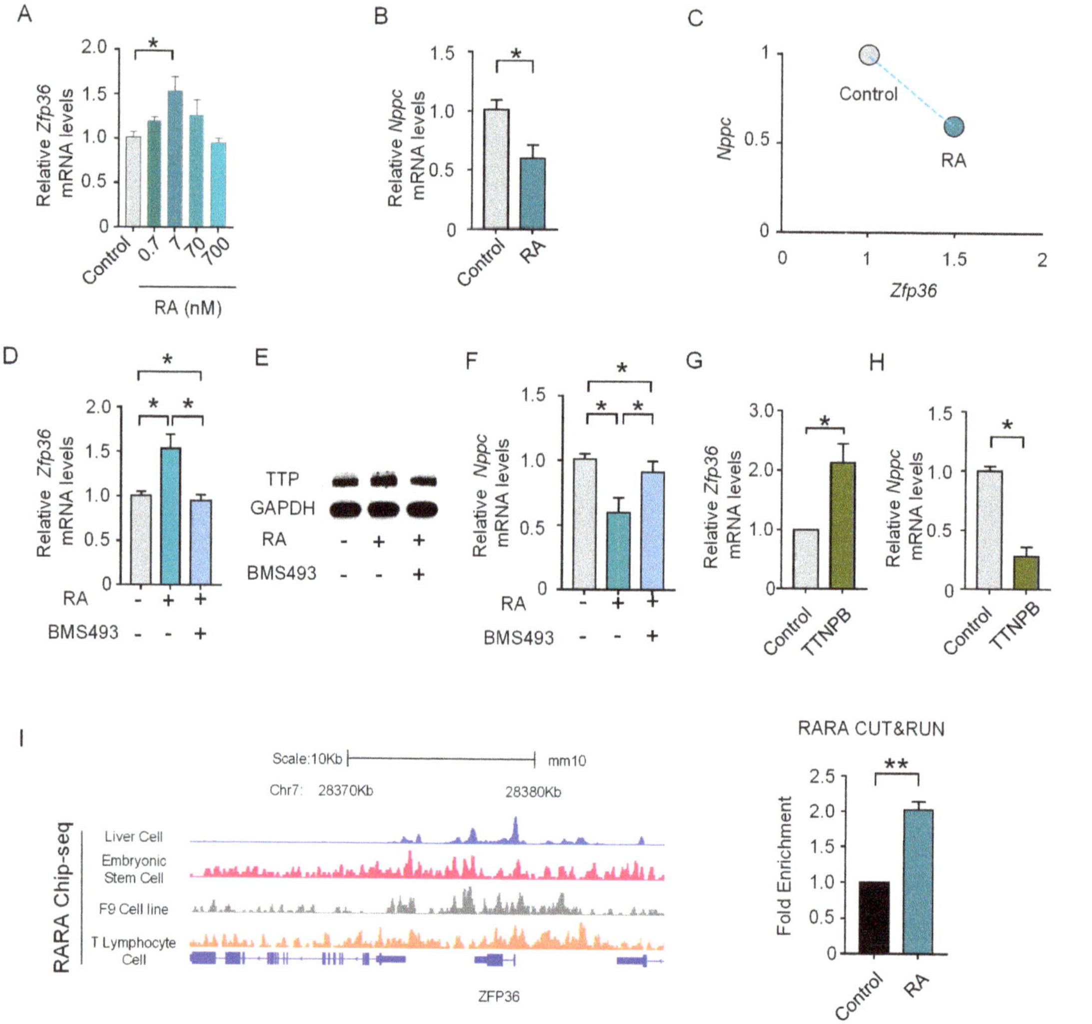

Figure 5. RA promotes the expression of *Zfp36* in MGC via RAR. (**A**) The mRNA levels of *Zfp36* in MGCs cultured with or without different concentrations of RA. * $p < 0.05$ (**B**) The mRNA levels of *Nppc* in MGCs cultured with or without 7 nM RA. * $p < 0.05$ (**C**) The negative correlation between *Zfp36* and *Nppc* expression levels in MGCs cultured with or without 7 nM RA. (**D**) The mRNA levels of *Zfp36* in MGCs cultured with RA alone or together with 1 μM BMS493. * $p < 0.05$ (**E**) The Western blotting analysis of TTP in MGCs cultured with RA alone or together with BMS493. * $p < 0.05$ (**F**) The mRNA levels of *Nppc* in MGCs cultured with RA alone or together with 1 μM BMS493. * $p < 0.05$ (**G**,**H**) The mRNA levels of *Zfp36* (**F**) and *Nppc* (**G**) in MGCs cultured with or without 0.1 μM TTNPB. * $p < 0.05$ (**I**) The IGV screenshots (left) of examples of RARA Chip-seq analysis in different cells collected by the Cistrome data browser and the RARA binding sites are highlighted by the grey area. The right panel showed the RARA CUT&RUN results by detecting the highlighted area. ** $p < 0.01$.

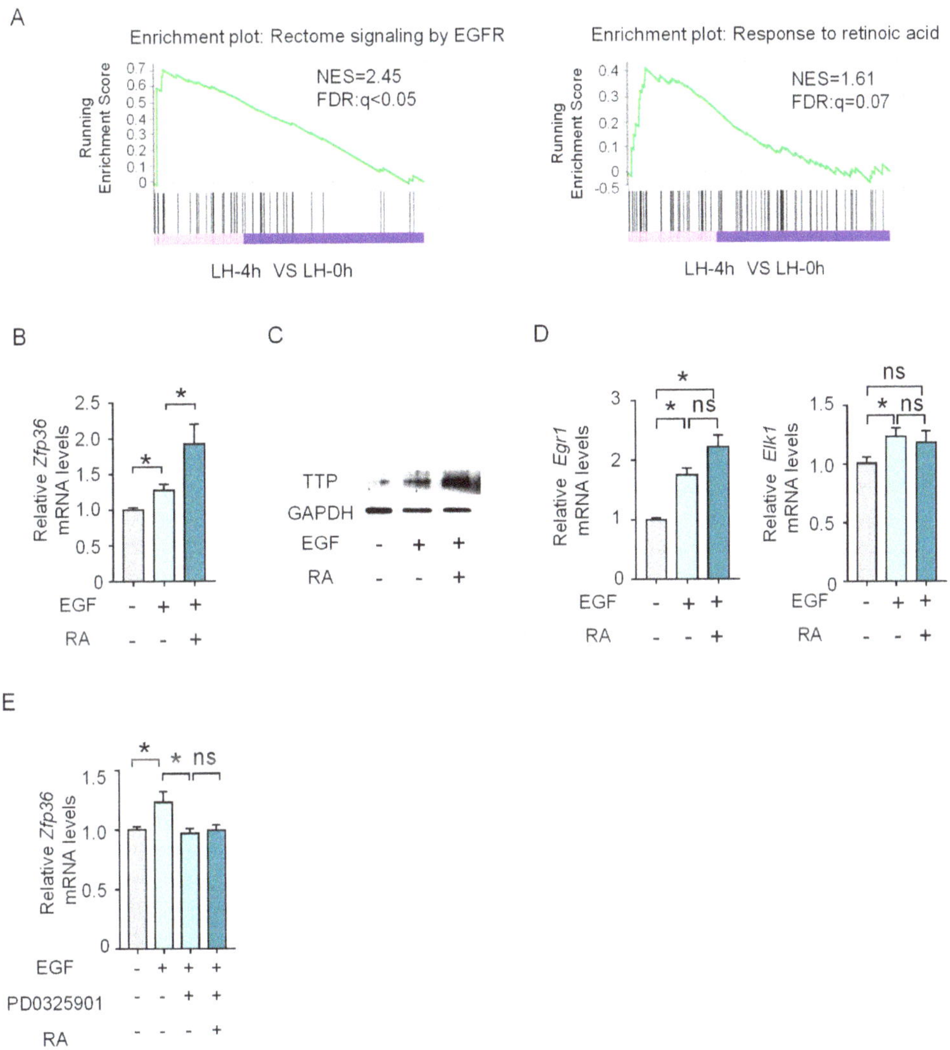

Figure 6. RA signaling cooperates with parallel EGF signaling to upregulate Zfp36 through the ERK-dependent pathway. (**A**) GSEA analysis showing enrichment of Rectome signaling by EGFR and Response to retinoic acid, Normalized enrichment score (NES) and false discovery rate (FDR) q value are shown. (**B**) The mRNA expression levels of *Zfp36* in MGCs cultured with EGF alone or together with RA. * $p < 0.05$ (**C**) The Western blotting analysis of TTP in MGCs cultured with EGF alone or together with RA. * $p < 0.05$ (**D**) The mRNA expression levels of *Egr1* and *Elk1* in MGCs cultured with EGF alone or together with RA. * $p < 0.05$, ns, not significant ($p > 0.05$). (**E**) The mRNA expression levels of *Zfp36* in MGCs cultured with EGF alone or various combinations with RA and PD0325901. * $p < 0.05$, ns, not significant ($p > 0.05$).

4. Discussion

It has been clear that RA regulates myriads of physiological functions and processes. Among these, RA acts as a key driver of entry into meiosis in both male and female germ cells [36]. In contrast, however, whether RA regulates the exit from oocyte meiotic arrest has not been definitely determined, despite the presence of some suggestive evidence. Results based on an in vitro study showed RA enhanced mouse oocyte maturation in vitro and improved fertilization and development rates in a dose-dependent manner [37]. In addition, exogenous injections of VA into gilts after the second estrous cycle dramatically enhanced the GVBD rate of oocytes [38].

In the present study, using well-established in vivo and in vitro models, we identify that RA plays a critical role in ensuring successful oocyte meiotic resumption in an MGC-dependent manner (Figure 7). Of note, although it has been frequently reported in previous studies [39], our study further highlights the importance of MGCs in mediating LH-induced oocyte resumption. We proved, in vitro and in vivo, that MGCs are indispensable for ensuring the prompting effect of RA on oocyte meiotic resumption.

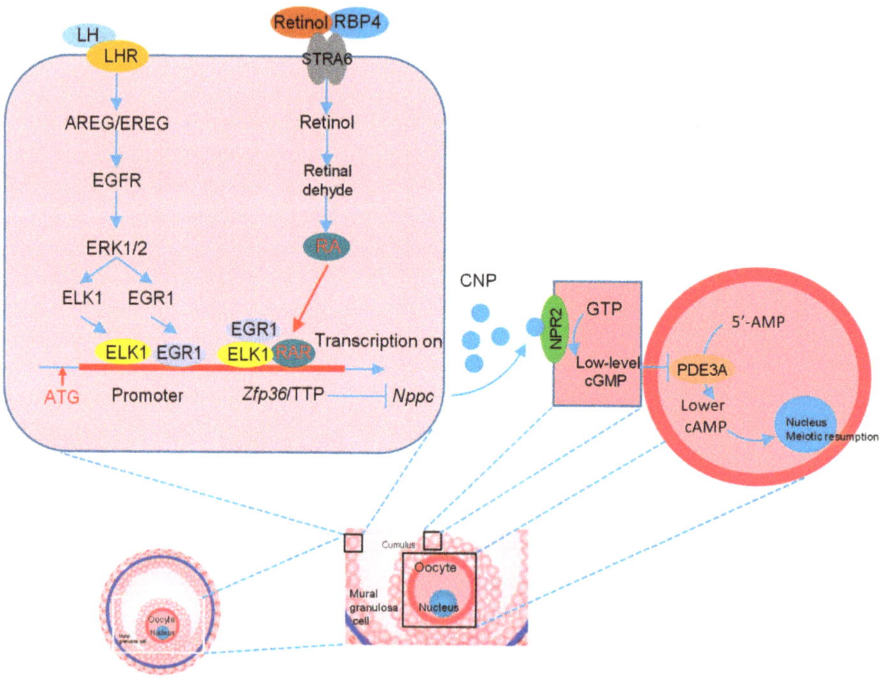

Figure 7. Working model illustrating the mechanism that underlies the role of RA in regulating LH-induced oocyte meiotic resumption. The left indicates the previously reported pathway responsible for Zfp36 upregulation, and the right indicates the pathway identified in the present study.

Our study has also focused on the mechanism underlying the role of RA in prompting oocyte meiotic resumption. Current knowledge mainly focused on the LH-induced signaling cascades in MGCs, e.g., LHR and EGFR activation, CNP-NPR2 inactivation and decrease in cGMP [13,40], while our study showed a remarkable metabolic change in MGCs and its functional role in regulating oocyte meiotic resumption. The spatiotemporal coincidence of LH-induced EGF signaling and RA signaling in MGCs also suggests the synergistic cooperation between these two signalings. We also found RA signaling acts in parallel with EGF signaling. This is in line with the previous notion that both EGF-dependent and -independent intrafollicular pathways are important for inducing oocyte meiotic resumption [21,41].

With regards to the receptor that mediates the prompting effect of RA signaling on meiotic resumption, our experiments based on both atRA supplementation and receptor inhibition suggested RAR is the main receptor. In general, it is well documented RAR is the preferred receptor for all-trans retinoic acid (atRA), and RXR is preferred for 9-cis RA. There are three subtypes of RAR, i.e., Rarα, Rarβ and Rarγ, and our bioinformatic analysis showed the binding motifs of RARα and RARγ within the promoter region of Zfp36, suggesting that both RARα and RARγ may participate in the transcriptional regulation of Zfp36.

The present study, together with our previous work [19], indicates that Zfp36 is the co-target of these two signalings. *Zfp36* encodes tristetraprolin (TTP), a ubiquitously expressed mRNA-destabilizing protein. We have reported that Zfp36/TTP is one of the regulatory components responsible for the LH-induced rapid decrease in *Nppc* mRNA/CNP, a critical prerequisite for oocyte meiotic resumption. We have also revealed the mechanism underlying the transient upregulation of Zfp36 in response to the LH. ELK1 and EGR1 are the essential transcriptional factors that are responsible for the LH-induced rapid upregulation of Zfp36, and their function is dependent on the EGF-ERK pathway [42]. In the present study, we have coupled RA signaling to EGF signaling and proposed that the transcriptional factors in these two parallel pathways may cooperate together to upregulate Zfp36 in response to LH.

In summary, our data provide direct evidence that intrafollicular RA signaling is important for LH-induced oocyte meiotic resumption. In response to LH surge, RA signaling is activated in MGCs. We also identify that RA signaling participates in LH-induced Zpf36 via RAR, which may cooperate with ELK1 and EGR1. Our results not only advance current knowledge of LH-induced Zfp36 upregulation and Nppc decrease but also further emphasize again the importance of MGCs in regulating oocyte meiotic resumption. In addition, these findings extend our understanding of the role of RA in regulating oocyte meiosis: it not only governs the initiation of oocyte meiosis during female embryonic development but also participates in the resumption of oocyte meiotic arrest when LH restarts meiosis.

Supplementary Materials: The following supporting information can be downloaded at: https://www.mdpi.com/article/10.3390/genes14040946/s1, Figure S1: Screening experiments of suitable concentration for RA, RAR antagonist (BMS493), RAR agonist (TTNPB) in follicile culture or MGCs culture; Table S1: All the primers used in this research; Table S2: Vitamin A deficient diet ingredients table.

Author Contributions: F.W., Q.Y., L.A. and J.T. conceived and designed the experiments. F.W., Y.T., Y.C., R.Y., Z.W. and X.W. performed the experiments. J.T. and L.A. improved the experimental design and the manuscript. F.W., L.A. and W.W. analyzed the data, wrote, and finalized the manuscript. All authors approved the paper. All authors have read and agreed to the published version of the manuscript.

Funding: This research was funded by grants from the Major Science and Technology Program of Inner Mongolia Autonomous Region (2021ZD0012), the National Key R&D Program (2021YFD1300901, 2017YFD0501901, and 2017YFD0501905), the National Natural Science Foundation of China (No. 31930103 and 3167246) and the Beijing Innovation Consortium of Livestock Research System (grant number BAIC05-2022).

Institutional Review Board Statement: The animal study protocol was approved by the Institutional Animal Care and Use Committee of China Agricultural University. The ethical clearance approval number is AW11107102-1-2.

Informed Consent Statement: Not applicable.

Data Availability Statement: The data presented in this study are available on request from the corresponding author.

Conflicts of Interest: The authors declare no competing financial interests.

References

1. Das, B.C.; Thapa, P.; Karki, R.; Das, S.; Mahapatra, S.; Liu, T.; Torregroza, I.; Wallace, D.P.; Kambhampati, S.; Van Veldhuizen, P.; et al. Retinoic acid signaling pathways in development and diseases. *Bioorg. Med. Chem.* **2014**, *22*, 673–683. [CrossRef] [PubMed]
2. Kawai, T.; Yanaka, N.; Richards, J.S.; Shimada, M. De Novo-Synthesized Retinoic Acid in Ovarian Antral Follicles Enhances FSH-Mediated Ovarian Follicular Cell Differentiation and Female Fertility. *Endocrinology* **2016**, *157*, 2160–2172. [CrossRef] [PubMed]
3. Kin Ting Kam, R.; Deng, Y.; Chen, Y.; Zhao, H. Retinoic acid synthesis and functions in early embryonic development. *Cell Biosci.* **2012**, *2*, 11. [CrossRef]
4. Bowles, J.; Koopman, P. Retinoic acid, meiosis and germ cell fate in mammals. *Development* **2007**, *134*, 3401–3411. [CrossRef]
5. Koubova, J.; Menke, D.B.; Zhou, Q.; Capel, B.; Griswold, M.D.; Page, D.C. Retinoic acid regulates sex-specific timing of meiotic initiation in mice. *Proc. Natl. Acad. Sci. USA* **2006**, *103*, 2474–2479. [CrossRef]

6. Menke, D.B.; Koubova, J.; Page, D.C. Sexual differentiation of germ cells in XX mouse gonads occurs in an anterior-to-posterior wave. *Dev. Biol.* **2003**, *262*, 303–312. [CrossRef] [PubMed]
7. Abouzaripour, M.; Fathi, F.; Daneshi, E.; Mortezaee, K.; Rezaie, M.J.; Abdi, M. Combined Effect of Retinoic Acid and Basic Fibroblast Growth Factor on Maturation of Mouse Oocyte and Subsequent Fertilization and Development. *Int. J. Fertil. Steril.* **2018**, *12*, 68–71. [CrossRef]
8. Ara Ahmed, J.; Dutta, D.; Nashiruddullah, N. Comparative efficacy of antioxidant retinol, melatonin, and zinc during in vitro maturation of bovine oocytes under induced heat stress. *Turk. J. Vet. Anim. Sci.* **2016**, *40*, 365–373. [CrossRef]
9. Lilibeth, A.; Cajuday, A.A.H.D. Effect of Retinoic Acid on the Development of Water Buffalo Embryos In Vitro. *Philipp. J. Vet. Anim. Sci.* **2012**, *2*, 107–116.
10. Tahaei, L.S.; Eimani, H.; Yazdi, P.E.; Ebrahimi, B.; Fathi, R. Effects of retinoic acid on maturation of immature mouse oocytes in the presence and absence of a granulosa cell co-culture system. *J. Assist. Reprod. Genet.* **2011**, *28*, 553–558. [CrossRef]
11. Liang, S.; Kang, J.; Jin, H.; Liu, X.; Li, J.; Li, S.; Lu, Y.; Wang, W.; Yin, X. The influence of 9-cis-retinoic acid on nuclear and cytoplasmic maturation and gene expression in canine oocytes during in vitro maturation. *Theriogenology* **2012**, *77*, 1198–1205. [CrossRef] [PubMed]
12. Vahedi, V.; Zeinoaldini, S.; Kohram, H.; Farahavar, A. Retinoic acid effects on nuclear maturation of bovine oocytes in vitro. *Afr. J. Biotechnol.* **2009**, *8*, 3974–3978.
13. Richards, J.S.; Ascoli, M. Endocrine, Paracrine, and Autocrine Signaling Pathways That Regulate Ovulation. *Trends Endocrinol. Metab.* **2018**, *29*, 313–325. [CrossRef]
14. Gómez, E.; Royo, L.; Duque, P.; Carneiro, G.; Hidalgo, C.; Goyache, F.; Lorenzo, P.L.; Alvarez, I.; Facal, N.; Díez, C. 9-cis-retinoic acid during in vitro maturation improves development of the bovine oocyte and increases midkine but not IGF-I expression in cumulus-granulosa cells. *Mol. Reprod. Dev.* **2003**, *66*, 247–255. [CrossRef]
15. Almiñana, C.; Gil, M.A.; Cuello, C.; Caballero, I.; Roca, J.; Vazquez, J.M.; Gomez, E.; Martinez, E.A. In vitro maturation of porcine oocytes with retinoids improves embryonic development. *Reprod. Fertil. Dev.* **2008**, *20*, 483. [CrossRef]
16. Zhang, M.; Su, Y.; Sugiura, K.; Xia, G.; Eppig, J.J. Granulosa Cell Ligand NPPC and Its Receptor NPR2 Maintain Meiotic Arrest in Mouse Oocytes. *Science* **2010**, *330*, 366–369. [CrossRef]
17. Norris, R.P.; Ratzan, W.J.; Freudzon, M.; Mehlmann, L.M.; Krall, J.; Movsesian, M.A.; Wang, H.; Ke, H.; Nikolaev, V.O.; Jaffe, L.A. Cyclic GMP from the surrounding somatic cells regulates cyclic AMP and meiosis in the mouse oocyte. *Development* **2009**, *136*, 1869–1878. [CrossRef] [PubMed]
18. Robinson, J.W.; Zhang, M.; Shuhaibar, L.C.; Norris, R.P.; Geerts, A.; Wunder, F.; Eppig, J.J.; Potter, L.R.; Jaffe, L.A. Luteinizing hormone reduces the activity of the NPR2 guanylyl cyclase in mouse ovarian follicles, contributing to the cyclic GMP decrease that promotes resumption of meiosis in oocytes. *Dev. Biol.* **2012**, *366*, 308–316. [CrossRef]
19. Xi, G.; An, L.; Wang, W.; Hao, J.; Yang, Q.; Ma, L.; Lu, J.; Wang, Y.; Wang, W.; Zhao, W.; et al. The mRNA-destabilizing protein Tristetraprolin targets "meiosis arrester" Nppc mRNA in mammalian preovulatory follicles. *Proc. Natl. Acad. Sci. USA* **2021**, *118*, e2018345118. [CrossRef]
20. Shirafuta, Y.; Tamura, I.; Ohkawa, Y.; Maekawa, R.; Doi-Tanaka, Y.; Takagi, H.; Mihara, Y.; Shinagawa, M.; Taketani, T.; Sato, S.; et al. Integrated Analysis of Transcriptome and Histone Modifications in Granulosa Cells During Ovulation in Female Mice. *Endocrinology* **2021**, *162*, bqab128. [CrossRef]
21. Liu, X.; Xie, F.; Zamah, A.M.; Cao, B.; Conti, M. Multiple Pathways Mediate Luteinizing Hormone Regulation of cGMP Signaling in the Mouse Ovarian Follicle1. *Biol. Reprod.* **2014**, *91*, 1–11. [CrossRef] [PubMed]
22. Campos-Junior, P.H.; Silva, C.A.; Grazia, J.G.; Soares, M.B.; Santos, R.R.; Viana, J.H. Use of ultrasound biomicroscopy to evaluate induced ovarian follicular growth and ovulation in mice. *Lab. Anim.* **2011**, *45*, 254–258. [CrossRef]
23. Crispo, M.; Meikle, M.N.; Schlapp, G.; Menchaca, A. Ovarian superstimulatory response and embryo development using a new recombinant glycoprotein with eCG-like activity in mice. *Theriogenology* **2021**, *164*, 31–35. [CrossRef] [PubMed]
24. Lima, P.F.; Oliveira, M.A.L.; Santos, M.H.B.; Reichenbach, H.D.; Weppert, M.; Paula-Lopes, F.F.; Neto, C.C.C.; Gonçalves, P.B.D. Effect of retinoids and growth factor on in vitro bovine embryos produced under chemically defined conditions. *Anim. Reprod. Sci.* **2006**, *95*, 184–192. [CrossRef] [PubMed]
25. Sato, Y.; Cheng, Y.; Kawamura, K.; Takae, S.; Hsueh, A.J.W. C-Type Natriuretic Peptide Stimulates Ovarian Follicle Development. *Mol. Endocrinol.* **2012**, *26*, 1158–1166. [CrossRef]
26. Simona Marc Zarcula, G.G.C.M.; Tulcan, H.C.G.O.; Andreea Ciobota, I.H.C.M. Morphological aspects of cumulus –oocyte complexes in different species. In Proceedings of the 53th Annual Meeting of Veterinary Medicine—"Towards a Global Health", Iasi, Romania, October 2014; pp. 91–97.
27. Choi, H.; Lee, J.; Yoon, J.D.; Hwang, S.U.; Cai, L.; Kim, M.; Kim, G.; Oh, D.; Kim, E.; Hyun, S.H. The effect of copper supplementation on in vitro maturation of porcine cumulus-oocyte complexes and subsequent developmental competence after parthenogenetic activation. *Theriogenology* **2021**, *164*, 84–92. [CrossRef]
28. Ming, T.; Nielsen, H.; Chen, Z. Maturation arrest of human oocytes at germinal vesicle stage. *J. Hum. Reprod. Sci.* **2010**, *3*, 153. [CrossRef]
29. Tombes, R.M.; Simerly, C.; Borisy, G.G.; Schatten, G. Meiosis, egg activation, and nuclear envelope breakdown are differentially reliant on Ca^{2+}, whereas germinal vesicle breakdown is Ca^{2+} independent in the mouse oocyte. *J. Cell Biol.* **1992**, *117*, 799–811. [CrossRef]

30. Hainer, S.J.; Bošković, A.; Mccannell, K.N.; Rando, O.J.; Fazzio, T.G. Profiling of Pluripotency Factors in Single Cells and Early Embryos. *Cell* **2019**, *177*, 1319–1329. [CrossRef]
31. Skene, P.J.; Henikoff, J.G.; Henikoff, S. Targeted in situ genome-wide profiling with high efficiency for low cell numbers. *Nat. Protoc.* **2018**, *13*, 1006–1019. [CrossRef]
32. Hainer, S.J.; Fazzio, T.G. High-Resolution Chromatin Profiling Using CUT&RUN. *Curr. Protoc. Mol. Biol.* **2018**, *126*, e85. [CrossRef]
33. White, J.C.; Highland, M.; Clagett-Dame, M. Abnormal development of the sinuatrial venous valve and posterior hindbrain may contribute to late fetal resorption of vitamin A-deficient rat embryos. *Teratology* **2000**, *62*, 374–384. [CrossRef]
34. See, A.W.; Kaiser, M.E.; White, J.C.; Clagett-Dame, M. A nutritional model of late embryonic vitamin A deficiency produces defects in organogenesis at a high penetrance and reveals new roles for the vitamin in skeletal development. *Dev. Biol.* **2008**, *316*, 171–190. [CrossRef] [PubMed]
35. Florkowska, M.; Tymoszuk, P.; Balwierz, A.; Skucha, A.; Kochan, J.; Wawro, M.; Stalinska, K.; Kasza, A. EGF activates TTP expression by activation of ELK-1 and EGR-1 transcription factors. *BMC Mol. Biol.* **2012**, *13*, 8. [CrossRef] [PubMed]
36. Koubova, J.; Hu, Y.C.; Bhattacharyya, T.; Soh, Y.Q.; Gill, M.E.; Goodheart, M.L.; Hogarth, C.A.; Griswold, M.D.; Page, D.C. Retinoic acid activates two pathways required for meiosis in mice. *PLoS Genet.* **2014**, *10*, e1004541. [CrossRef] [PubMed]
37. Nasiri, E.; Mahmoudi, R.; Bahadori, M.H.; Amiri, I. The Effect of Retinoic Acid on in vitro Maturation and Fertilization Rate of Mouse Germinal Vesicle Stage Oocytes. *Cell J.* **2011**, *13*, 19–24.
38. Whaley, S.L.; Hedgpeth, V.S.; Farin, C.E.; Martus, N.S.; Jayes, F.C.; Britt, J.H. Influence of vitamin A injection before mating on oocyte development, follicular hormones, and ovulation in gilts fed high-energy diets. *J. Anim. Sci.* **2000**, *78*, 1598–1607. [CrossRef]
39. Jaffe, L.A.; Egbert, J.R. Regulation of Mammalian Oocyte Meiosis by Intercellular Communication Within the Ovarian Follicle. *Annu. Rev. Physiol.* **2017**, *79*, 237–260. [CrossRef]
40. He, M.; Zhang, T.; Yang, Y.; Wang, C. Mechanisms of Oocyte Maturation and Related Epigenetic Regulation. *Front. Cell. Dev. Biol.* **2021**, *9*, 654028. [CrossRef]
41. Egbert, J.R.; Yee, S.; Jaffe, L.A. Luteinizing hormone signaling phosphorylates and activates the cyclic GMP phosphodiesterase PDE5 in mouse ovarian follicles, contributing an adsditional component to the hormonally induced decrease in cyclic GMP that reinitiates meiosis. *Dev. Biol.* **2018**, *435*, 6–14. [CrossRef]
42. Hsieh, M.; Conti, M. G-protein-coupled receptor signaling and the EGF network in endocrine systems. *Trends Endocrinol. Metab.* **2005**, *16*, 320–326. [CrossRef] [PubMed]

genes

MDPI

Article

Melatonin Protects the Apoptosis of Sheep Granulosa Cells by Suppressing Oxidative Stress via MAP3K8 and FOS Pathway

Bo Zhai [1], Xu Li [1], Zhongli Zhao [1], Yang Cao [1], Xinxin Liu [1], Zheng Liu [1], Huihai Ma [1,*] and Wenfa Lu [2,*]

1 Institute of Animal Science, Jilin Academy of Agricultural Science, Changchun 136100, China; 4513046@163.com (Xu Li)
2 College of Animal Science and Technology, Jilin Agricultural University, Changchun 130118, China
* Correspondence: mahuihai@cjaas.com (H.M.); lvwenfa@jlau.edu.cn (W.L.); Tel.: +86-137-5629-2797 (H.M.); +86-138-4410-2196 (W.L.)

Abstract: Melatonin is not only a highly effective active oxygen scavenger but also an important reproductive hormone. Melatonin has a regulatory effect on animal reproduction, especially on the ovaries. It can affect the proliferation and apoptosis of cells in follicles. However, the mechanisms of the dual antioxidation and anti-apoptosis effects of melatonin on granulosa cells are still not clear, especially in sheep. Therefore, we investigated the mechanisms of the protective effect of melatonin against oxidative damage in granulosa cells. At a concentration of 250 μmol/L, H_2O_2 promoted granulosa cell apoptosis; however, 10 ng/mL melatonin effectively alleviated the pro-apoptotic effect of H_2O_2. Furthermore, through the application of high-throughput sequencing technology, we identified 109 significantly differentially expressed genes (35 upregulated and 74 downregulated genes) involved in the protective effect of melatonin against apoptosis. The expression levels of nine related genes, i.e., *ATF3, FIBIN, FOS, HSPA6, MAP3K8, FOSB, PET117, DLX2*, and *TRIB1*, changed significantly. *MAP3K8* and *FOS* gene overexpression impacted the protective effect of melatonin in granulosa cells; the two genes exhibited an upstream and downstream regulatory relationship. Our findings indicated that melatonin alleviated H_2O_2-induced apoptosis in sheep granulosa cells through the MAP3K8-FOS pathway.

Keywords: melatonin; apoptosis; sheep; granulosa cells; oxidative stress

Citation: Zhai, B.; Li, X.; Zhao, Z.; Cao, Y.; Liu, X.; Liu, Z.; Ma, H.; Lu, W. Melatonin Protects the Apoptosis of Sheep Granulosa Cells by Suppressing Oxidative Stress via MAP3K8 and FOS Pathway. *Genes* **2023**, *14*, 1067. https://doi.org/10.3390/genes14051067

Academic Editors: Qiuyue Liu and Bao Yuan

Received: 2 April 2023
Revised: 4 May 2023
Accepted: 8 May 2023
Published: 11 May 2023

1. Introduction

Melatonin (MT) is a neuroendocrine hormone initially discovered in the bovine pineal gland. It is not only a highly effective reactive oxygen species (ROS) scavenger but also an important reproductive hormone. As a natural fat-soluble antioxidative factor, melatonin scavenges different types of free radicals, such as hydroxyl radicals (-OH) and hydrogen peroxide (H_2O_2) [1]. Melatonin can maintain mitochondrial function by removing or reducing oxidative stress in mitochondria, thereby reducing the occurrence of subsequent apoptosis. The effects of melatonin on the reproductive performance of female animals mainly manifest in sexual maturation, hormone secretion, follicular development, luteal function, ovulation, pregnancy, and childbirth [2]. In addition to its synthesis in the pineal gland, melatonin is also synthesized in the cumulus oocyte complex and granulosa cells and participates in follicular development [3]. The high concentration of melatonin in follicular fluid is very important for inhibiting follicular atresia. Melatonin can protect oocytes from free radical damage during ovulation, allowing the full development of preovulatory follicles.

Granulosa cell apoptosis is a key influencing factor for follicular atresia [4]. Melatonin exerts anti-apoptotic effects in the process of follicular atresia in a variety of animals, inhibits granulosa cell apoptosis, and promotes follicular development [5–11]. Melatonin can not only directly affect the growth status of granulosa cells and the production of steroid hormones but also indirectly alleviate the damage caused by extracellular stimuli [11–14].

Melatonin can affect the proliferation and apoptosis of bovine granulosa cells and stimulate the secretion of progesterone through melatonin 2 (MT2) and melatonin 1 (MT1) receptors in vitro [15]. It can also regulate the proliferation and apoptosis of chicken granulosa cells through mTOR signaling pathway activation [16]. Melatonin has a significant reparative effect on the exogenous damage to granulosa cells. When granulosa cells are treated with bisphenol A [17] and BMP6 (Recombinant Human Bone Morphogenetic Protein 6) [18], melatonin effectively resolves the induced granulosa cell proliferation and steroid hormones production. Melatonin affects the proliferation and apoptosis of sheep granulosa cells under heat stress [19].

Studies on the regulation of granulosa cells by melatonin mainly focus on swine, cattle, rats and other animals. However, the protective effect of melatonin from oxidative stress in sheep granulosa cells has not been reported. This study on the ability of melatonin to alleviate H_2O_2-induced apoptosis in sheep granulosa cells through the MAP3K8-FOS pathway provides a theoretical reference for the role of melatonin in the regulation of sheep reproductive functions.

2. Materials and Methods

2.1. Ethics Statement

The collection of test samples in this research was approved by the Animal Care and Use Committee of Jilin Agricultural University.

2.2. Isolation and Culture of Granulosa Cells

Follicles (3–5 mm) isolated from the ovary of small tailed han sheep (six years old) were soaked in 75% alcohol and placed in precooled Dulbecco's phosphate-buffered saline (DPBS). The follicles were cut open in petri dishes (containing DPBS), and the inner wall of the follicle was gently scraped to obtain parietal granulosa cells, which were collected in a 15-mL centrifuge tube. After centrifugation at 1000× *g* r/min for 5 min, the supernatant was discarded, and the pellet was resuspended in a complete medium (DMEM/F12), centrifuged and washed twice. A cell suspension in the complete medium was spread on a culture dish, and the dishes were incubated in a cell incubator (37 °C, 5% CO_2) for subsequent experiments.

2.3. Cell Viability Assay

The cells were seeded in 96-well plates (3 replicates for each group) at a seeding density of 1×10^4 cells/well. After reaching confluency and attaching to the wells, the cells were cultured in a serum-free medium with different concentrations of H_2O_2 (0, 100, 200, 250, 300, 400, 500 umol/L) and melatonin (0.1, 1, 10, 100, 1000 ng/L), and MTT (3-(4,5-Dimethylthiazol-2-yl)-2,5-diphenyltetrazolium bromide), diluted with 5% PBS was added to the wells at 20 µL/well and removed after 4 h. After an appropriate amount of DMSO (Dimethyl sulfoxide) was added, cell viability was detected at 490 nm.

2.4. Flow Cytometric Analysis of Apoptotic cells

The granulosa cells from the groups with the control, H_2O_2 treatment and H_2O_2+Melatonin treatment, were transferred to 1.5 mL centrifuge tubes, and 200 µL of 1× Binding Buffer was added to each tube in accordance with the BD Apoptosis Detection Kit. Propidium iodide (PI) was added to one tube of cells, and fluorescein isothiocyanate (FITC) was added to another tube of cells. Five microliters each of FITC and PI were added to the cells in each tube of different treatments and mixed by inversion, and the cells were incubated in a cell culture incubator for 20 min. Then, 200 µL of 1× Binding Buffer was added to each tube, and the cells were filtered through a nylon mesh sieve. The percentage of apoptotic cells was determined by flow cytometry.

2.5. Real Time Quantitative Polymerase Chain Reaction (qRT-PCR) Analysis

The total RNA was isolated from the cells, using the TRIzol™ reagent (Invitrogen, Waltham, MA, USA), and 1 μg was employed to generate cDNA, using a PrimeScript RT reagent Kit (Takara, Kusatsu, Japan). Next, qRT-PCR was performed using SYBR® Premix Ex Taq™ II (Takara, Kusatsu, Japan) on a Real-Time PCR Detection System (Santa, TX, USA). The primer sequences are shown in Table 1. The relative expression of genes was calculated, using the $2^{-\Delta\Delta Ct}$ method [20].

Table 1. Primer information for quantitative real-time polymerase chain reaction.

Genes	Primer Sequence (5′–3′)	Size (bp)
FOSB	F: TCCGCCGAGTCTCAGTATCTGTC R: TGGAGGTCCTGGCTGGTTGTG	146
TRIB1	F: GACGGGAGCCTTCGGAGAGAC R: GTCACTGTCCTCCTGGTACTCTGG	140
DLX2	F: TTGACAGTCTGGTGGCTGATATGC R: GCTTGTGGAGGCTGCTGCTG	129
PET117	F: GAGACGCTCCAAGGTGGTGTTG R: GTCCTGCCGCTGCTTCAGATG	85
MAP3K8	F: ACAACGAGGAGGAGCGGTCTG R: ACGCTTCGCAGTGTTGGATATGT	131
HSPA6	F: AAGGAGACGGCGGAGGCTTAC R:GCTGCGAGTCGTTGAAGTAGGC	82
FOS	F: CACCGACCTGCCTGCAAGATC R: GCTCCATGCTGCCGACACTC	193
FIBIN	F:CAGGCTCAACGAGGACTTCTTGG R:CATGGCTCCTGATGGTGAGTGC	125
ATF3	F: GAAGTCAGTGCCTCTGCCATCG R: GGCGAACCTCAGCTCTTCCTTG	108
Bcl2	F: CGCATCGTGGCCTTCTTT R: CGGTTCAGGTACTCGGTCATC	113
Bax	F: CGAGTGGCGGCTGAAAT R: GGTCTGCCATGTGGGTGTC	237
β-actin	F: GCAAAGACCTCTACGCCAAC R: GGGCAGTGATCTCTTTCTGC	90

2.6. High-Throughput Sequencing Analysis

One milliliter of TRIzol (Invitrogen, USA) was added to sheep GCs with a different treatment for extracting the total RNA, and the quality was analyzed using an Agilent Bioanalyzer 2100. The total RNA extracted from the different treatment GCs was stored at −80 °C and used for the high-throughput sequencing. After the concentration was measured using a Qubit 2.0 Fluorometer, a library was constructed by Novogene Bioinformatics Technology Co., Ltd. (Beijing, China) and sequenced using the Illumina HiSeq 2500 platform.

The raw image file, which had now passed through base identification and error filtering, finally resulted in the original sequencing fragment that could be used for analysis, which we call Reads. Sequencing raw reads may contain unqualified reads such as low overall quality, sequencing primers, and low-end quality, which are likely to have a certain impact on the quality of analysis, so they must be filtered to obtain clean reads that can be used for data analysis, and Seqtk (https://github.com/lh3/seqtk accessed on 15 September 2020) should be used for filtering. To render the gene expression levels of different genes and different samples comparable, reads are converted into FPKM (Fragments Per Kilobase of exon model per Million mapped reads) for gene expression standardization. We first used Stringtie (version:1.3.0) to count the number of fragments of each gene after Hisat2 alignment, and then normalized the number of fragments using the TMM (trimmed mean of M values) method, and finally used a perl script to calculate the FPKM value of each gene. Subsequently, edgeR was adopted for genetic differences analysis between samples, and multiple hypothesis test correction was carried out after the p-value was obtained, and the threshold of the p-value was determined by controlling the FDR (False Discovery Rate), and the corrected p-value was q-value. At the same time, we calculated the difference expression multiple, i.e., fold change, based on the FPKM value.

Gene Ontology (GO) analysis was applied to explore the functions of differentially expressed genes identified in this study. GO analysis organizes genes into hierarchical categories and can uncover gene regulatory networks on the basis of biological processes and molecular functions (http://www.geneontology.org accessed on 30 March 2023). Specifically, the two-sided Fisher's exact test was used to classify the GO category, and the GO annotation list was by chance greater than expected. An FDR was calculated to correct the p-value. We computed p-values for GOs enriched among differentially expressed genes (the recommended p-value cutoff is 0.05). Pathway analysis was employed to place differentially expressed genes according to the Kyoto Encyclopedia of Genes and Genomes (KEGG, http://www.genome.jp/kegg/ accessed on 30 March 2023). Fisher's exact tests were also used to identify pathways, and the threshold of significance was defined by the p-value. The enrichment was calculated in a similar manner as the GO analysis.

2.7. Western Blot Analysis

The total protein was extracted from sheep ovine granulosa cells using RIPA cell lysate. After protein quantification, 30 μg of total protein was subjected to SDS-PAGE gel electrophoresis, transferred to a membrane, and blocked. The membrane was incubated with diluted primary antibodies for Bcl2 (1:500), Bax (1:1000), Caspase-3 (1:1000), and β-actin (1:10,000) at 4 °C overnight. After incubation with the secondary antibody at room temperature for 1 h, ECL, the membrane was incubated in a chromogenic solution for 30 s, and the protein bands were visualized using a Chemi Scope imaging system (CLiNX Science Instruments, Shanghai, China). The grayscale values of the protein bands were analyzed using ImageJ, and the relative protein expression levels were calculated using β-actin as an internal reference.

2.8. Overexpression Vector Construction

Firstly, the candidate gene and recombinant expression vector were amplified and identified, and the PCR product was linked to the pcDNA3.1 (−) eukaryotic expression vector after digestion. The mRNA of the recombinant expression vector was tested.

2.9. Statistical Analysis

SPSS 22.0 (2019)software was used for qRT-PCR, Western blot, MTT and Flow cytometric results data analysis, and the Hisat (version:2.0.4) analysis was used for the sequence data. GraphPad Prism 7.0 (2020) software was used for graphing. The independent sample t test and the analysis of the variance were used to identify significant differences between groups (Control, Melation, Melation+H_2O_2). GraphPad Prism 7.0 software was used for graphing. Experimental data are expressed as "mean ± standard error", and all experiments were divided into three groups with three samples for each group. A significant difference is indicated by $p < 0.05$.

3. Results

3.1. Melatonin Protects against H_2O_2-Induced Apoptosis in Sheep Granulosa Cells

To determine the concentration for exogenous treatment and detect the effects of treatment on granulosa cell proliferation and apoptosis, the MTT assay and double-staining flow cytometry were used. After 3 h of treatment with different concentrations (0–500 μM) of H_2O_2, the cell viability decreased with the increase in the concentration of H_2O_2 in a dose-dependent manner; the median lethal dose (LD50) was 250 μM. After treatment with the serum-free medium containing different concentrations of melatonin (1–1000 ng/mL) for 24 h, the cell viability in the 0.1 ng/mL and 1 ng/mL treatment groups increased significantly ($p < 0.05$), and cell viability in the 10 ng/mL treatment group demonstrated an extremely significant increase ($p < 0.01$). When concentrations were higher than 10 ng/mL, cell viability decreased (Figure 1).

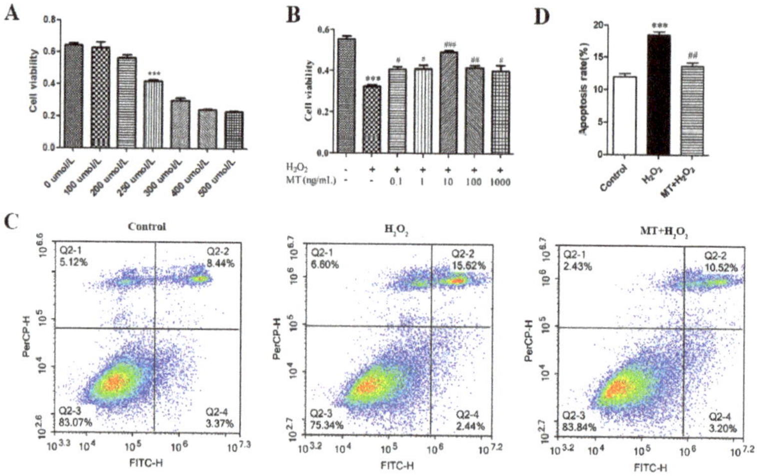

Figure 1. Melatonin represses H_2O_2 −induced apoptosis in cultured granulosa cells. (**A**) The effects of different concentrations (0, 100, 200, 250, 300, 400, 500 uM) of H_2O_2 on the viability of sheep granulosa cells. (**B**) The effects of different concentrations (0, 0.1, 1,10, 10, 100 and 1000 ng/mL) of melatonin on the viability of sheep granulosa cells with 250 uM H_2O_2 treatment. (**C**) Apoptosis levels measured by flow cytometry assay. (**D**) Apoptosis rate of flow cytometry assay. The results were acquired from three independent experiments and are presented as the mean ± SD. For *** $p < 0.01$ vs. the negative control group, the difference is extremely significant; for # $p < 0.05$ (##,### $p < 0.01$,) vs. the H_2O_2 group, the difference is significant or extremely significant.

The cells were cultured in a serum-free medium containing 10 ng/mL melatonin for 24 h and treated with 250 μmol/L H_2O_2 for 3 h.Double-staining flow cytometry indicated that the apoptotic rate for the group treated with melatonin was significantly reduced ($p < 0.01$), that the addition of melatonin effectively protected against sheep granulosa cell apoptosis after H_2O_2 treatment.

3.2. High-Throughput Sequencing Analysis

After we excluded the adaptors and N-containing and low-quality reads from raw reads obtained by high-throughput sequencing, a total of 57,693,778 clean reads were obtained for the blank group (N), accounting for 92.83%. A total of 64,057,360 clean reads were obtained for the H_2O_2 treatment group, accounting for 93.38%, and 61,994,733 clean reads were obtained for the melatonin+H_2O_2 (M+H) treatment group, accounting for 92.80%. The obtained clean reads were compared with reads in the Oar-v3.1 RefSeq database, and 27,055 annotated transcripts were obtained.

In the process of differential transcript analysis, the FPKM (fragments per kilobase of transcript per million fragments mapped) of cells in the three groups was normalized, DESeq2 software was used to set FPKM ≥ 1 and Q value < 0.05, and 492, 369, and 361 differentially expressed genes were obtained in the three groups (H/N, (M+H)/N, (M+H)/H) of sheep granulosa cells, respectively. The following parameters were set for the differentially expressed genes: FPKM ≥ 1, H/N-FPKM/(M+H)/N-FPKM/(M+H)/H-FPKM > 1, $p < 0.05$, and 98, 48, 109 differently expressed mRNA were obtained. Among them, 13 mRNAs were significantly upregulated in the H/N group and downregulated in the (M+H)/H group; 4 mRNAs were significantly downregulated in the H/N group and upregulated in the (M+H)/H group (Table 2). The differentially expressed genes in the (M+H)/H group caused by melatonin mainly reflect changes in genes that protect against H_2O_2 damage (Figure 2).

Table 2. 17 genes in the H/N comparisons that are regulated in the opposite way in the (M+H)/H comparisons.

Gene	Full Genetic Name	*p*-Value	Error Discovery Rate	Log2	Up/Down
UNC13C	unc-13 homolog C	4.03×10^{-16}	1.69×10^{-12}	2.035441	UP
STAR	steroidogenic acute regulatory protein	1.13×10^{-11}	1.88×10^{-08}	1.465299	UP
MMP9	matrix metallopeptidase 9	6.34×10^{-08}	4.42×10^{-05}	1.785441	UP
PET117	collagen type II alpha 1 chain	6.12×10^{-08}	4.42×10^{-05}	4.656607	UP
HSPA6	heat shock protein family A (Hsp70) 6member 6	2.68×10^{-29}	4.49×10^{-25}	−2.76246	DOWN
FGL2	fibrinogen like 2	2.12×10^{-25}	1.77×10^{-21}	−2.2515	DOWN
TRIB1	tribbles pseudokinase 1	4.25×10^{-17}	2.37×10^{-13}	−1.75357	DOWN
FOS	Fos proto-oncogene, AP-1 transcription factor subunit	1.94×10^{-15}	5.43×10^{-12}	−1.57275	DOWN
FOSB	Fos proto-oncogene B, AP-1 transcription factor subunit	5.24×10^{-13}	1.25×10^{-09}	−2.60732	DOWN
MAK3P8	mitogen-activated protein kinase kinase kinase 8	1.32×10^{-12}	2.77×10^{-09}	−3.29209	DOWN
SFRP4	secreted frizzled related protein 4	5.21×10^{-12}	9.70×10^{-09}	−1.6262	DOWN
JUN	Jun proto-oncogene, AP-1 transcription factor subunit	4.70×10^{-10}	7.15×10^{-07}	−1.1781	DOWN
ATF3	Activating transcription factor 3	5.95×10^{-08}	4.42×10^{-05}	−1.347907	DOWN
FIBIN	fin bud initiation factor homolog (zebrafish)	2.54×10^{-09}	3.28×10^{-06}	−1.1054	DOWN
STC1	stanniocalcin 1	1.12×10^{-07}	7.49×10^{-05}	−1.283589	DOWN
TGM2	transglutaminase 2	1.43×10^{-07}	9.19×10^{-05}	−1.092309	DOWN
DLX2	distal-less homeobox 2	4.64×10^{-09}	5.55×10^{-06}	−2.56006	DOWN

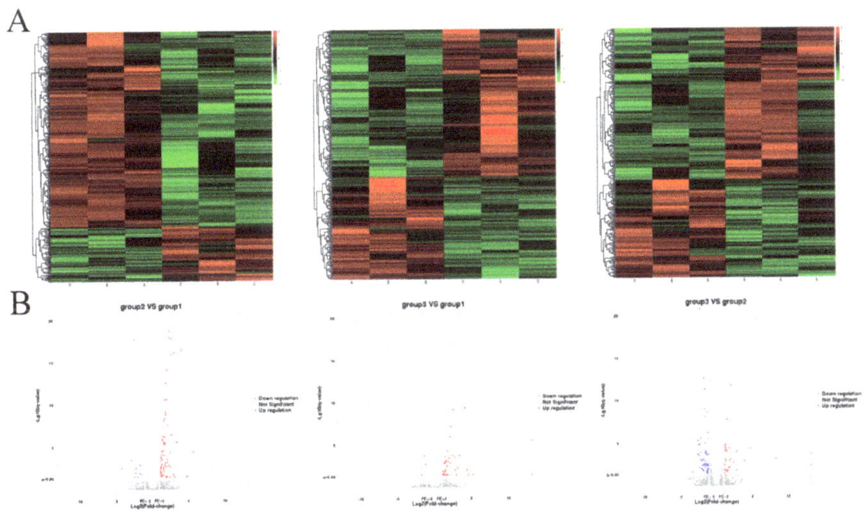

Figure 2. Hierarchical Clustering of Differentially Expressed Genes. (**A**) Differential gene heatmap. C and H represent the granulosa cells' samples of control and H_2O_2 treatment, respectively. M represents the granulosa cells' samples of H_2O_2+Melatonin treatment. The red colors represent the genes with higher expression, and the green colors represent the genes with lower expression. Colored bars indicate the expression level. (**B**) Differential gene volcano plot. The volcano plot of differentially expressed genes in C, H and M. Red dots represent the up-regulated genes, and blue dots show down−regulated genes. Gray dots represent genes that did not show obvious changes amongdifferential samples.

3.3. Functional Enrichment and Signal Pathway Analysis

A GO functional enrichment analysis of 98, 48, and 109 differentially expressed genes obtained from the comparison of the H/N, (M+H)/N and (M+H)/H groups was performed using GOseq software, and three large categories were obtained. Among the 49 enriched functions in the H/N group, biological processes accounted for 46.9%, cell components accounted for 30.6%, and molecular functions accounted for 32.7%. Among the 47 enriched functions in the (M+H)/N group, biological processes accounted for 46.8%, cell components accounted for 46.8%, and molecular functions accounted for 23.4%. Among the 46 enriched functions in the (M+H)/H group, biological processes accounted for 50.0%, cell components accounted for 30.4%, and molecular functions accounted for 19.6%.

A KEGG signaling pathway analysis was performed using KOBAS software on the differentially expressed genes obtained by comparing the H/N, (M+H)/N and (M+H)/H groups, and 24, 14 and 16 pathways were identified, respectively. Among them, the signal transduction pathway was the most enriched, with 18 in the (M+H)/H group. In addition, five genes were also found to be involved in the growth and development of granulosa cells (Figure 3).

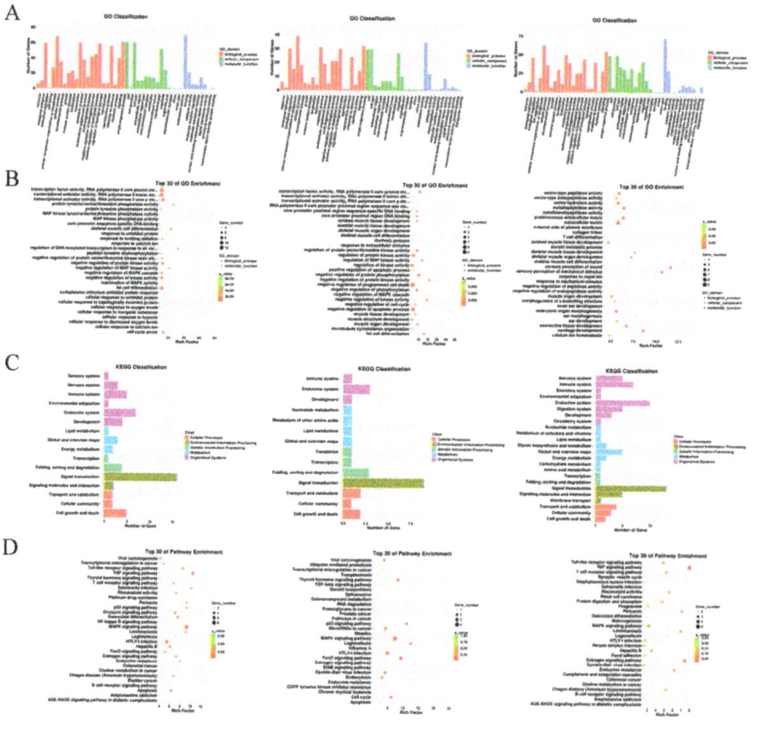

Figure 3. GO and KEGG analysis. GO-enriched bar graph of differentially expressed genes in granulosa cells with the control, H_2O_2 treatment and H_2O_2+Melatonin treatment. (**A**) The red bar indicates biological processes; green indicates cellular components; blue indicates molecular functions. (**B**) The scatter plot shows 30 GO terms in granulosa cells with the control, H_2O_2 treatment and H_2O_2+Melatonin treatment. (**C**) The ordinate indicates the GO terms, and the abscissa indicates the Rich factor. The size of the point indicates how many differentially expressed genes are in the GO terms, and the color of the point corresponds to the *p*.adj value range. The top 30 KEGG enrichment scatter plot indicates differentially expressed genes in granulosa cells with control, H_2O_2 treatment and H_2O_2+Melatonin treatment. (**D**) The ordinate indicates the KEGG terms, and the abscissa indicates the Rich factor. The size of the point indicates how many differentially expressed genes are in the KEGG pathways, and the color of the point corresponds to the *p*.adj value range.

3.4. RT–qPCR Validation

Through functional analysis, we identified nine genes that may be closely related to the protective effect of melatonin against oxidative damage in sheep granulosa cells. Among them, ATF3, FIBIN, FOS, HSPA6, MAP3K8, FOSB, DLX2, and TRIB1 were upregulated in the H group and downregulated in the M+T group; PET117 was downregulated in the H group and upregulated in M+T group. The RT–qPCR results indicated that although the expression levels of genes in each group were significantly different, the expression trend was consistent with the high-throughput sequencing results (Figure 4).

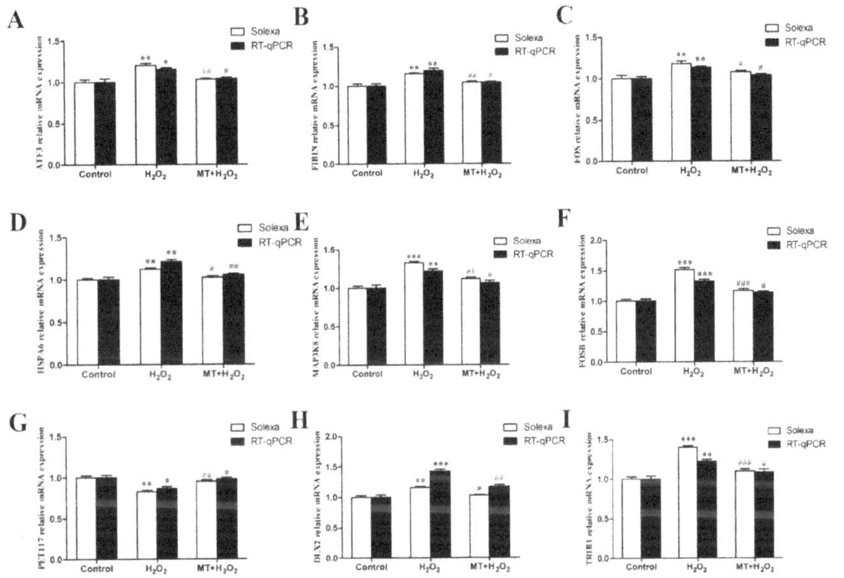

Figure 4. Comparison of the genes' mRNA expression levels between high-throughput sequencing and qRT-PCR. (**A–I**): ATF3, FIBIN, FOS, HSPA6, MAP3K8, FOSB, PET117,DLX2 and TRIB1. The white bar represents high-throughput sequencing and the black bar represents qRT-PCR. The results were acquired from three independent experiments and are presented as the mean ± SD. For * $p < 0.05$ (***,** $p < 0.01$) vs. the control group, the difference is significant or extremely significant; for F # $p < 0.05$ (###,## $p < 0.01$) vs. the H_2O_2 group, the difference is significant or extremely significant.

3.5. Effect of Candidate Genes on Sheep Granulosa Cell Apoptosis

Combined with KEGG analysis and the significance of gene expression differences, the MAPSK8 and FOS genes were screened as candidate genes for subsequent studies. After an overexpression vector of the candidate gene was successfully constructed, pcDNA3.1-MAP3K8 (referred to as MAP3K8), pcDNA3.1-FOS (referred to as FOS) and pcDNA3.1 (referred to as NC) were transfected into sheep granulosa cells, and 24 h later, cells were treated with H_2O_2 for 3 h, followed by the detection of apoptosis-related indicators. Compared with that in the control group, the viability of granulosa cells significantly decreased in both the MAP3K8 overexpression and FOS overexpression groups ($p < 0.01$), and the granulosa cell apoptosis rate significantly increased ($p < 0.01$). The expression levels of the proapoptotic genes Bax and Caspase-3 significantly decreased, and the expression level of Bcl2 significantly decreased ($p < 0.05$); the protein expression level was consistent with the mRNA level. Therefore, MAP3K8 and FOS gene overexpression can inhibit the viability of granulosa cells and promote granulosa cell apoptosis (Figure 5A–H).

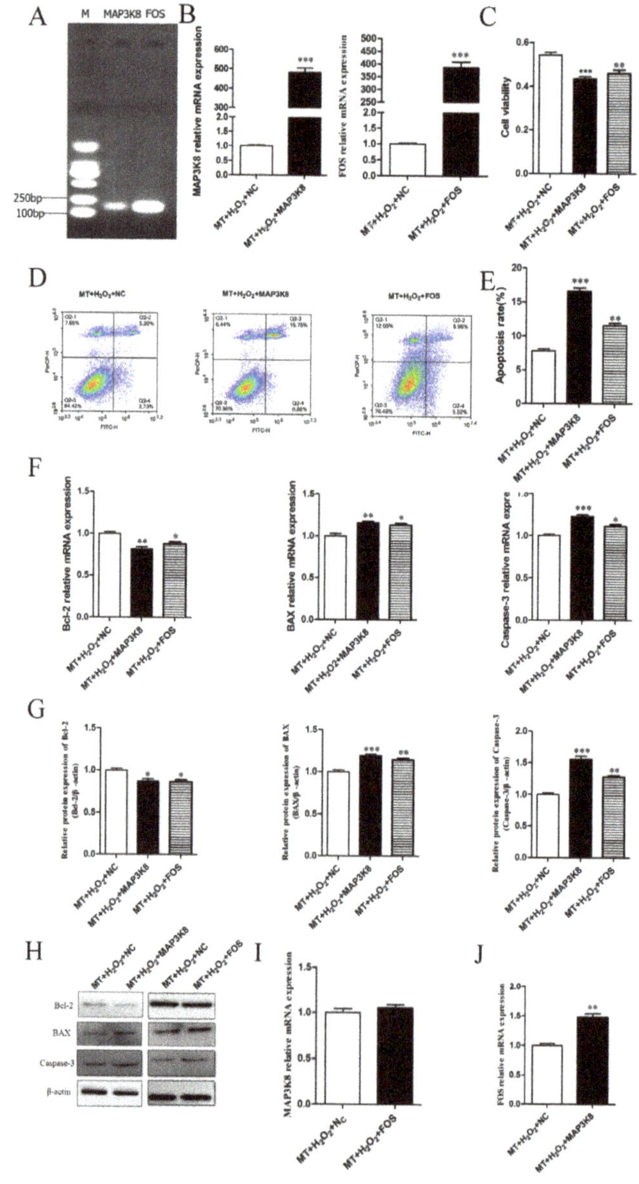

Figure 5. Melatonin inhibits H_2O_2-induced apoptosis of granulosa cells via MAP3K8 and FOS. (**A**,**B**) Construction of the overexpression vector of MAP3K8 and FOS. The expression of MAP3K8 and FOSB was determined by qRT-PCR; (**C**) Granulosa cells were transfected with MAP3K8 and FOS or negative control (NC). The viability of the different groups was measured by MTT; (**D**,**E**) Apoptosis levels measured by a flow cytometry assay. (**F**) The mRNA expression of Bax, Caspase-3 and Bcl-2 was determined by qRT-PCR analysis. (**G**,**H**) The protein expression of Bax, Caspase-3 and Bcl-2 was determined by western blot analysis. (**I**,**J**) Upstream and downstream relationship identification between MAP3K8 and FOS was measured by qRT-PCR. The results were acquired from three independent experiments and are presented as the mean ± SD. For * $p < 0.05$, the difference is significant; for ***,** $p < 0.01$, the difference is extremely significant.

As seen in Figure 5I–J, the increase in MAP3K8 expression significantly increased the expression of FOS ($p < 0.01$). Furthermore, the increase in FOS expression had no significant effect on the expression of MAP3K8 ($p > 0.05$), indicating that MAP3K8 was an upstream regulator of FOS expression.

4. Discussion

This study proves that melatonin inhibits the H_2O_2-induced apoptosis of sheep granulosa cells by suppressing oxidative stress via activation of the MAP3K8-FOS pathway. To the best of our knowledge, this is the first report revealing the inhibitory effect of melatonin on H_2O_2-induced apoptosis in sheep granulosa cells and the underlying mechanism of this effect.

Melatonin has been found in the follicles of many mammals, such as cattle [21] and swine [22]. It may participate in the protection of follicular cells from oxidative damage and play an important regulatory role in the maturation of oocytes. As a potent antioxidant, melatonin itself and its metabolites can protect the body from oxidative damage. Melatonin may act through PKA and PKB simultaneously in human granulosa cells to prevent follicular atresia and early luteinization at the antral stage [23]. Studies on the in vitro culture of porcine granulosa cells showed that, with the extension of the culture time, the endogenous melatonin content gradually decreased and the addition of melatonin to the culture medium effectively delayed the apoptosis of follicular granulosa cells and inhibited the apoptosis of granulosa cells and follicle atresia [8]. Melatonin attenuates palmitic acid-induced apoptosis in mouse granulosa cells through endoplasmic reticulum stress [24]. Melatonin attenuates deoxynivalenol-induced apoptosis in human granulosa cells [25] and affects the proliferation and apoptosis of sheep granulosa cells under heat stress [19]. The addition of melatonin and H_2O_2 affects the growth status of cells, and screening at different concentrations can optimize the effect. In the current study, the addition of 10 ng/mL melatonin effectively protected against sheep granulosa cell apoptosis caused by H_2O_2 (250 μM) damage. Therefore, melatonin can protect granulosa cells from apoptosis caused by a variety of exogenous stresses.

After the normal growth state of granulosa cells is stimulated by the external environment, the conformation of proapoptotic proteins changes, allowing them to transfer from the cytosol to the outer mitochondrial membrane and interact with the antiapoptotic proteins on and inside the membrane, eventually leading to apoptosis [26,27]. High-throughput sequencing identified 581 differentially expressed genes in melatonin-treated hen granulosa cells. These genes were mainly involved in biological processes such as cell proliferation and apoptosis. KEGG analysis revealed that these genes were mainly involved in the mTOR, PI3K-Akt, FoxO, and MAPK signaling pathways [28]. In the current study, 255 differentially expressed genes were selected from melatonin-treated and melatonin- and H_2O_2-treated granulosa cells by high-throughput sequencing; the expression levels of 17 genes changed significantly after melatonin treatment. This change indicates that melatonin plays a significant role in damage repair. The results of functional and KEGG analyses indicated that these differentially expressed genes were mainly involved in the MAPK, TNF, Toll-like receptor, and T-cell receptor-signaling pathways. We all know that a variety of genes are involved in cell damage and apoptosis in the MAPK and TNF pathways, and the addition of melatonin changes the gene expression in the above pathways, indicating that the regulatory pathway of melatonin in protecting the apoptosis function of granulosa cells is likely to be regulated by signals such as MAPK and TNF. In a previous study on the ameliorative effects of melatonin as an adjuvant for fertility preservation, it was shown that there was a significant downregulation in the mRNA expression of TNF-α, NF-Kβ, mTOR and p38-MAPK [29]. Therefore, melatonin is involved in many influencing factors and gene regulatory pathways in the regulation of granulosa cell apoptosis and can protect granulosa cells from apoptosis and alleviate the damage caused by exogenous stimuli through a variety of regulatory methods.

Melatonin can inhibit mitochondrial autophagy and inhibit FOXO1-mediated autophagy to protect mouse granulosa cells from oxidative damage [30]. Melatonin can enhance PDK1/Akt or PINK1/Parkin signaling in granulosa cells in polycystic ovary syndrome (PCOS) patients to alter the expression levels of the SIRT1 gene, thereby regulating mitochondrial function [31,32]. Melatonin can reduce hypoxia-induced apoptosis in granulosa cells by reducing ROS and activating the MTNR1B-PKA-Caspase8/9 pathway [33]. As important regulatory factors in the MAPK pathway, MAP3K8 and FOS are closely related to cell proliferation and apoptosis. MAP3K8 can inhibit monocyte apoptosis [34]. In myeloma tumor cells, MAP3K8 acts as a mitogen in mitosis, induces MAP3K, and inhibits tumor cell apoptosis [35]. Treatment with MAP3K8 inhibitors alone induced DNA fragmentation and poly (ADP-ribose) polymerase (PARP) cleavage of the two markers, indicating that apoptosis was induced in 8505C monolayer cells [36]. FOS alleviated apoptosis and rescued impaired GLP-1 release in TNF-α-treated L cells [37]. Furthermore, miR-29c-3p regulated FOS to repress EMT and cell proliferation and facilitate apoptosis [38]. In this study, MAP3K8 and FOS gene overexpression significantly increased the granulosa cell apoptosis rate, indicating that the two genes are upstream and downstream regulators. Therefore, it is proposed that melatonin can alleviate the apoptosis caused by H_2O_2 oxidative damage to granulosa cells by downregulating the expression of the MAP3K8 and FOS genes.

In summary, because MAP3K8 and FOS are important regulatory genes in the MAPK signaling pathway, we inferred that melatonin alleviates the proapoptotic injury induced by H_2O_2 in sheep granulosa cells through MAP3K8-FOS in the MAPK signaling pathway. The results of this study provide another theoretical basis for the mechanism of action of melatonin in alleviating oxidative stress in granulosa cells.

Author Contributions: Conceptualization, B.Z. and X.L. (Xu Li); methodology, Y.C. and X.L.(Xu Li); Investigation, B.Z., Z.Z., Z.L., X.L. (Xu Li) and X.L. (Xinxin Liu); Resources, H.M.; validation, X.L.(Xu Li)., X.L. (Xinxin Liu) and Z.Z.; Formal analysis, X.L. (Xu Li).; writing—original draft preparation, B.Z. and X.L. (Xu Li).; writing—review and editing, Y.C. and Z.Z.; supervision, W.L.; funding acquisition, H.M. All authors have read and agreed to the published version of the manuscript.

Funding: This research was funded by the National Key R & D Program of China (2021YFD1300904); Major Key of Jilin Province Innovation Project (CXGC2021ZD101).

Institutional Review Board Statement: The study was conducted according to the guidelines of the Declaration of China, and approved by the Animal Care and Use Committee of Jilin Agricultural University.

Data Availability Statement: Not applicable.

Acknowledgments: The study acknowledges the materials used for experiments by the Team of Wenfa Lu in Jilin Agricultural University.

Conflicts of Interest: The authors declare that they have no known competing financial interests or personal relationships that could have appeared to influence the work reported in this paper.

References

1. Reiter, R.J.; Paredes, S.D.; Manchester, L.C.; Tan, D.X. Reducing oxidative/nitrosative stress: A newlydiscovered genre for melatonin. *Crit. Rev. Biochem. Mol. Biol.* **2009**, *44*, 175–200. [CrossRef] [PubMed]
2. Reiter, R.J.; Tamura, H.; Tan, D.X.; Xu, X.Y. Melatonin and the circadian system: Contributions to successful female reproduction. *Fertil. Steril.* **2014**, *102*, 321–328. [CrossRef] [PubMed]
3. Tamura, H.; Takasaki, A.; Taketani, T.; Tanabe, M.; Kizuka, F.; Lee, L.; Tamura, I.; Maekawa, R.; Asada, H.; Yamagata, Y.; et al. Melatonin as a free radical scavenger in the ovarian follicle. *Endocr. J.* **2013**, *60*, 1–13. [CrossRef] [PubMed]
4. Hussein, M.R.; Bedaiwy, M.A.; Falcone, T. Analysis of apoptotic cell death, Bcl-2, and p53 protein expression in freshly fixed and cryopreserved ovarian tissue after exposure to warm ischemia. *Fertil. Steril.* **2006**, *85*, 1082–1092. [CrossRef]
5. Barberino, R.S.; Menezes, V.G.; Ribeiro, A.E.; Palheta, R.C., Jr.; Jiang, X.; Smitz, J.E.; Matos, M.H.T. Melatonin protects against cisplatin-induced ovarian damage in mice via the MT1 receptor and antioxidant activity. *Biol. Reprod.* **2017**, *96*, 1244–1255. [CrossRef]
6. Talpur, H.S.; Worku, T.; Rehman, Z.U.; Dad, R.; Bhattarai, D.; Bano, I.; Farmanullah; Liang, A.; He, C.; Yang, L. Knockdown of melatonin receptor 1 and induction of follicle-stimulating hormoneon the regulation of mouse granulosa cell function. *Reprod. Biol.* **2017**, *17*, 380–388. [CrossRef]

7. Lin, T.; Lee, J.E.; Kang, J.W.; Oqani, R.K.; Cho, E.S.; Kim, S.B.; Jin, D., II. Melatonin supplementation during prolonged in vitro maturation improves the quality and development of poor-quality porcine oocytes via anti-oxidative andanti-apoptotic effects. *Mol. Reprod. Dev.* **2018**, *85*, 665–681. [CrossRef]
8. He, Y.; Deng, H.; Jiang, Z.; Li, Q.; Shi, M.; Chen, H.; Han, Z. Effects of melatonin on follicular atresia and granulosa cell apoptosis in the porcine. *Mol. Reprod. Dev.* **2016**, *83*, 692–700. [CrossRef]
9. Wang, Y.; Zeng, S. Melatonin promotes ubiquitination of phosphorylated pro-apoptotic protein Bcl2-interacting mediator of cell death-extra long (BimEL) in porcine granulosa cells. *Int. J. Mol. Sci.* **2018**, *19*, 3431–3446. [CrossRef]
10. Fang, Y.; Deng, S.; Zhang, J.; Liu, H.; Li, Y.; Zhang, X.; Liu, Y. Melatonin-mediated development of ovine cumulus cells, perhaps by regulation of DNA methylation. *Molecules* **2018**, *23*, 494–506. [CrossRef]
11. Riaz, H.; Yousuf, M.R.; Liang, A.; Hua, G.H.; Yang, L. Effect of melatonin on regulation of apoptosis and steroidogenesis in cultured buffalo granulosa cells. *Anim. Sci. J.* **2019**, *90*, 473–480. [CrossRef] [PubMed]
12. Zhao, S.-Q.; Zhang, Y.; Gao, Y.; Yang, X.-P.; Yang, Z.; Yang, Z.-J. The in vitro effects of melatonin and Cry gene on the secretion of estradiol from camel ovarian granulosa cells. *Domest. Anim. Endocrinol.* **2021**, *74*, 106497.
13. Liu, Y.; Yang, Y.; Li, W.; Ao, H.; Zhang, Y.; Zhou, R.; Li, K. Effects of melatonin on the synthesis of estradiol and gene expression in pig granulosa cells. *J. Pineal Res.* **2019**, *66*, e12546. [CrossRef] [PubMed]
14. Perego, M.C.; Bellitto, N.; Maylem, E.R.S.; Caloni, F.; Spicer, L.J. Effects of selected hormones and their combination on progesterone and estradiol production and proliferation of feline granulosa cells cultured in vitro. *Theriogenology* **2021**, *168*, 1–12. [CrossRef] [PubMed]
15. Wang, S.J.; Liu, W.J.; Wu, C.J.; Ma, F.H.; Ahmad, S.; Liu, B.R.; Han, L.; Jiang, X.P.; Zhang, S.J.; Yang, L.G. Melatonin suppresses apoptosis and stimulates progesterone production by bovine granulosa cells via its receptors (MT1 and MT2). *Theriogenology* **2012**, *78*, 1517–1526. [CrossRef]
16. Hao, E.-Y.; Wang, D.-H.; Chang, L.-Y.; Huang, C.-X.; Chen, H.; Yue, Q.-X.; Zhou, R.-Y.; Huang, R.-L. Melatonin regulates chicken granulosa cell proliferation and apoptosis by activating the mTOR signaling pathway via its receptors. *Poult. Sci.* **2020**, *99*, 6147–6162. [CrossRef]
17. Wu, G.; Song, D.; Wei, Q.; Xing, J.; Shi, X.; Shi, F. Melatonin mitigates bisphenol A-induced estradiol production and proliferation by porcine ovarian granulosa cells in vitro. *Anim. Reprod. Sci.* **2018**, *192*, 91–98. [CrossRef]
18. Nakamura, E.; Otsuka, F.; Terasaka, T.; Inagaki, K.; Hosoya, T.; Tsukamoto-Yamauchi, N.; Toma, K.; Makino, H. Melatonin counteracts BMP-6 regulation of steroidogenesis by rat granulosa cells. *J. Steroid Biochem. Mol. Biol.* **2014**, *143*, 233–239. [CrossRef]
19. Fu, Y.; He, C.-J.; Ji, P.-Y.; Zhuo, Z.-Y.; Tian, X.-Z.; Wang, F.; Tan, D.-X.; Liu, G.-S. Effects of melatonin on the proliferation and apoptosis of sheep granulosa cells under thermal stress. *Gene* **2014**, *15*, 21090–21104. [CrossRef]
20. Li, X.; Xu, G.; Li, Z.; Liu, H.; Ma, X.; Yang, L.; Zhang, P.; Zhao, J.; Wang, J.; Lu, W. Gonadotropin-Inhibiting Hormone Promotes Apoptosis of Bovine Ovary Granulosa Cells. *Life Sci.* **2021**, *270*, 119063. [CrossRef]
21. Tian, X.; Wang, F.; He, C.; Zhang, L.; Tan, D.; Reiter, R.J.; Xu, J.; Ji, P.; Liu, G. Beneficial effects of melatonin on bovine oocytes maturation: A mechanistic approach. *J. Pineal Res.* **2014**, *57*, 239–247. [CrossRef] [PubMed]
22. Shi, J.-M.; Tian, X.-Z.; Zhou, G.-B.; Wang, L.; Gao, C.; Zhu, S.-E.; Zeng, S.-M.; Tian, J.-H.; Liu, G.-S. Melatonin exists in porcine follicular fluid and improves in vitro maturation and parthenogenetic development of porcine oocytes. *J. Pineal Res.* **2010**, *47*, 318–323. [CrossRef] [PubMed]
23. Asma, A.; Marc-André, S. Melatonin Signaling Pathways Implicated in Metabolic Processes in Human Granulosa Cells (KGN). *Int. J. Mol. Sci.* **2022**, *23*, 2988. [CrossRef]
24. Chen, Z.; Lei, L.; Wen, D.; Yang, L. Melatonin attenuates palmitic acid-induced mouse granulosa cells apoptosis via endoplasmic reticulum stress. *J. Ovarian Res.* **2019**, *12*, 43. [CrossRef] [PubMed]
25. Xue, R.; Li, S.; Zou, H.; Ji, D.; Lv, M.; Zhou, P.; Wei, Z.; Zhang, Z.; Cao, Y. Melatonin alleviates deoxynivalenol-induced apoptosis of human granulosa cells by reducing mutually accentuated FOXO1 and ER stress. *Biol. Reprod.* **2021**, *105*, 554–566. [CrossRef]
26. Basini, G.; Simona, B.; Santini, S.E.; Grasselli, F. Reactive oxygen species and anti-oxidant defences in swine follicular fluids. *Reprod. Fertil. Dev.* **2008**, *20*, 269–274. [CrossRef]
27. Agarwal, A.; Gupta, S.; Sharma, R.K. Role of oxidative stress in female reproduction. *Reprod. Biol. Endocrinol.* **2005**, *3*, 28–49. [CrossRef]
28. Hao, E. Research on the Regulatory Mechanism of mTOR Signaling Pathway Mediated Melatonin to Delay Ovary Aginginlate-Phase Laying Hen. Ph.D. Thesis, HeBei Agricultural University, Baoding, China, 2020.
29. Al-Shahat, A.; Hulail, M.A.E.; Soliman, N.M.M.; Khamis, T.; Fericean, L.M.; Arisha, A.H.; Moawad, R.S. Melatonin Mitigates Cisplatin-Induced Ovarian Dysfunction via Altering Steroidogenesis, Inflammation, Apoptosis, Oxidative Stress, and PTEN/PI3K/Akt/mTOR/AMPK Signaling Pathway in Female Rats. *Pharmaceutics* **2022**, *14*, 2769. [CrossRef]
30. Shen, M.; Cao, Y.; Jiang, Y.; Wei, Y.; Liu, H. Melatonin protects mouse granulosa cells against oxidative damage by inhibiting FOXO1-mediated autophagy: Implication of an antioxidation-independent mechanism. *Redox Biol.* **2018**, *18*, 138–157. [CrossRef]
31. Zheng, B.; Meng, J.; Zhu, Y.; Ding, M.; Zhang, Y.; Zhou, J. Melatonin enhances SIRT1 to ameliorate mitochondrial membrane damage by activating PDK1/Akt in granulosa cells of PCOS. *J. Ovarian Res.* **2021**, *14*, 152. [CrossRef]
32. Yi, S.; Zheng, B.; Zhu, Y.; Cai, Y.; Sun, H.; Zhou, J. Melatonin ameliorates excessive PINK1/Parkin-mediated mitophagy by enhancing SIRT1 expression in granulosa cells of PCOS. *Am. J. Physiol. Endocrinol. Metab.* **2020**, *319*, E91–E101. [CrossRef] [PubMed]

33. Tao, J.-L.; Zhang, X.; Zhou, J.-Q.; Li, C.-Y.; Yang, M.-H.; Liu, Z.-J.; Zhang, L.-L.; Deng, S.-L.; Zhang, L.; Shen, M.; et al. Melatonin Alleviates Hypoxia-Induced Apoptosis of Granulosa Cells by Reducing ROS and Activating MTNR1B-PKA-Caspase8/9 Pathway. *Antioxidants* **2021**, *10*, 184. [CrossRef] [PubMed]
34. Sanz-Garcia, C.; Sánchez, Á.; Contreras-Jurado, C.; Cales, C.; Barranquero, C.; Muñoz, M.; Merino, R.; Escudero, P.; Sanz, M.-J.; Osada, J.; et al. MAP3K8 Modulates Monocyte State and Atherogenesis in ApoE-/- Mice. *Arter. Thromb. Vasc. Biol.* **2017**, *37*, 237–246. [CrossRef] [PubMed]
35. Hebron, E.; Hope, C.; Kim, J.; Jensen, J.L.; Flanagan, C.; Bhatia, N.; Maroulakou, I.; Mitsiades, C.; Miyamoto, S.; Callander, N.; et al. MAP3K8 kinase regulates myeloma growth by cell-autonomous and non-autonomous mechanisms involving myeloma-associated monocytes/macrophages. *Br. J. Haematol.* **2013**, *160*, 779–784. [CrossRef] [PubMed]
36. Gianì, F.; Russo, G.; Pennisi, M.; Sciacca, L.; Frasca, F.; Pappalardo, F. Computational modeling reveals MAP3K8 as mediator of resistance to vemurafenib in thyroid cancer stem cells. *Bioinformatics* **2019**, *35*, 2267–2275. [CrossRef] [PubMed]
37. Wongkrasant, P.; Pongkorpsakol, P.; Chitwattananont, S.; Satianrapapong, W.; Tuangkijkul, N.; Muanprasat, C. 4 Fructo-oligosaccharides alleviate inflammation- associated apoptosis of GLP-1 secreting L cells via inhibition of iNOS and cleaved caspase-3 expression. *J. Pharm. Sci.* **2020**, *143*, 65–73. [CrossRef]
38. Yao, L.; Yang, L.; Song, H.; Liu, T.; Yan, H. MicroRNA miR-29c-3p modulates FOS expression to repress EMT and cell proliferation while induces apoptosis in TGF-β2-treated lens epithelial cells regulated by lncRNA KCNQ1OT1. *Biomed. Pharmacother.* **2020**, *129*, 110290. [CrossRef]

genes

Article

Identification and Functional Analysis of Transcriptome Profiles, Long Non-Coding RNAs, Single-Nucleotide Polymorphisms, and Alternative Splicing from the Oocyte to the Preimplantation Stage of Sheep by Single-Cell RNA Sequencing

Zijing Zhang [1], Qiaoting Shi [1], Xiaoting Zhu [1], Lei Jin [1], Limin Lang [1], Shijie Lyu [1], Xiaoling Xin [1], Yongzhen Huang [2], Xiang Yu [3], Zhiming Li [4], Sujuan Chen [5], Zhaoxue Xu [1], Wei Zhang [6,*] and Eryao Wang [1,*]

1 Institute of Animal Husbandry and Veterinary Science, Henan Academy of Agricultural Sciences, No. 116 Hua Yuan Road, Zhengzhou 450002, China; vincezhang163@163.com (Z.Z.); sqtsw@126.com (Q.S.); xtzhu2019@outlook.com (X.Z.); langlimin666@outlook.com (L.L.); sjlyu@outlook.com (S.L.); xuzhaoxue11@163.com (Z.X.)
2 College of Animal Science and Technology, Northwest A&F University, Yangling 712100, China; hyzsci@nwafu.edu.cn
3 Henan Animal Health Supervision Institute, Zhengzhou 450003, China; 15838282287@163.com
4 Henan Provincial Animal Husbandry General Station, Zhengzhou 450008, China; hna1920@163.com
5 Synthetic Biology Engineering Lab of Henan Province, School of Sciences and Technology, Xinxiang Medical University, Xinxiang 453003, China; chensujuan101@163.com
6 College of Animal Science and Veterinary Medicine, Henan Institute of Science and Technology, Xinxiang 453003, China
* Correspondence: yuntaishan2003@aliyun.com (W.Z.); eryaowang6@outlook.com (E.W.)

Abstract: Numerous dynamic and complicated processes characterize development from the oocyte to the embryo. However, given the importance of functional transcriptome profiles, long non-coding RNAs, single-nucleotide polymorphisms, and alternative splicing during embryonic development, the effect that these features have on the blastomeres of 2-, 4-, 8-, 16-cell, and morula stages of development has not been studied. Here, we carried out experiments to identify and functionally analyze the transcriptome profiles, long non-coding RNAs, single-nucleotide polymorphisms (SNPs), and alternative splicing (AS) of cells from sheep from the oocyte to the blastocyst developmental stages. We found between the oocyte and zygote groups significantly down-regulated genes and the second-largest change in gene expression occurred between the 8- and 16-cell stages. We used various methods to construct a profile to characterize cellular and molecular features and systematically analyze the related GO and KEGG profile of cells of all stages from the oocyte to the blastocyst. This large-scale, single-cell atlas provides key cellular information and will likely assist clinical studies in improving preimplantation genetic diagnosis.

Keywords: transcriptome profiles; long non-coding RNAs; single-nucleotide polymorphisms; alternative splicing; sheep; preimplantation; single-cell RNA sequencing

Citation: Zhang, Z.; Shi, Q.; Zhu, X.; Jin, L.; Lang, L.; Lyu, S.; Xin, X.; Huang, Y.; Yu, X.; Li, Z.; et al. Identification and Functional Analysis of Transcriptome Profiles, Long Non-Coding RNAs, Single-Nucleotide Polymorphisms, and Alternative Splicing from the Oocyte to the Preimplantation Stage of Sheep by Single-Cell RNA Sequencing. *Genes* **2023**, *14*, 1145. https://doi.org/10.3390/genes14061145

Academic Editor: Bao Yuan

Received: 3 April 2023
Revised: 16 May 2023
Accepted: 18 May 2023
Published: 25 May 2023

1. Introduction

From the development of the oocyte to the embryo, many dynamic and complicated processes take place as the zygote undergoes several rapid rounds of division and produces a mass of cells within the zona pellucida [1,2]. During these dynamic stages, differential gene expression in individual cells is a key determinant of cellular differentiation, function, and physiology [3,4]. In recent years, several studies have documented the key developmental processes underlying the formation of blastomeres of 4- and 8-cell embryos of mice [5], cattle [6], and goats [7]. However, the exact mechanism and developmental patterns underlying how the blastomeres of the 2-, 4-, 8-, 16-cell, morula, and blastocyst stages undergo asymmetric division are still unclear. Understanding the molecular mechanism underlying cleavage-stage development is critically important for improving preimplantation genetic diagnosis.

Single-cell RNA sequencing (scRNA-seq) provides an alternative method for studying the cellular heterogeneity of human preimplantation embryos and embryonic stem cells [8,9], mouse oocytes [10], porcine oocyte maturation [11], and Haimen white goat's oocyte maturation by generating a readout of the abundance of a transcript within a cell [12]. Indeed, scRNA-seq applied to mammalian gametes has generated new insights into the composition of transcripts and disease-related functional abnormalities. However, scRNA-seq studies have not thoroughly characterized how the blastomeres of the 2-, 4-, 8-, 16-cell, and especially the morula and blastocyst stages, undergo asymmetric division. Here, we aimed to transcriptionally profile nucleated cells present during the blastomeres of 2-, 4-, 8-, 16-cell and morula stage undergoing asymmetric division to provide a broad profile of the composition of transcripts in the cell and their transcriptomes.

Long non-coding RNAs (lncRNAs) are a diverse group of RNAs that are longer than an arbitrary limit of 200 nt and do not encode proteins [13]. Nevertheless, lnc RNAs can be located in exonic, intronic, and intergenic regions and can regulate gene expression by interacting with other biological macromolecules, such as RNA, DNA, proteins, and other factors, to promote normal cell function [14,15]. Compared with the characteristics of protein-coding genes, lncRNAs tend to be less conserved across species and often show lower expression levels and high tissue specificity [16]. During the development of embryos, some lncRNAs, such as Xist, Tsix, and H19, have significant regulatory functions and can potentially determine the cell's fate and direction of differentiation during embryogenesis to form different organs or special tissues that contain various cells expressing specific genes [17–19].

An increasing number of studies has reported several thousands of annotated lncRNA loci in mammalian oocytes and early embryos [1,20]. For example, a total of 2733 novel lncRNAs was found to be expressed in specific developmental stages among 8701 lncRNAs using single-cell sequencing analysis of 124 individual cells from human embryonic stem cell (ES) and human preimplantation embryos at different passages. In addition, 5204 novel lncRNAs were obtained from in vivo and somatic cell nuclear transfer mouse preimplantation embryo, suggesting that several lncRNAs and their association with known protein-coding genes might be involved in mouse embryonic development and cell reprogramming [8,21]. Another study reported that approximately a quarter of the 1600 lncRNA loci expressed during the murine oocyte-to-embryo transition employed promoters from a long terminal repeat retrotransposon class, which exhibited either maternal or zygotic expression and showed signatures of massive expansion in the evolution of the rodent genome [16]. In bovine early embryos, some lncRNAs play a role in the translational control of target mRNA and are thus important for managing maternal reserves, especially before embryonic genome activation, as these reserves contain the embryonic program [22]. Despite the fact that various functional lncRNAs are important during embryonic development, the effect of lncRNAs on the blastomeres of the 2-, 4-, 8-, 16-cell, and morula stages of development has not been studied; nevertheless, this subject requires more attention and discussion by scientists. LncRNAs play an important role in biological processes, including epigenetic regulation, dosage compensation, cell cycle, cell differentiation, proliferation, apoptosis through gene imprinting, chromatin remodeling, transcriptional activation, transcriptional interference, and nuclear splicing regulation.

Here, we conducted an scRNA-seq survey of 24 sheep cells during their development from the oocyte to the blastocyst stages. We then conducted experiments involved in the identification and functional analysis of transcriptome profiles, lncRNAs, single-nucleotide polymorphisms (SNPs), and alternative splicing (AS). Using these different methods, we constructed a profile to characterize cellular and molecular features and systematically analyze the related GO (Gene Ontology) and KEGG (Kyoto Encyclopedia of Genes and Genomes) profile of cells during sheep development. This large-scale single-cell atlas provides key cellular information and will likely aid clinical studies in the development of more efficient preimplantation genetic diagnosis.

2. Materials and Methods

2.1. Animals and Sample Collection

All work with animals was completed in accordance and with the approval of the Henan Academy of Agricultural Sciences institutional animal care and use committee. Mature sheep were obtained from Hui yuan Sheep Industry Co., Ltd. (Puyang, Henan province, China). We used 15-month-old female sheep (40 kg) for our study. The animals were provided with grass and drinking water and had access to an animal exercise pen. All animals were healthy, showed a normal appetite, and had smooth wool. Artificial insemination using semen from one of three fertile rams was conducted at 12- and 24-h post-standing heat (Day 0). Donor animals were anesthetized and in vivo developed oocytes and embryos at the 2- to 16-cell stages were collected by oviductal flushing 2–4 days after estrus. Early morulae, compact morulae, and blastocysts were collected by routine non-surgical uterine flushing on days 5, 6, and 7. All oocytes and embryos were examined and staged under light microscopy. Only morphologically intact embryos meeting the standards of Grade 1 by the International Embryo Transfer Society were included in this study. Embryos were washed twice in D-PBS before being frozen and stored individually in RNAlater (Ambion, Grand Island, NY, USA) in liquid nitrogen.

2.2. RNA Isolation, Library Preparation, and Sequencing

First, total RNA was isolated from the sheep sample (three biological replicates per sample combined) using single cell RNA kit (Single Cell RNA Purification Kit NGB) and extracted (Norgen Biotek, CA, USA) following the manufacturer's procedure. The single cell RNA quantity and purity were analyzed with the Bioanalyzer 2100 and RNA 6000 Nano LabChip Kit (Agilent, CA, USA) with RIN number > 7.0. Approximately 10 μg of total RNA representing a specific adipose type was subjected to isolate Poly (A) mRNA with poly-T oligo-attached magnetic beads (Invitrogen, Carlsbad, CA, USA). Following purification, all amplifications were carried out in parallel with positive and no-template controls for quality assurance using the SMARTer Universal low Input RNA kit (Clontech) for cDNA library. Briefly, RNA was converted to cDNA, and then adapters for Illumina sequencing (with specific barcodes) were added through PCR using only a limited number of cycles. The PCR products were purified (Protocol C), and then ribosomal cDNA was depleted. The cDNA fragments were further amplified with primers universal to all libraries. Lastly, the PCR products were purified once more to yield the final cDNA library. Then, the mRNA was fragmented into small pieces using divalent cations under elevated temperature. Then the cleaved RNA fragments were reverse-transcribed to create the final cDNA library in accordance with the protocol for the mRNASeqsample preparation kit (Illumina, San Diego, CA, USA); the average insert size for the paired-end-libraries was 300 bp (±50 bp). Following this, we performed the paired-end sequencing on an IlluminaHiseq4000 at the (LC Sciences, Houston, TX, USA) using the vendor's recommended protocol.

2.3. Quality Control and Assembly of Transcriptome Data

Raw data in FASTQ format were first processed through in-house perl scripts. Clean reads were obtained by removing reads containing adapters, reads containing poly-N, and low-quality raw reads. Furthermore, Q20, Q30, and GC contents of the clean data were calculated. All of the downstream sequencing analyses were based on high-quality clean reads. For the RNA-seq data, all clean reads from each sample were aligned to the sheep reference genome (https://www.ncbi.nlm.nih.gov/genome/83?genome_assembly_id=351950, accessed on 1 October 2019) using Tophat v2.0.949. The distribution of known gene types was analyzed by HTSeq software. The mapped reads of each sample were then assembled by both Scripture (beta2) and Cufflinks (v2.1.1) using a reference-based approach.

2.4. RNA-Seq Reads Mapping

We aligned reads to the UCSC (http://genome.ucsc.edu/, accessed on 1 October 2019) sheep reference genome using the Tophat package, which initially removes a portion of the reads based on quality information accompanying each read and then maps the reads to the reference genome. Tophat allows multiple alignments to be read (up to 20 by default) and a maximum of two mismatches when mapping the reads to the reference. Tophat builds a database of potential splice junctions and confirms these by comparing the previously unmapped reads against the database of putative junctions.

2.5. Estimation of Transcript Abundance and Differential Expression and Principal Component Analysis (PCA)

The mapped reads of each sample were assembled using StringTie. All transcriptomes from the samples were then merged to reconstruct a comprehensive transcriptome using perl scripts. After the final transcriptome was generated, StringTie and Ballgown were used to estimate the expression levels of all transcripts. StringTie was used to determine the expression level of mRNAs by calculating FPKM (fragments per kilobase million). Differentially expressed mRNAs and genes were identified if log2 (fold change) > 1 or log2 (fold change) < −1 and by statistical significance (*p*-value < 0.05) through the R package (1.12.0) Ballgown and then used to generate a PCA plot, showing the relationship of gene expression between the different stages as a previous study [9].

2.6. WGCNA Analysis

The co-expression network analysis was performed using WGCNA (version:1.61) [23]. First, the soft threshold for network construction was selected. The soft threshold constrains the adjacency matrix to assume a continuous value between 0 and 1 so that the constructed network conforms to the power-law distribution and is closer to the real biological network state. Second, the scale-free network was constructed using the blockw iseModules function, followed by the module partition analysis to identify gene co-expression modules, which groups genes with similar patterns of expression. The modules were defined by cutting the clustering tree into branches using a dynamic tree-cutting algorithm and assigning different colors for visualization to the modules [23]. The module eigengene (ME) of each module was then calculated. ME represents the expression level for each module. Next, the correlation between ME and the clinical trait in each module was calculated, followed by the determination of gene significance.

2.7. Transcript Assembly and Identification of Candidate lncRNAs

First, Cutadapt was used to remove the reads that were contaminated with adaptors [24], low-quality bases, and undetermined bases. Sequence quality was then verified using FastQC (http://www.bioinformatics.babraham.ac.uk/projects/fastqc/, accessed on 1 October 2019). We used Bowtie2 (FastQC) and Tophat2 to map read to the genome [25]. The mapped reads of each sample were assembled using StringTie [26]. All transcriptomes from the samples were then merged to reconstruct a comprehensive transcriptome using perl scripts. After the transcriptome was generated, StringTie and Ballgown were used to estimate the expression levels of all transcripts. Transcripts that overlapped with known mRNAs and transcripts shorter than 200 bp were discarded. We utilized CPC and CNCI to predict transcripts with coding potential [27]. All transcripts with CPC scores < −1 and CNCI scores < 0 were removed. The remaining transcripts were considered lncRNAs.

2.8. Classification of lncRNAs

The annotated lncRNAs were subdivided into the following four categories based on their locations relative to the nearest protein-coding genes: (i) lncRNAs that do not overlap protein-coding genes (lincRNAs); (ii) lncRNAs located entirely within a protein-coding locus (intragenic lncRNAs); (iii) lncRNAs partially overlapping a protein-coding gene (overlapping lncRNAs); (iv) lncRNAs that overlapped exons of a protein-coding transcript on the opposite strand (antisense lncRNAs). Perl scripts were developed to classify these four categories.

2.9. Quantification and Differential Expression Analysis

The relative abundances of both candidate lncRNAs and coding genes in each sample were computed by calculating the FPKM using Cufflinks (v2.1.1). Differentially expressed lncRNAs in comparison groups were identified using the Cuffdiff program. For biological replicates, transcripts with adjusted p-values < 0.05 were considered differentially expressed between the two groups.

2.10. Predictions of Cis and Trans-Target Genes

To explore the function of lncRNAs, we first predicted the cis and trans-target genes of lncRNAs. We searched for coding, cis-target genes 10 k and 100 k upstream and downstream, respectively, of candidate lncRNAs, and then analyzed their functions. We calculated the expressed correlation between lncRNAs and coding genes with custom scripts and then analyzed their functions through functional enrichment analysis. The trans-target genes and lncRNAs were identified by their expression levels.

2.11. SNP Analysis

To further characterize the SNPs, we categorized them as genic or intergenic. Approximately half of the SNPs (47%) were located in genic regions, and the rest was located in intergenic regions. In addition, nonsynonymous SNPs in the exon region were analyzed to determine whether their amino acid character had changed (e.g., hydrophobic to basic or stop codons), given that compositional changes of the amino acids in proteins can result in changes in structural conformation or enzymatic activities and thus generate phenotypic diversity or critical functional variations. First, 20 amino acids were clustered into several character groups. Non-synonymous SNPs that caused amino acid changes from one group to another were searched. Common SNPs in the eight groups representing each line were then classified.

2.12. AS Data Collection

Data on 96 melanoma cases with clinicopathological information were obtained to explore changes in AS events. To analyze the AS profiles for each patient, the SpliceSeq tool (version 2.1), a java application, was used to evaluate the splicing patterns of mRNA in the melanoma cohort. The percentage spliced in value was calculated to quantify alternative splicing events and ranged between 0 and 1 for seven types of AS events, including exon skip (ES), alternate terminator (AT), and mutually exclusive (ME).

2.13. GO and KEGG Enrichment Analysis

GO enrichment analysis of the aforementioned groups was conducted using the GOseq R package while correcting for biases in gene length. In addition, KOBAS software and the KEGG database (http://www.genome.jp/kegg/, accessed on 1 October 2019) were used to analyze the statistical enrichment of target genes of differentially expressed lncRNAs in KEGG pathways. Lower p-values corresponded to more relevant pathways, and corrected p-values < 0.05 were considered significantly enriched by DEGs.

2.14. Statistical Analysis

Statistical analysis was performed using SPSS13.0 software. Proportional data were compared using chi-squared analysis or Fisher's exact tests, and differences were considered significant when $p < 0.05$. The percentage of blastocyst formed represented the number of blastocysts formed divided by the total number of embryos cultured. The percentage of high-quality blastocyst formed represented the number of high-quality blastocysts divided by the total number of blastocysts cultured.

3. Results

3.1. Transcriptome Profiles

To establish single-cell transcriptome profiles during the blastomeres of the 2-, 4-, 8-, 16-cell, and morula stage as they underwent asymmetric division, we used previously published protocols; single cells were subjected to RNA-seq sample preparation with several steps modified [8]. Overall, scRNA-seq was performed on 24 cells samples using the Illumina HiSeq2000 sequencer (Table 1). We generated 384 Gb of sequencing data from the 24 cells samples, with, on average, 10.2 million reads per cell with read lengths of 100 bp. To determine whether these gene expression profiles correlated with developmental stages, we analyzed RNA-seq data from blastocyst cells of the oocytes using a heat map. The greatest changes in gene expression were observed between the morula and blastocyst stages, which may be explained by the major morula–blastocyst transition and are largely representative of the expression patterns occurring during sheep preimplantation development (Figure 1A). Weighted gene co-expression network analysis (WGCNA) is commonly used to reveal correlations between genes in different samples. Previous WGCNAs have been used to reveal developmentally important genes of the bovine sham nuclear-transfer (shNT) blastocysts [28]. Following the approach of a previous bioinformatics study [23], the soft threshold for network construction was set to 30 (Figure 1B). Meanwhile, the fitting degree of the scale-free topological model was 0.85. Thus, this network conformed to the power-law distribution and was closer to the real biological network state. A total of 26 modules (Figure 1C) was obtained in the current study. The differentially expressed genes (DEGs) in grey were not included in any module; thus, subsequent analyses were no longer performed on DEGs in grey. Mutually exclusive (ME) was in accordance with the expression pattern of DEGs in each module. The pattern of similar within-stage and different between-stage expression patterns were also supported by principal component analysis (PCA) (Figure 1D). Notably, differences in the expression patterns between oocytes, 4-cell, 8-cell, and blastocyst stages confirmed the key timing of the transcriptome profiles throughout the various stages of sheep development.

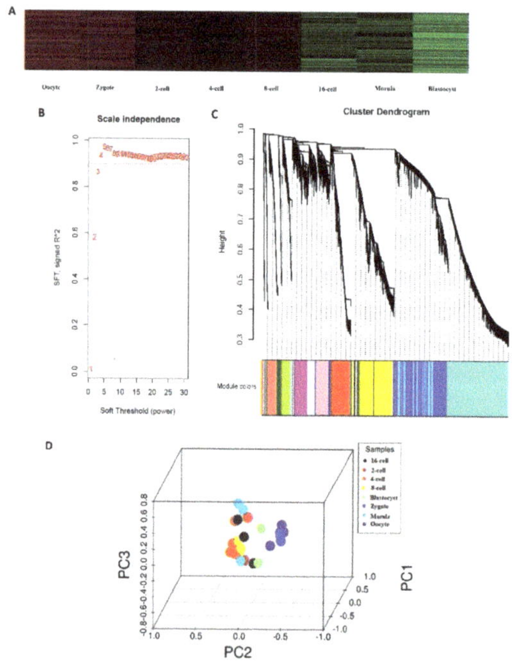

Figure 1. Expression profiles during sheep development from the oocyte to the preimplantation stage. (**A**) Heat map of gene expression profiles correlated with developmental stages. (**B**) Determination of

the soft threshold with the WGCNA algorithm. The approximate scale-free fit index can be attained at the soft-thresholding power of 18. (**C**) Clustering dendrograms showing 26 modules containing a group of highly connected genes. Each color represents a certain gene module. (**D**) Principal component analysis. PC1, PC2, and PC3 represent the top three dimensions of the genes showing differential expression among preimplantation blastomeres.

Table 1. Numbers of embryos and cells analyzed by single-cell RNA-Seq analysis.

Sample	No. of Sample	No. of Single Cells
Oocyte	3	3
Zygote	3	3
2-cell	3	3
4-cell	3	3
8-cell	3	3
16-cell	3	3
Morula	3	3
Blastocyst	3	3
Total	24	24

3.2. Differentially Expressed Genes

Although the total numbers of genes expressed in different stages from the oocyte to the blastocyst varied little, the identities of the genes expressed during early development were dramatically different (Table 2). A total of 2127 unique genes were identified to be differentially expressed among all of the developmental stages ($p < 0.05$). Consistent with the Pearson correlations between all detected genes, the majority of the DEGs (1948) were identified between the morula and blastocyst cell stages, indicating that the greatest changes in gene expression occurred during this transition. Among these genes, 1092 and 856 were down- and up-regulated, respectively (Figure 2). For example, between the oocyte and zygote groups, significantly down-regulated genes included *BTF3* (basic transcription factor 3), *TLR1* (toll-like receptor 1), and *SPINT2* (serine peptidase inhibitor, Kunitz type 2), while significantly up-regulated genes included *PEX12* (peroxisomal biogenesis factor 12) and *PGK1* (phosphoglycerate kinase 1). The second-largest change in gene expression occurred between the 8- and 16-cell stages. A total of 1211 DEGs were detected, and 724 and 487 were down- and up-regulated, respectively (Figure 2).

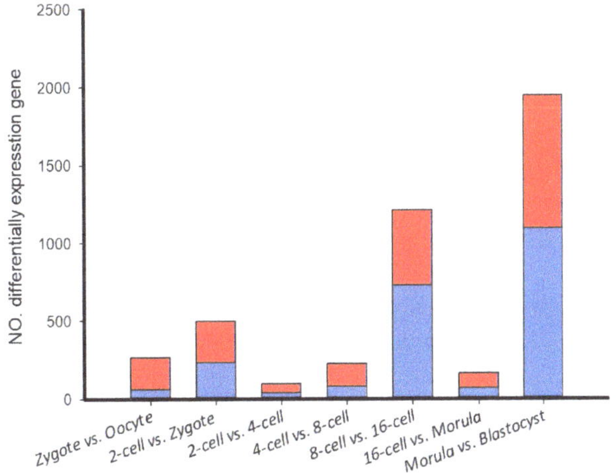

Down-regulated	61	232	40	80	724	67	1092
Up-regulated	207	266	58	146	487	97	856

Figure 2. Differentially expressed genes in sheep development from the oocyte to the blastocyst stages.

Table 2. The numbers of genes detected in sheep from oocyte to blastocyst development.

Stage	No. of Genes (FPKM > 0.1)
Oocyte	24,729
Zygote	25,455
2-cell	26,714
4-cell	28,948
8-cell	25,920
16-cell	24,341
Morula	22,242
Blastocyst	31,157

FPKM: fragments per kilobase of exon per million fragments mapped.

3.3. Gene Ontology and Kyoto Encyclopedia of Genes and Genomes Analyses of DEGs

Between the oocytes and the zygote, 61 and 207 of the 268 DEGs were down- and up-regulated, respectively. These genes represent a rapid degradation of the maternally stored transcripts. Gene ontology (GO) analysis indicated that there was a significant over-representation of elements involved in nuclear speck and cytoplasm. The Kyoto Encyclopedia of Genes and Genomes (KEGG) analysis showed that most DEGs were primarily involved in the spliceosome, ribosome biogenesis in eukaryotes, and Epstein–Barr virus infection pathway Table S1-a, Figure 3A(1,2). Between the zygote and the 2-cell stage, 266 and 232 of the 498 DEGs were down- and up-regulated, respectively. GO analysis indicated that there was a significant over-representation of elements involved in translation, in the structural constituent of ribosome, inRNA binding, and in extracellular exosome. The KEGG analysis showed that most DEGs were primarily involved in RNA transport and the ribosome pathway (Table S1-b, Figure 3B(1,2)). Between the 2-cell and the 4-cell stage, 58 and 40 of the 98 DEGs were down and up-regulated, respectively. GO analysis indicated that there was a significant over-representation of elements involved in the positive regulation of cell proliferation, negative regulation of transcription, DNA-templated, and negative regulation of cell differentiation. The KEGG analysis showed that most DEGs were primarily involved in the phagosome pathway (Table S1-c, Figure 3C(1,2)). Between the 4-cell and the 8-cell stage, 146 and 80 of the 226 DEGs were down- and up-regulated, respectively, which was inconsistent with the findings of a previous study that used scRNA-seq to profile human preimplantation embryos and embryonic stem cells [23,29]. GO analysis indicated that there was a significant over-representation of elements involved in the nucleus. The KEGG analysis showed that most DEGs were primarily involved in the cell cycle pathway (Table S1-d, Figure 3D(1,2)). Between the 8-cell and the 16-cell stage, 487 and 724 of the 1211 DEGs were down- and up-regulated, respectively. GO analysis indicated that there was a significant over-representation of elements involved in RNA binding and mitochondrion. The KEGG analysis showed that most of the DEGs were primarily involved in the ribosome pathway (Table S1-e, Figure 3E(1,2)). Between the 16-cell and the morula stage, 97 and 67 of the 174 DEGs were down and up-regulated, respectively. GO analysis indicated that there was a significant over-representation of elements involved in transferase activity. The KEGG analysis showed that most of the DEGs were primarily involved in the DNA replication pathway (Table S1-f, Figure 3F(1,2)). The majority of the DEGs, 856 and 1092 of the 1948 DEGs were down- and up-regulated, respectively, between the morula and blastocyst stages. GO analysis indicated that there was a significant over-representation of elements involved in extracellular exosome and cytosol. The KEGG analysis showed that most of the DEGs were primarily involved in the transcriptional misregulation in cancer, lysosome, and protein processing in the endoplasmic reticulum pathway (Table S1-g, Figure 3G(1,2)).

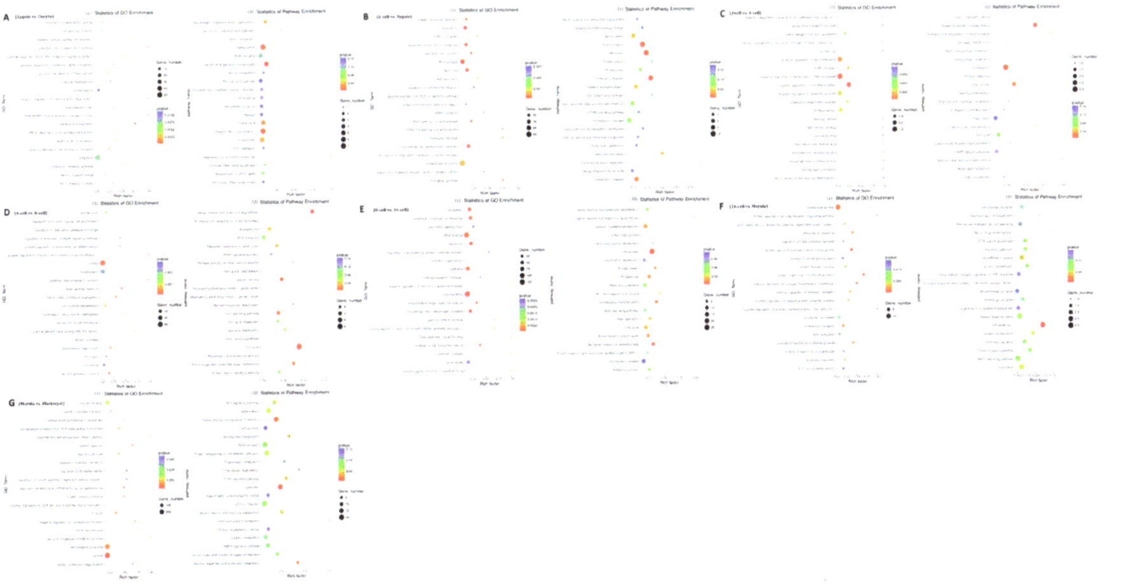

Figure 3. Gene ontology (GO) and Kyoto Encyclopedia of Genes and Genomes (KEGG) analyses of differentially expressed genes in sheep development from the oocyte to the blastocyst stages. GO and KEGG analyses between (**A**) the oocyte and zygote stages, (**B**) the zygote and 2-cell stages, (**C**) the 2-cell and 4-cell stages, (**D**) the 4-cell and 8-cell stages, (**E**) the 8-cell and 16-cell stages, (**F**) the 16-cell and morula stages, and (**G**) the morula and blastocyst stages.

3.4. Genome-Wide Discovery and Identification of lncRNAs

To identify new lncRNAs involved in sheep development from the oocyte to the blastocyst stages, cell samples from oocytes, zygote, and blastomeres of the 2-, 4-, 8-, 16-cell, morula, and blastocyst stages were collected. Transcriptome sequencing was then performed using the Illumina HiSeqTM 4000 platform. An overview of the analysis pipeline is shown in Figure 4. After removing adaptor reads, reads containing poly-N > 10%, and low-quality reads, clean reads were obtained for each sample and used in the following analyses. The GC content of each sample was between 43.5 and 45%. Thus, approximately 75.28–91.45% of the total clean reads could be mapped to the mouse reference genome sequence using Tophat v2.0.9 (Table 3). The different gene subtypes of the above-mapped reads are shown in Figure 1B and are based on genomic overlap with existing annotations using the HTseq program. A total of 274,470 assembled transcripts were produced using both Scripture (beta2) and Cufflinks (v2.1.1).

Table 3. Summary of read filter and alignment.

Sample	Raw Reads	Clean Reads	Clean Bases	Error Rate (%)	GC Content (%)
oocyte	102,383,290	86,741,418	13.01 G	0.04	43.50
Zygote	111,921,426	87,081,144	13.06 G	0.02	44.50
2-cell	123,273,652	95,679,634	14.35 G	0.04	44
4-cell	108,522,186	96,419,102	14.46 G	0.03	44
8-cell	142,178,172	123,561,104	18.53 G	0.02	45
16-cell	133,023,048	90,477,914	13.57 G	0.04	43.5
morula	122,556,646	90,506,796	13.58 G	0.02	44.5
blastula	141,488,370	105,354,406	15.80 G	0.03	44

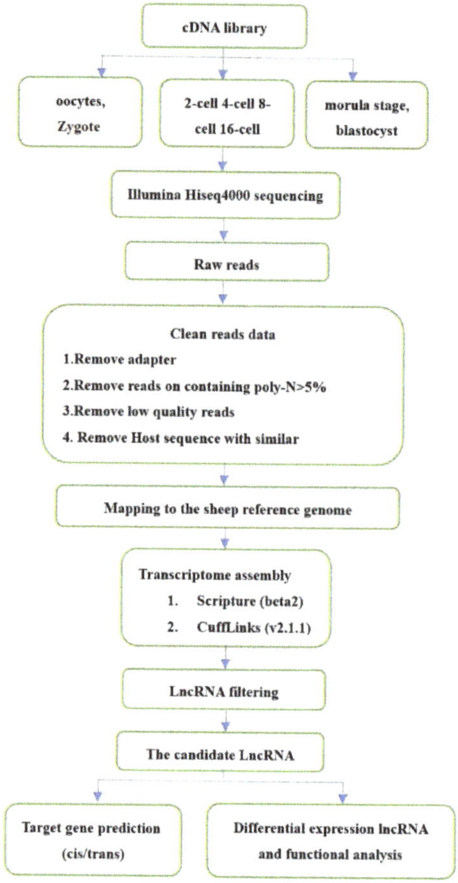

Figure 4. Pipeline for identification of lncRNAs.

3.5. Features of lncRNAs

The fragments per kilobase of exon per million fragments mapped (FPKM) values and numbers of lncRNAs demonstrated that lncRNAs in cell samples from the oocyte, fertilized egg, 2-, 4-, 8-, 16-cell, morula, and blastocyst stages were expressed at lower levels compared with the levels of protein-coding genes expressed (Figure 5A). However, the lncRNA transcript length was mostly ≥1000 bp, which was not significantly different compared with transcript lengths observed for protein-coding genes (Figure 5B). In addition, significant differences in the distributions of exon number between protein-coding genes and lncRNAs were also detected, and 82.41% of all lncRNAs only contained two exons (Figure 5C). Furthermore, most of the lncRNAs contained relatively shorter open reading frames (ORFs) compared with protein-coding genes (Figure 5D).

3.6. Differentially Expressed lncRNAs

A total of 42 differentially expressed lncRNAs ($p < 0.05$; 52 transcript isoforms) were detected (Figure 6). The most noticeable changes in lncRNA expression occurred between the morula and blastocyst stages, in which 19 (23 transcript isoforms) were significantly up-regulated and six (10 transcript isoforms) were down-regulated (Figure 6). Overall, differentially expressed lncRNA transcripts were fewer in number in sheep compared with the implantation and inter-implantation sites of pregnant mice [30].

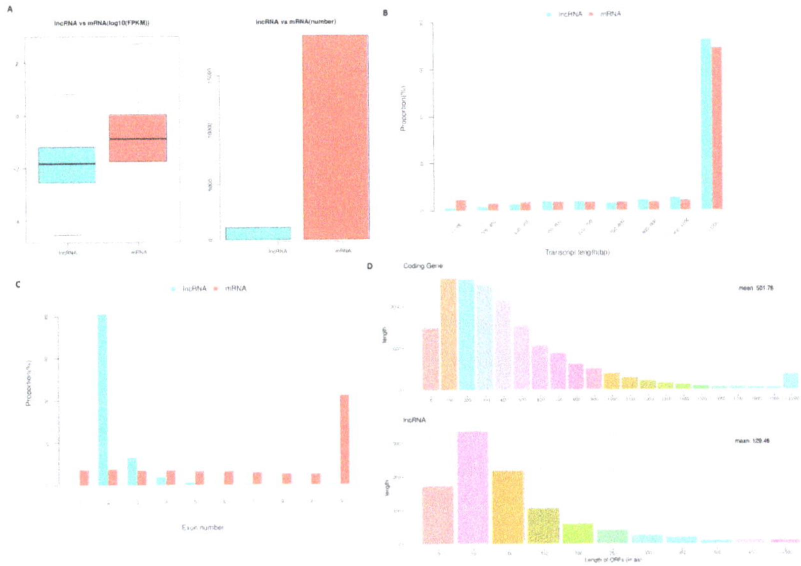

Figure 5. LncRNA characteristics in sheep development from the oocyte to the blastocyst stages. (**A**) Comparison of the expression level between lncRNA and protein-coding genes in terms of fragments per kilobase of exon per million fragments mapped (FPKM). The FPKM distribution of lncRNAs in mouse uterus was lower than that of protein-coding genes. (**B**) Distribution of transcript lengths in the lncRNAs and protein-coding genes. The average size of lncRNA transcripts was generally shorter than that of protein-coding genes. (**C**) The number of exons in lncRNAs and protein-coding genes. A total of 88.41% of the lncRNAs contained two to four exons, while the majority of protein-coding genes contained more than 10 exons. (**D**) The number of ORFs identified in the lncRNAs and protein-coding genes using Estscan. As expected, the ORFs of lncRNAs were substantially shorter than the ORFs in protein-coding genes.

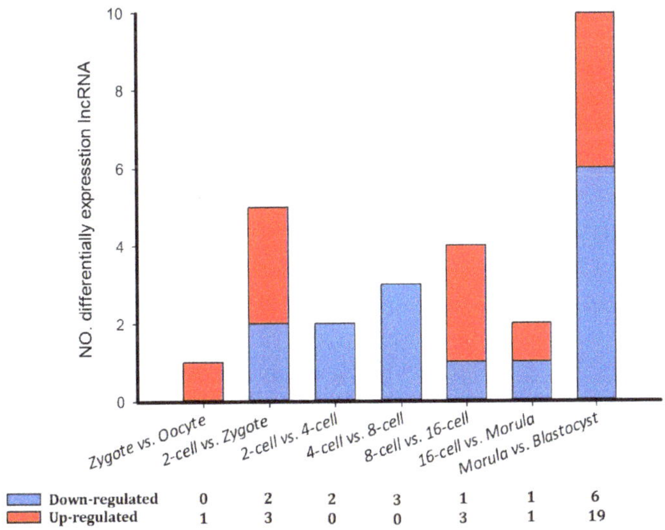

Figure 6. The number of differentially expressed lncRNAs in eight comparison groups in sheep development from the oocyte to the blastocyst stages.

3.7. *Cis- and Trans-Target Genes of lncRNAs*

To investigate the function of lncRNAs, we first predicted the putative lncRNA cis- and trans-regulatory target genes. Protein-coding genes located within 100 kb upstream and downstream of lncRNAs were considered cis-targets (Table S2-a). The trans-target genes located far from lncRNAs are shown in Table S2-b.

3.8. *Functional Analysis of Differentially Expressed lncRNAs*

To predict the function of lncRNAs during the different aforementioned stages, GO and KEGG analyses of the cis and trans-target genes of the lncRNAs in the eight comparison groups were performed. GO analysis of the cis-targets revealed only one significantly enriched GO term (mitochondrion) in the oocyte vs. zygote stages. The KEGG analysis revealed that the significantly enriched pathways in the oocyte vs. zygote stages were "Spliceosome" and "Carbon metabolism" ($p < 0.05$, Figure 7A(1,2) and Table S3-1). The significantly enriched GO terms of the cis-targets in the zygote vs. 2-cell stages, which represented biological processes and molecular functions, were associated primarily with nucleus, mitochondrion, extracellular exosome, cytosol, and cytoplasm. The KEGG analysis revealed that the significantly enriched cis pathways in the oocyte vs. zygote stages were "Non-alcoholic fatty liver disease," "Huntington's disease," and "Alzheimer's disease" (corrected p-value < 0.05, Figure 7B(1,2) and Table S3-2). There was only one significantly enriched GO term detected in the 2-cell vs. 4-cell stages: intracellular signal transduction. The KEGG analysis revealed that the significantly enriched cis pathways were the "MAPK signaling pathway" and "Glycerophospholipid metabolism" ($p < 0.05$, Figure 7C and Table S3-3). Furthermore, no GO terms were significantly enriched in the 4-cell vs. 8-cell stages. The KEGG analysis revealed that the only significantly enriched cis pathway was "cell cycle" (corrected p-value < 0.05, Figure 7C(1,2) and Table S3-4). In the 8-cell vs. 16-cell stages, there were additional significantly enriched GO terms: nucleus, membrane, and others. The KEGG analysis revealed that the only significantly enriched cis pathway was "Ribosome" (corrected p-value < 0.05, Figure 7E(1,2) and Table S3-5). Seven significantly enriched GO terms were detected in the 16-cell vs. morula stages: nucleotide binding, membrane, and others. The KEGG analysis revealed that the significantly enriched cis pathways were "RNA transport" and "AMPK signaling pathway" ($p < 0.05$, Figure 7F(1,2) and Table S3-6). In the morula vs. blastula stages, there were additional significantly enriched GO terms: RNA binding, nucleus, and others. The KEGG analysis revealed that the significantly enriched cis pathways were "Huntington's disease" and "Oxidative phosphorylation" (corrected p-value < 0.05, Figure 7G(1,2) and Table S3-7).

3.9. *Distribution of Different SNP and Indel Types in Sheep from Oocyte to Blastocyst Development*

Single-nucleotide polymorphisms (SNPs) are the most common form of genetic variation in humans and drive phenotypic variation. Due to evolutionary conservation, SNPs and indels (insertion and deletions) are depleted in functionally important sequence elements [31,32]. Here, using the SAMtools/Popoolation software package. A total of 4,352,847 putative SNPs and 297,411 INDEL was predicted, with an average of 181,368 SNPs and 12,392 INDEL per sample, respectively. After removing redundant SNPs among all samples, we had 678,433 and 8454 INDEL from SAMtools/Popoolation2 (Table S4). Then, concerning all the putative SNPs in sheep from oocyte to pre-implantation development, there were 79,516, 70,858, 99,129, 115,926, 72,615, 78,782, 66,420, 224,557 intergenic SNPs. In addition, 47,852, 40,178, 60,970, 67,126, 37,762, 64,925, 39,054, 201,642 SNPs from different stages in sheep from oocyte to pre-implantation development were genic, and defined exactly as in the 5′UTR, 3′UTR, and upstream and downstream of protein-coding genes. Furthermore, in these three SNP datasets, there were large percentages of intergenic (including upstream/downstream) SNPs (37–49%). There were 79,516, 70,858, 99,129, 115,926, 72,615, 78,782, 66,420, 22,4557 intergenic SNPs. In addition, 47,852, 40,178, 60,970, 67,126, 37,762, 64,925, 39,054, 201,642 SNPs from different stages in sheep from oocyte to pre-implantation development were genic, and of these genic SNPs (Table 4).

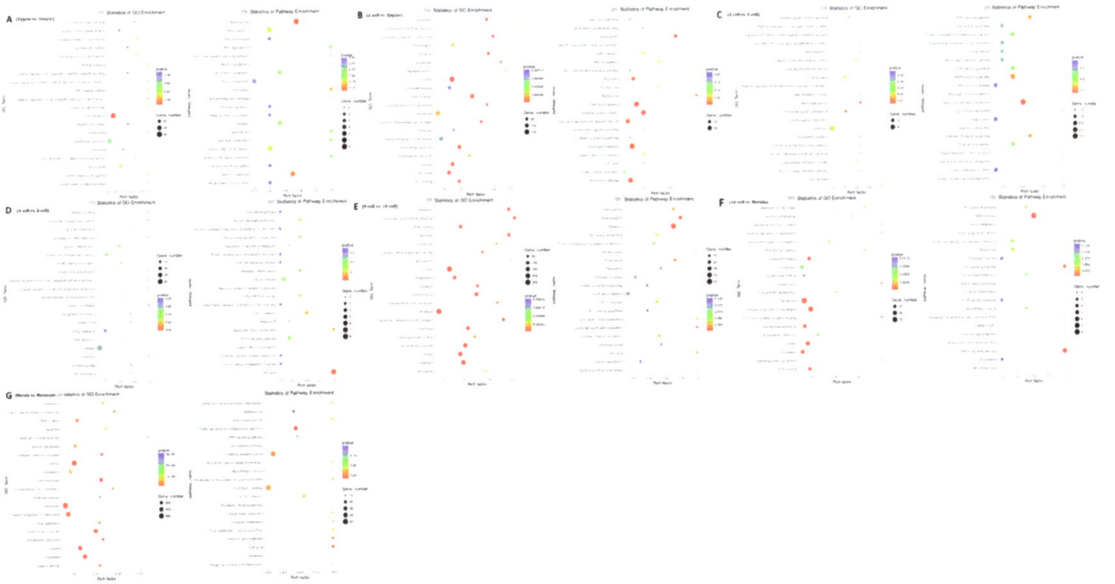

Figure 7. GO and KEGG analysis with the cis and trans-target genes of lncRNAs. GO and KEGG analyses between (**A**) oocyte and zygote stages, (**B**) zygote and 2-cell stages, (**C**) 2-cell and 4-cell stags, (**D**) 4-cell and 8-cell stages, (**E**) 8-cell and 16-cell stages, (**F**) 16-cell and morula stages, and (**G**) morula and blastocyst stages.

3.10. SNP and Indel Functional Annotation

Functional annotation of SNPs with allelic imbalances was performed using the Blast2GO suite [32–35]. The SNP-flanking sequences were searched against the NCBI nr-protein database using BLASTx. Associated genes and GO terms were then obtained. In the biological processes' category, SNP genes were associated with various cellular processes that were primarily involved in development-related mechanisms, including the regulation of the MyD88-dependent toll-like receptor signaling pathway, regulation of metabolic and oxidation-reduction processes, and protein translation. In the molecular function category, SNP-containing genes were associated with binding phosphoprotein, nucleic acid, and actin. In addition, a significant number of the genes were associated with transferase, motor, oxidoreductase, and structural molecule activities. In the cellular component category, many of the genes were associated with the cytoplasmic compartment, membranes, myosin complex, and extracellular region compartment (Supplemental Tables S4 and S5), which is consistent with the findings of previous studies [31].

Additionally, KEGG pathway mapping was used to assign functions to the SNP-containing transcripts. A search of transcripts against the KEGG database yielded 1043 transcripts (13.15%) with significant hits to 632 KEGG Orthologies belonging to different pathways (Supplemental Tables S5 and S6).

Table 4. Summary of SNP and indel classification for different sets.

Functional Class	SNP								INDEL							
	Oocyte	Zygote	2-Cell	4-Cell	8-Cell	16-Cell	Morula	Blastocyst	Oocyte	Zygote	2-Cell	4-Cell	8-Cell	16-Cell	Morula	Blastocyst
Intergenic	79,516	70,858	99,129	115,926	72,615	78,782	66,420	224,557	5650	5260	7088	8129	6163	4939	4765	15,130
Intronic	47,852	40,178	60,970	67,126	37,762	64,925	39,054	201,642	3104	2831	3973	4172	2793	3580	2566	15,373
exonic	28,898	24,940	33,951	35,201	26,715	32,414	27,569	66,325	336	298	438	411	415	326	440	512
3′UTR	3991	4763	5879	5761	4773	4822	3991	8356	502	564	666	651	598	547	577	862
5′UTR	1075	904	1142	1269	964	1265	1075	2145	34	32	35	42	35	41	44	60
upstream	4343	3612	5330	5836	3937	5551	4343	9568	243	233	359	408	285	306	299	588
downstream	14,572	16,348	19,799	20,220	16,904	16,251	14,572	27,967	1845	1814	2268	2209	2072	1749	2010	2886
splicing	632	375	421	494	390	390	632	525	47	43	53	52	47	44	53	71
Total number	184,380	162,027	226,690	251,891	164,117	204,485	157,743	541,244	11,763	11,076	14,882	16,076	12,408	11,534	10,755	35,487

3.11. Overview of AS Events in Sheep from Oocyte to Blastocyst Development

Splicing events were comprehensively analyzed for 24 single cells based on relevant RNA-seq data. In the oocyte stage, there were in total 101,835 AS events detected in 54,266 genes, comprising 7003 alternative exon ends (AE) events detected in 2603 genes, 1862 retention of single (IR) events in 723 genes, 146 multiple (MIR) intron events in 56 genes, 5260 multi-exon SKIP (MSKIP) events in 1400 genes, 17,236 Skipped exon (SKIP) events in 4310 genes, 33,409 Alternative 5′ first exon (TSS) events in 20,899 genes, 26,563 alternative transcription termination site (TTS) events in 20,900 genes, 2063 approximate AE (XAE) events in 729 genes, 1055 approximate IR (XIR) events in 359 genes, 76 approximate MIR (XMIR) events in 31 genes, 1969 approximate MSKIP (XMSKIP) events in 631 genes, and 5193 approximate SKIP (XSKIP) events in 1625 genes (Figure 8A). Next, splicing events were comprehensively analyzed in fertilized egg stage, 2-cell, 4-cell, 8 cell, 16 cell, morula and blastocyst shown in Figure 8B–H. Of those stags, in morula and blastula, only one type of AS event was detected in most genes, although there were some exceptions; generally, it was demonstrated that three or more splicing events could be attributed to one gene, with a maximum of five types of AS events observable for a single gene. However, TSS was the predominant type of event in all the histologic STS subtypes, which revealed that TSS was the most common splicing event in sheep from oocyte to blastocyst development.

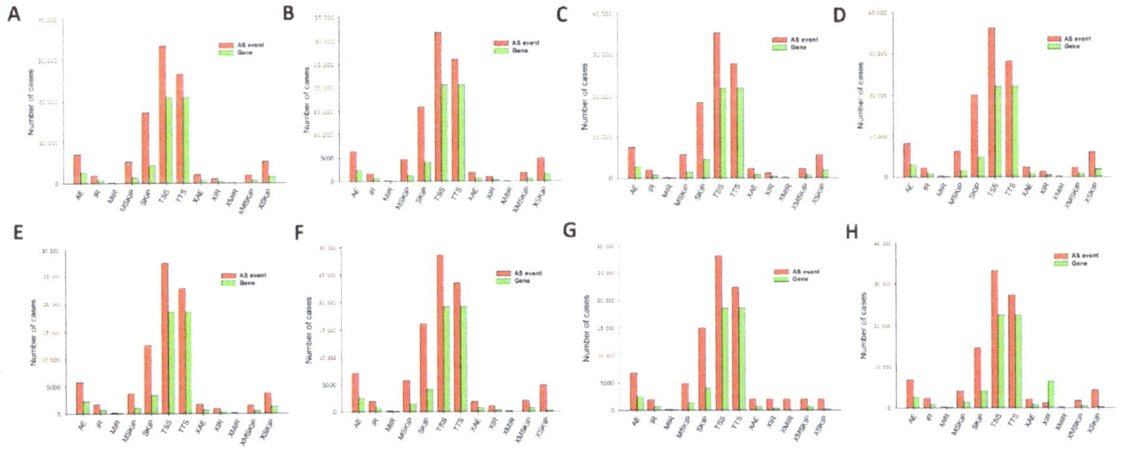

Figure 8. The number of alternative splicing events and associated genes in sheep development from the oocyte to the blastocyst stages. TSS was the most frequent of the eleven types of events. (**A**) Oocyte stage. (**B**) Zygote stage. (**C**) 2-Cell stage. (**D**) 4-Cell stage. (**E**) 8-Cell stage. (**F**) 16-Cell stage. (**G**) Morula stage. (**H**) Blastocyst stage. *AE* alternative exon ends, *AD* alternate donor, *AP* alternate promoter, *AT* alternate terminator, *ES* exon skip, *ME* mutually exclusive exon, *RI* retained intron.

3.12. Associated AS Events

The main AS events were SKIP, TSS, and TTS (Figure 1). Therefore, we analyzed associated AS events and genes by UpSet plot. SKIP was detected in 2566 genes at every stage (Figure 9A). Furthermore, TSS and TTS were detected in 11,441 genes at every stage (Figure 9B,C). More than one AS event could occur in one gene, and up to three types of splicing events were identified in one gene. Thus, one gene might have two or more AS events that were markedly related to the PFI of PTC patients. The UpSet plot revealed that TSS was the most common prognosis-related event.

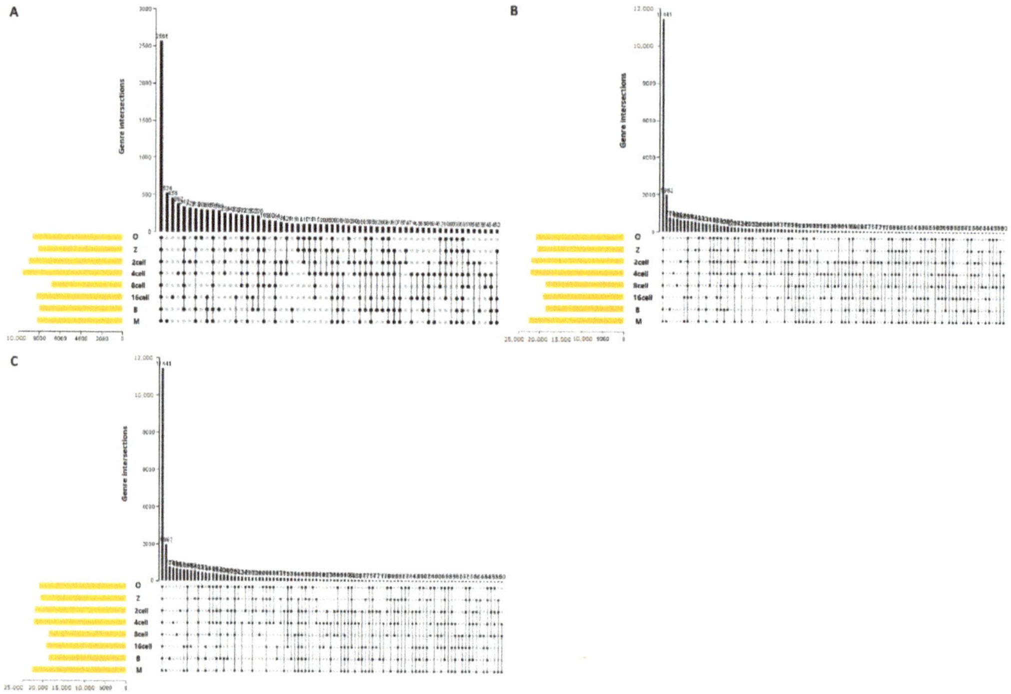

Figure 9. UpSet plot of the three main alternative splicing events at different stages in sheep development from the oocyte to the blastocyst stages. UpSet plot of the interactions of the alternative splicing events associated with genes. (**A**) UpSet plot of the interactions of SKIP. (**B**) UpSet plot of the interactions of TSS. (**C**) UpSet plot of the interactions of TTS. *TTS* alternative transcription termination site.

3.13. Molecular Characteristics of the Most Important AS Events

Based on UpSet plot, there were 13,663 genes with one or more AS events at different stages (Figure 10A). We then carried out GO and KEGG analysis. The functional annotations revealed that "regulation of transcription, DNA templated (884 genes)," "transport (609 genes)," and "proteolysis (397 genes)" were the three most significant biological process terms. "Membrane (4323 genes)," "nucleus (3966 genes)," and "integral component of membrane (3055 genes)" were the three most significant cellular component terms. For molecular function, "ATP binding (1498 genes)," "nucleotide binding (1441 genes)," and "nucleic acid binding (1412 genes)" were the three most enriched categories (Figure 10B).

Furthermore, we found that the related pathways were metabolism, environmental information processing, and human diseases. The "metabolism" and human diseases correlated pathways were mostly genes involved in purine metabolism (131 genes), pyrimidine metabolism (93 genes), and pathways in cancer (181 genes) (Figure 10C). In our study, there were 13,663 genes with one more AS event suggesting that AS may play an important role at different developmental stages. Furthermore, "ATP binding (1498 genes)," "nucleotide binding (1441 genes)," and "nucleic acid binding (1412 genes)" were the three most enriched categories, suggesting that alternative protein isoforms with distinct functions are expressed. Thus, defects in splicing control might be able to induce losses in function and severe phenotypes and require further study.

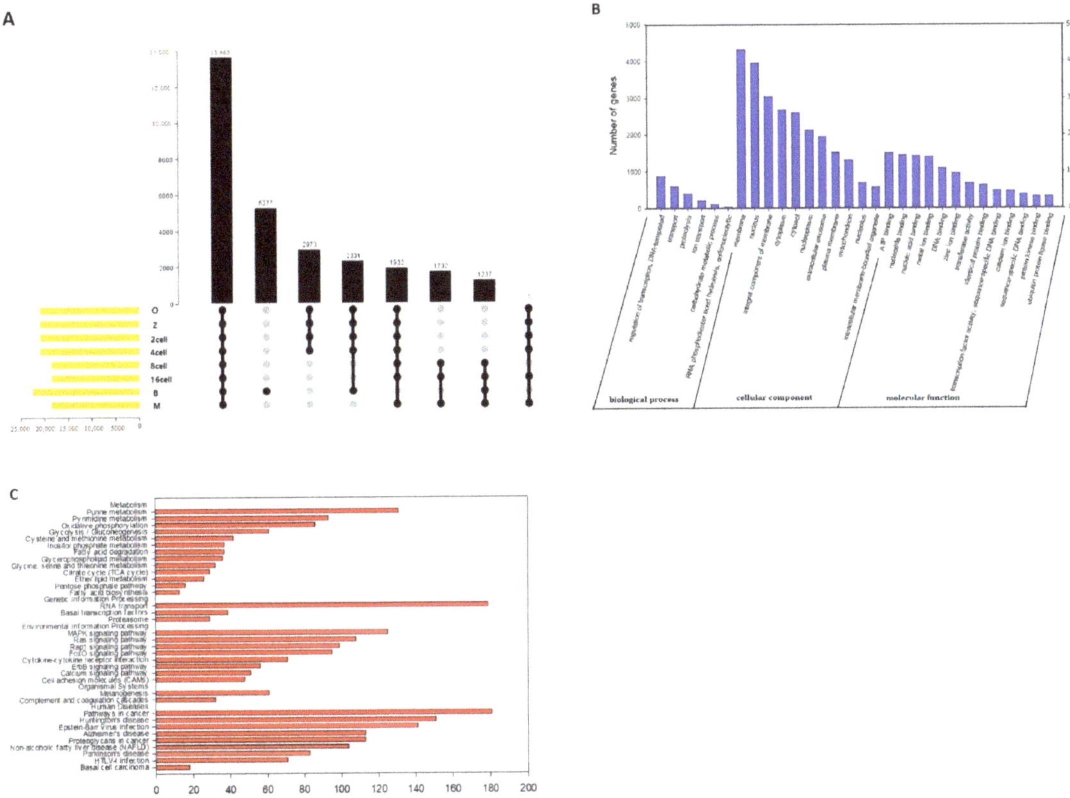

Figure 10. Molecular characteristics of the main alternative splicing events in sheep development from the oocyte to the blastocyst stages. (**A**) UpSet plot of the main genes at different stages in sheep development from the oocyte to the blastocyst stages. (**B**) GO analysis. (**C**) KEGG analysis.

4. Discussion

In recent years, much research has focused on the study of the development of blastomeres of the 4- and 8-cell embryos of mice [5], cattle [6], and goats [7]. However, the exact mechanism and the developmental patterns underlying how the blastomeres of the 2-, 4-, 8-, 16-cell, morula, and blastocyst stages undergo asymmetric division are still unclear. For the first time, we used scRNA-seq to study the transcriptome profiles during sheep development from the oocyte to the blastocyst stages. Our data showed that from the 4-cell to the 8-cell stage, there were no noticeable changes in the transcriptome profile as has been shown for the 4- and 8-cell embryos of mice [6,7,36] (Figure 1A). However, the first major change was noted between the 8-cell and 16-cell stages, which is similar to the pattern observed in the bovine embryonic genome. The greatest changes in gene expression were observed between the morula and blastocyst stages, which may be explained by a major morula–blastocyst transition that results in expression patterns characteristic of sheep preimplantation development [11,37]. Furthermore, 1092 and 856 DEGs were down- and up-regulated, respectively (Figure 2). *BTF3* (basic transcription factor 3), *TLR1*(toll-like receptor 1), and *SPINT2* (serine peptidase inhibitor, Kunitz type 2) were markedly up-regulated, while *PEX12* (peroxisomal biogenesis factor 12) and *PGK1* (phosphoglycerate kinase 1) were significantly down-regulated between the oocyte and zygote stages. In others reports, BTF3 as one of the important transcription factors was proved in gastric cancer development and in progression by enhance transcription [38]. TLR4 enhances blastocyst attachment to endometrial cells in mice via miR-Let-7a suggesting that inflammatory responses are beneficial in the fetomaternal interface and supposedly accelerate the chances for successful implantation. In our study, TLR1 was markedly up expressed from oocyte to blastocyst development [39]. Therefore, the role of inflammatory responses is interesting and needs further study. GO and KEGG analyses of DEGs were then conducted. At different stages, GO analysis indicated that there was a significant over-representation of elements involved in nuclear speck and cytoplasm. The KEGG analysis revealed that most of the DEGs were primarily involved in the spliceosome, ribosome biogenesis in eukaryotes, and the Epstein–Barr virus infection pathway, RNA transport, ribosome pathway, transferase activity, and others. The above pathways have also been reported in the development of bovine and monkey embryos [11,40,41]. Therefore, the transcriptome profiles in sheep from oocyte to blastocyst development showed the same pattern.

We then characterized the features of the lncRNAs and their target genes. The lncRNAs in cell samples from the oocyte, fertilized egg, 2-, 4-, 8-, 16-cell, morula, and blastocyst stages had lower expression levels compared with the expression levels observed in protein-coding genes. However, the transcript lengths of lncRNAs were mostly $\geq$1000 bp, and the mean transcript length of lncRNAs was not significantly different relative to the mean transcript length of protein-coding genes (Figure 5B). In addition, there was a total of 42 differentially expressed lncRNAs (52 transcript isoforms) (Figure 6). The major changes in lncRNA expression occurred between the morula and blastocyst stages, of which 19 (23 transcript isoforms) lncRNAs were significantly up-regulated, such as lnc MSTRG.2676, lnc MSTRG.3585, and six (10 transcript isoforms) lncRNAs were down-regulated, such as lnc MSTRG.8262 and lncMSTRG.11966 (Figure 6). Overall, there were fewer differentially expressed lncRNA transcripts during sheep development compared with the number of differentially expressed transcripts detected in implantation and inter-implantations sites in day-5 pregnant mice [30]. GO and KEGG analyses of the cis- and trans-target genes of the lncRNAs in the eight comparison groups were performed. GO analysis of the cis lncRNA targets revealed the following significantly enriched GO terms: RNA binding, nucleus, and others. The KEGG analysis revealed that the significantly enriched cis-pathways were "Huntington's disease" and "Oxidative phosphorylation" between the morula and blastula groups (corrected p-value < 0.05, Figure 7G and Table S3-7).

We then studied the distribution of different SNP and indel types (Table 4). Approximately 10% intergenic and 30% non-coding SNPs have been reported in humans from RNA-seq data [42]. The high percentages of intergenic SNPs in humans may be partially explained by the incomplete annotation of protein-coding genes and exons in the current version of the rainbow trout reference genome sequence [35]. We then conducted GO analysis of the associated genes. In the biological process's category, SNP genes were associated with various cellular processes that were primarily involved in development-related mechanisms, including regulation of the MyD88-dependent toll-like receptor signaling pathway, regulation of metabolic and oxidation-reduction processes, and protein translation. In the molecular function category, SNP-containing genes were associated with binding phosphoprotein, nucleic acid, and actin. In addition, many genes were associated with transferase, motor, oxidoreductase, and structural molecule activities. In the cellular component category, many of the genes were associated with the cytoplasmic compartment, membranes, myosin complex, and extracellular region compartment (Tables S5 and S6), which is consistent with the findings of previous studies [31].

Finally, we also detected AS events. There are relatively few studies that have reported the numbers of AS events during development from the oocyte to the blastocyst. In this study, splicing events were comprehensively analyzed for 24 single cells based on relevant RNA-seq data. For example, there was a total of 101,835 AS events that were detected in 54,266 genes, comprising 7003 AE events detected in 2603 genes in the oocyte stage (Figure 8A). AS can control the transcriptional identity of the maternal transcriptome by the RNA-binding protein, which is essential for the development of fertilized-competent oocytes [43]. Therefore, our comprehensive analysis of AS suggests that AS plays an important role in sheep development. GO and KEGG analyses showed that there were

13,663 genes with one or more AS events at different stages during sheep development based on the UpSet plot (Figure 10A). The most significant biological process terms were "regulation of transcription, DNA templated," "transport", and "proteolysis". The three most significant cellular component terms were "membrane," "nucleus," and "integral component of membrane". The significant terms for molecular function were "ATP binding," "nucleotide binding", and "nucleic acid binding" (Figure 10B). In a previous study, AS has been shown to play an important role in the protein-coding genes of mouse oocytes and zygotes by RNA-binding protein, and the correct combination of exons through AS ensures that gene isoforms are expressed for the specific context in which they are required [44,45]. Furthermore, the "metabolism", "genetic information processing" and "human diseases" correlated pathways were mostly genes involved in purine metabolism, RNA transport, and pathways in cancer (Figure 10C). In metabolomic analyses of fetal germ cells in mice on embryonic day (E)13.5 and E18.5, purine metabolism was involved in demonstrating sex- and developmental stage-dependent changes in these processes [46]. RNA transport in our study was mostly genes involved which also as one of the important AS events is frequent across developmental stages and tissues in embryonic day 8.5, 9.5. and 11.5 mouse embryos and placenta [47]. Furthermore, we detected 13,663 genes with one or more AS events, suggesting that AS may play an important role in sheep development. Furthermore, "ATP binding (1498 genes)," "nucleotide binding (1441 genes)," and "nucleic acid binding (1412 genes)" were the three most enriched categories, suggesting that alternative protein isoforms with distinct functions were expressed. Thus, defects in splicing control might be able to induce losses in function and severe phenotypes and thus require further study [48].

5. Conclusions

Here, we conducted a scRNA-seq survey of cells from sheep from the oocyte to the blastocyst developmental stages. We then carried out experiments to identify and functionally analyze the transcriptome profiles, long non-coding RNAs, single-nucleotide polymorphisms (SNPs), and alternative splicing (AS). We found between the oocyte and zygote groups significantly down-regulated genes and the second-largest change in gene expression occurred between the 8- and 16-cell stages. We used various methods to construct a profile to characterize cellular and molecular features and systematically analyze the related GO and KEGG profile of cells of all stages from the oocyte to the blastocyst. This large-scale, single-cell atlas provides key cellular information and will likely assist clinical studies in improving preimplantation genetic diagnosis.

Supplementary Materials: The following supporting information can be downloaded at: https://www.mdpi.com/article/10.3390/genes14061145/s1, Table S1: the GO and KEGG analysis between different group of DEGs, Table S2: Cis- and Trans-Target Genes of lncRNAs, Table S3: the GO and KEGG analysis between different group of Expressed lncRNAs, Table S4: INDEL in different samples, Table S5 and S6: the GO and KEGG analysis between different group of INDEL.

Author Contributions: Conceived the study: Z.Z.; Designed and supervised the experiments: Z.Z. and E.W.; Performed the experiments: Q.S., X.Z., L.J., L.L., S.C., S.L., X.X., Y.H., X.Y., Z.L. and Z.X. Analyzed and interpreted data: Z.Z. and W.Z.; Wrote the paper: W.Z. All authors have read and agreed to the published version of the manuscript.

Funding: This work was supported for the Henan Beef Cattle Industrial Technology System (HARS-22-13-S), The National Beef Cattle Industrial Technology System (No. CARS-37), Scientific and Technological Key Project of Henan Province (222102110069, 222102110018, 212102110008), the Science-Technology Foundation for Outstanding Young Scientists of Henan Academy of Agricultural Sciences (2022YQ20), the Key Scientific and Technological Special Projects of Henan Province (201300111200), Major Scientific and Technological Special Project of Henan Province (221100110200), the Breeding and production of cattle and sheep by scientific and technological innovation team of Henan Academy of Agricultural Sciences (2023TD28).

Institutional Review Board Statement: All work with animals was completed in accordance and with the approval of the Henan Academy of Agricultural Sciences institutional animal care and use committee (2019008).

Informed Consent Statement: Not applicable.

Data Availability Statement: Not applicable.

Conflicts of Interest: The authors declare that they have no conflict of interest.

References

1. Zhang, K.; Huang, K.; Luo, Y.; Li, S. Identification and functional analysis of long non-coding RNAs in mouse cleavage stageembryonic development based on single cell transcriptome data. *BMC Genom.* **2014**, *15*, 845. [CrossRef]
2. Daughtry, B.L.; Rosenkrantz, J.L.; Lazar, N.H.; Fei, S.S.; Redmayne, N.; Torkenczy, K.A.; Adey, A.; Yan, M.; Gao, L.; Park, B.; et al. Single-cell sequencing of primate preimplantation embryos reveals chromosome elimination via cellular fragmentation and blastomere exclusion. *Genome Res.* **2019**, *29*, 367–382. [CrossRef]
3. Cheung, V.G.; Spielman, R.S. The genetics of variation in gene expression. *Nat. Genet.* **2002**, *32*, S522–S525. [CrossRef]
4. Cheung, V.G.; Bruzel, A.; Burdick, J.T.; Morley, M.; Devlin, J.L.; Spielman, R.S. Monozygotic twins reveal germline contribution to allelic expression differences. *Am. J. Hum. Genet.* **2008**, *82*, 1357–1360. [CrossRef]
5. Zhang, X.; Li, T.; Zhang, L.; Jiang, L.; Cui, T.; Yuan, X.; Wang, C.; Liu, Z.; Zhang, Y.; Li, W.; et al. Individual blastomeres of 4- and 8-cell embryos have ability to develop into a full organism in mouse. *J. Genet. Genom.* **2018**, *45*, 677–680. [CrossRef]
6. Daigneault, B.W.; Rajput, S.; Smith, G.W.; Ross, P.J. Required for Expanded Bovine Blastocyst Formation. *Sci. Rep.* **2018**, *8*, 7753. [CrossRef]
7. Deng, M.; Liu, Z.; Ren, C.; Zhang, G.; Pang, J.; Zhang, Y.; Wang, F.; Wan, Y. Long noncoding RNAs exchange during zygotic genome activation in goat. *Biol. Reprod.* **2018**, *99*, 707–717. [CrossRef]
8. Yan, L.; Yang, M.; Guo, H.; Yang, L.; Wu, J.; Li, R.; Liu, P.; Lian, Y.; Zheng, X.; Yan, J.; et al. Single-cell RNA-Seq. profiling of human preimplantation embryos and embryonic stem cells. *Nat. Struct. Mol. Biol.* **2013**, *20*, 1131–1139. [CrossRef]
9. Blakeley, P.; Fogarty, N.M.; del Valle, I.; Wamaitha, S.E.; Hu, T.X.; Elder, K.; Snell, P.; Christie, L.; Robson, P.; Niakan, K.K. Defining the three cell lineages of the human blastocyst by single-cell RNA-seq. *Development* **2015**, *142*, 3151–3165. [CrossRef]
10. Tang, F.; Lao, K.; Surani, M.A. Development and applications of single-cell transcriptome analysis. *Nat. Methods* **2011**, *8*, S6–S11. [CrossRef]
11. Liu, X.M.; Wang, Y.K.; Liu, Y.H.; Yu, X.X.; Wang, P.C.; Li, X.; Du, Z.Q.; Yang, C.X. Single-cell transcriptome sequencing reveals that cell division cycle 5-like protein is essential for porcine oocyte maturation. *J. Biol. Chem.* **2018**, *293*, 1767–1780. [CrossRef]
12. Yin, X.Y.; Cheng, G.H.; Guo, H.Y.; Wang, Q.; Li, Y.J.; Zhang, H. Single cell transcriptome profiling revealed differences in gene expression during oocyte maturation in Haimen white goats. *Genet. Mol. Res.* **2017**, *16*, 1–12. [CrossRef]
13. Svoboda, P. Long and small noncoding RNAs during oocyte-to-embryo transition in mammals. *Biochem. Soc. Trans.* **2017**, *45*, 1117–1124. [CrossRef]
14. Abbastabar, M.; Sarfi, M.; Golestani, A.; Khalili, E. lncRNA involvement in hepatocellular carcinoma metastasis ands prognosis. *EXCLI J.* **2018**, *17*, 900–913.
15. Youness, R.A.; Gad, M.Z. Long non-coding RNAs: Functional regulatory players in breast cancer. *Noncoding RNA Res.* **2019**, *4*, 36–44. [CrossRef]
16. Karlic, R.; Ganesh, S.; Franke, V.; Svobodova, E.; Urbanova, J.; Suzuki, Y.; Aoki, F.; Vlahovicek, K.; Svoboda, P. Long non-coding RNA exchange during the oocyte-to-embryo transition in mice. *DNA Res.* **2017**, *2*, 129–141.
17. Cabili, M.N.; Trapnell, C.; Goff, L.; Koziol, M.; Tazon-Vega, B.; Regev, A.; Rinn, J.L. Integrative annotation of human large intergenic noncoding RNAs reveals global properties and specific subclasses. *Genes Dev.* **2011**, *25*, 1915–1927. [CrossRef]
18. Rivera, R.M.; Ross, J.W. Epigenetics in fertilization and preimplantation embryo development. *Prog. Biophys. Mol. Biol.* **2013**, *113*, 423–432. [CrossRef]
19. Volders, P.J.; Verheggen, K.; Menschaert, G.; Vandepoele, K.; Martens, L.; Vandesompele, J.; Mestdagh, P. An update on LNCipedia: A database for annotated human lncRNA sequences. *Nucleic Acids Res.* **2015**, *43*, 4363–4364. [CrossRef]
20. Veselovska, L.; Smallwood, S.A.; Saadeh, H.; Stewart, K.R.; Krueger, F.; Maupetit-Méhouas, S.; Arnaud, P.; Tomizawa, S.; Andrews, S.; Kelsey, G. Deep sequencing and de novo assembly of the mouse oocyte transcriptome define the contribution of transcription to the DNA methylation landscape. *Genome Biol.* **2015**, *16*, 209. [CrossRef]
21. Wu, F.; Liu, Y.; Wu, Q.; Li, D.; Zhang, L.; Wu, X.; Wang, R.; Zhang, D.; Gao, S.; Li, W. Long non-coding RNAs potentially function synergistically in the cellular reprogramming of SCNT embryos. *BMC Genom.* **2018**, *19*, 631. [CrossRef]
22. Caballero, J.; Gilbert, I.; Fournier, E.; Gagné, D.; Scantland, S.; Macaulay, A.; Robert, C. Exploring the function of long non-coding RNA in the development of bovine early embryos. *Reprod. Fertil. Dev.* **2014**, *27*, 40–52. [CrossRef]
23. Langfelder, P.; Horvath, S. WGCNA: An R package for weighted correlation network analysis. *BMC Bioinform.* **2008**, *9*, 559. [CrossRef]
24. Martin, M. Cutadapt removes adapter sequences from high-throughput sequencing reads. *EMBnet. J.* **2011**, *17*, 11–12. [CrossRef]
25. Langmead, B.; Salzberg, S.L. Fast gapped-read alignment with Bowtie 2. *Nat. Methods* **2012**, *9*, 357–359. [CrossRef]

26. Pertea, M.; Pertea, G.M.; Antonescu, C.M.; Chang, T.C.; Mendell, J.T.; Salzberg, S.L. StringTie enables improved reconstruction of a transcriptome from RNA-seq reads. *Nat. Biotechnol.* **2015**, *33*, 290–295. [CrossRef]
27. Sun, L.; Luo, H.; Bu, D.; Zhao, G.; Yu, K.; Zhang, C.; Liu, Y.; Chen, R.; Zhao, Y. Utilizing sequence intrinsic composition to classify protein-coding and long non-coding transcripts. *Nucleic Acids Res.* **2013**, *41*, e166. [CrossRef]
28. Min, B.; Park, J.S.; Kang, Y.K. Determination of Oocyte-Manipulation, Zygote-Manipulation, and Genome-Reprogramming Effects on the Transcriptomes of Bovine Blastocysts. *Front. Genet.* **2018**, *9*, 143. [CrossRef]
29. Xie, D.; Chen, C.C.; Ptaszek, L.M.; Xiao, S.; Cao, X.; Fang, F.; Ng, H.H.; Lewin, H.A.; Cowan, C.; Zhong, S. Rewirable gene regulatory networks in the preimplantation embryonic development of three mammalian species. *Genome Res.* **2010**, *20*, 804–815. [CrossRef]
30. Wang, Q.; Wang, N.; Cai, R.; Zhao, F.; Xiong, Y.; Li, X.; Wang, A.; Lin, P.; Jin, Y. Genome-wide analysis and functional prediction of long non-coding RNAs in mouse uterus during the implantation window. *Oncotarget* **2017**, *8*, 84360–84372. [CrossRef]
31. Al-Tobasei, R.; Ali, A.; Leeds, T.D.; Liu, S.; Palti, Y.; Kenney, B.; Salem, M. Identification of SNPs associated with muscle yield and quality traits using allelic-imbalanceanalyses of pooled RNA-Seq samples in rainbow trout. *BMC Genom.* **2017**, *18*, 582. [CrossRef]
32. Neininger, K.; Marschall, T.; Helms, V. SNP and indel frequencies at transcription start sites and at canonical and alternative translation initiation sites in the human genome. *PLoS ONE* **2019**, *14*, e0214816. [CrossRef]
33. Raineri, E.; Ferretti, L.; Esteve-Codina, A.; Nevado, B.; Heath, S.; Pérez-Enciso, M. SNP calling by sequencing pooled samples. *BMC Bioinform.* **2012**, *13*, 239. [CrossRef]
34. Piskol, R.; Ramaswami, G.; Li, J.B. Reliable identification of genomic variants from RNA-seq data. *Am. J. Hum. Genet.* **2013**, *93*, 641–651. [CrossRef]
35. Berthelot, C.; Brunet, F.; Chalopin, D.; Juanchich, A.; Bernard, M.; Noel, B.; Bento, P.; Da Silva, C.; Labadie, K.; Alberti, A.; et al. The rainbow trout genome provides novel insights into evolution after whole-genome duplication in vertebrates. *Nat. Commun.* **2014**, *5*, 3657. [CrossRef]
36. Petropoulos, S.; Edsgärd, D.; Reinius, B.; Deng, Q.; Panula, S.P.; Codeluppi, S.; Plaza Reyes, A.; Linnarsson, S.; Sandberg, R.; Lanner, F. Single-Cell RNA-Seq Reveals Lineage and X Chromosome Dynamics in Human Preimplantation Embryos. *Cell* **2016**, *165*, 1012–1026. [CrossRef]
37. Macaulay, I.C.; Voet, T. Single cell genomics: Advances and future perspectives. *PLoS Genet.* **2014**, *10*, e1004126. [CrossRef]
38. Liu, Q.; Zhou, J.-P.; Li, B.; Huang, Z.-C.; Dong, H.-Y.; Li, G.-Y.; Zhou, K.; Nie, S.-L. Basic transcription factor 3 is involved in gastric cancer development and progression. *World J. Gastroenterol.* **2013**, *19*, 4495–4503. [CrossRef]
39. Hosseini, S.; Hosseini, S.; Salehi, M. Upregulation of Toll-like receptor 4 through anti-miR-Let-7a enhances blastocyst attachment to endometrial cells in mice. *J. Cell. Physiol.* **2020**, *235*, 9752–9762. [CrossRef]
40. Lavagi, I.; Krebs, S.; Simmet, K.; Beck, A.; Zakhartchenko, V.; Wolf, E.; Blum, H. Single-cell RNA sequencing reveals developmental heterogeneity of blastomeres during major genome activation in bovine embryos. *Sci. Rep.* **2018**, *8*, 4071. [CrossRef]
41. Razza, E.M.; Sudano, M.J.; Fontes, P.K.; Franchi, F.F.; Belaz, K.R.A.; Santos, P.H.; Castilho, A.C.S.; Rocha, D.F.O.; Eberlin, M.N.; Machado, M.F.; et al. Treatment with cyclic adenosine monophosphate modulators prior to in vitro maturation alters the lipid composition and transcript profile of bovine cumulus-oocyte complexes and blastocysts. *Reprod. Fertil. Dev.* **2018**, *30*, 1314–1328. [CrossRef]
42. Zhang, F.; Lupski, J.R. Non-coding genetic variants in human disease. *Hum. Mol. Genet.* **2015**, *24*, R102–R110. [CrossRef]
43. Do, D.V.; Strauss, B.; Cukuroglu, E.; Macaulay, I.; Wee, K.B.; Hu, T.X.; Igor, R.L.M.; Lee, C.; Harrison, A.; Butler, R.; et al. SRSF3 maintains transcriptome integrity in oocytes by regulation of alternative splicing and transposable elements. *Cell Discov.* **2018**, *4*, 33. [CrossRef]
44. Braunschweig, U.; Gueroussov, S.; Plocik, A.M.; Graveley, B.R.; Blencowe, B.J. Dynamic integration of splicing within gene regulatory pathways. *Cell* **2013**, *152*, 1252–1269. [CrossRef]
45. Daguenet, E.; Dujardin, G.; Valcarcel, J. The pathogenicity of splicing defects: Mechanistic insights into pre-mRNA processing inform novel therapeutic approaches. *EMBO Rep.* **2015**, *16*, 1640–1655. [CrossRef]
46. Hayashi, Y.; Mori, M.; Igarashi, K.; Tanaka, K.; Takehara, A.; Ito-Matsuoka, Y.; Kanai, A.; Yaegashi, N.; Soga, T.; Matsui, Y. Proteomic and metabolomic analyses uncover sex-specific regulatory pathways in mouse fetal germline differentiation. *Biol. Reprod.* **2020**, *103*, 717–735. [CrossRef]
47. Revil, T.; Gaffney, D.; Dias, C.; Majewski, J.; Jerome-Majewska, L.A. Alternative splicing is frequent during early embryonic development in mouse. *BMC Genom.* **2010**, *11*, 399. [CrossRef]
48. Yang, Q.; Wang, R.; Wei, B.; Peng, C.; Wang, L.; Hu, G.; Kong, D.; Du, C. Candidate Biomarkers and Molecular Mechanism Investigation for Glioblastoma MultiformeUtilizing WGCNA. *Biomed. Res. Int.* **2018**, *2018*, 4246703. [CrossRef]

genes

MDPI

Article

Transcriptomic Analysis of the Developing Testis and Spermatogenesis in Qianbei Ma Goats

Yue Zou [1,2,3], Xiang Chen [1,2,3,*], Xingzhou Tian [1,2,3], Wei Guo [1,2,3], Yong Ruan [1,2,3], Wen Tang [1,2,3], Kaibin Fu [1,2,3] and Taotao Ji [1,2,3]

1 Key Laboratory of Animal Genetics, Breeding and Reproduction in the Plateau Mountainous Region, Ministry of Education, Guizhou University, Guiyang 550025, China; zoekiwi@163.com (Y.Z.); tianxingzhou@yeah.net (X.T.); guow16@lzu.edu.cn (W.G.); yruan@gzu.edu.cn (Y.R.); went4863@gmail.com (W.T.); kbinfu@163.com (K.F.); taotaoji20@163.com (T.J.)

2 Key Laboratory of Animal Genetics, Breeding and Reproduction, Guiyang 550025, China

3 College of Animal Science, Guizhou University, Guiyang 550025, China

* Correspondence: xchen2@gzu.edu.cn

Abstract: Reproductive competence in male mammals depends on testicular function. Testicular development and spermatogenesis in goats involve highly complex physiological processes. In this study, six testes were, respectively, obtained from each age group, immature (1 month), sexually mature (6 months) and physically mature (12 months old) Qianbei Ma goats. RNA-Seq was performed to assess testicular mRNA expression in Qianbei Ma goats at different developmental stages. Totally, 18 libraries were constructed to screen genes and pathways involved in testis development and spermatogenesis. Totally, 9724 upregulated and 4153 downregulated DEGs were found between immature (I) and sexually mature (S) samples; 7 upregulated and 3 downregulated DEGs were found between sexually mature (S) and physically mature (P) samples, and about 4% of the DEGs underwent alternative splicing events between I and S. Select genes were assessed by qRT-PCR, corroborating RNA-Seq findings. The detected genes have key roles in multiple developmental stages of goat testicular development and spermatogenesis. Gene Ontology (GO) and Kyoto Encyclopedia of Genes and Genomes (KEGG) analyses were performed to determine differentially expressed genes (DEGs). GO analysis revealed DEGs between S and P contributed to "reproduction process", "channel activity" and "cell periphery part" between I and S, and in "ion transport process", "channel activity" and "transporter complex part". KEGG analysis suggested the involvement of "glycerolipid metabolism", "steroid hormone biosynthesis" and "MAPK signaling pathway" in testis development and spermatogenesis. Genes including *IGF1*, *TGFB1*, *TGFBR1* and *EGFR* may control the development of the testis from immature to sexually mature, which might be important candidate genes for the development of goat testis. The current study provides novel insights into goat testicular development and spermatogenesis.

Keywords: Qianbei Ma goat; testis development; RNA-Seq; mRNA expression

Citation: Zou, Y.; Chen, X.; Tian, X.; Guo, W.; Ruan, Y.; Tang, W.; Fu, K.; Ji, T. Transcriptomic Analysis of the Developing Testis and Spermatogenesis in Qianbei Ma Goats. *Genes* **2023**, *14*, 1334. https://doi.org/10.3390/genes14071334

Academic Editor: Bao Yuan

Received: 15 May 2023
Revised: 22 June 2023
Accepted: 23 June 2023
Published: 25 June 2023

1. Introduction

The Qianbei Ma goat is a unique goat breed in the mountainous area of Guizhou Plateau. It is large in size, strong in constitution and delicious in taste. It is a goat breed whose skin and meat are both valuable, and has many advantages such as strong adaptability, high fertility, stable genetic inheritance and strong disease resistance. However, the research on its reproductive performance mainly focuses on lambing and estrus in she-goat and the molecular biology understanding of testicular development or spermatogenesis in Qianbei Ma goat is very limited, deserving further investigation. The breeding performance of male animals is also important to the total fertility of the population. Additionally, the most important guarantee for the reproductive performance of rams is the normal development of testis and spermatogenesis. Testicular development and spermatogenesis are very complex and precise processes which are regulated by multiple genes.

The testis represents a critical organ of the male reproductive system in mammals, producing spermatozoids and androgens. Spermatogenesis constitutes a developmental event, which produces haploid spermatozoa from diploid spermatogonial stem cells during meiosis in the testis [1]. Spermatogenesis comprises three steps, including spermatogonial proliferation and differentiation, meiotic division of spermatocytes and spermatozoid maturation [2]. Aside from spermatogenic cells, spermatogenic mechanisms involve many somatic cells in the testicle, including Sertoli and Leydig cells. In the testis of mammals, spermatogenesis occurs in seminiferous tubules, where germ cells are associated with Sertoli cells. Genes associated with the latter cells also have critical functions in precise steps of spermatogenesis [3]. The proper development of the testicles ensures the spermatogenesis process.

RNA sequencing (RNA-Seq) allows an expression profiling of genes and might help map and quantify the transcriptome [4,5]. This approach offers many advantages compared to other transcriptomic tools, including high resolution/sensitivity, a broad dynamic range of gene expression, and the identification of new transcript sequences and splice isoforms of previously reported genes [6]. Ramsköld [7] evaluated multiple tissues from mammals by RNA-Seq and reported that most genes were specifically expressed in testicular samples. Additionally, considering RNA-Seq-based expression patterns, Djureinovic [8] categorized 20,050 putative human genes, which showed specific expression in the human testicle. Their evaluation revealed the testicular tissue had by far the largest quantity of tissue-specific genes. Using microarray analysis, Anand detected differentially expressed genes (DEGs) in testicular samples compared to other tissues, identifying 2868 upregulated transcripts and 2011 downregulated mRNAs [9]. The testicle appears to have a higher degree of metabolic activity relative to other normal tissues. Most current reports assessing the association of testicular development with spermatogenesis have been performed in the human or mouse species, and the goat is scarcely examined.

The present work aimed to perform transcriptome profiling of immature and mature Qianbei Ma goat testis specimens by RNA-Seq and bioinformatic analysis. The findings provide novel insights into the mechanisms regulating goat testicular development and spermatogenesis.

2. Materials and Methods

2.1. Ethics Statement

This study had approval from the Animal Ethics Committee of Guizhou University (Guiyang, China) (No. EAE-GZU-2021-P024, Guiyang, China; 30 March 2021).

2.2. Animal Handling and Sample Collection

Permission was granted to castrate eighteen healthy Qianbei Ma goats in Fuxing Husbandry Co., Ltd., Zunyi, Guizhou, China. Goat ages were obtained from goat farming records. There were six immature goats (1 month old, before sexual maturation, i.e., samples I1, I2, I3, I4, I5 and I6), six sexually mature goats (6 months old, after sexual maturation but before physical maturation, i.e., samples S1, S2, S3, S4, S5 and S6) and six physically mature goats (12 months old, after physical maturation, i.e., samples P1, P2, P3, P4, P5 and P6). We surgically collected the right testes from the eighteen goats by castration after anesthesia, followed by storage in RNA/DNA sample protector (Servicebio, Wuhan, China). The testis from each goat was cut longitudinally, and a small amount (3–5 g) of the parenchyma, including seminiferous tubules and Leydig cells, underwent snap freezing in liquid nitrogen and was transported to the lab for further studies. All castrated goats remained in Fuxing Husbandry Co., Ltd. (Guizhou, China) after our study, for fatten feeding.

2.3. RNA Quantitation and Quality

Total RNA extraction was carried out from the testicular tissue in groups I, S and P using TRIzol reagent (Servicebio, Wuhan, China) and RNeasy RNA purification kit (Servicebio, Wuhan, China) with DNase as directed by the manufacturer. A NanoDrop™ One spectrophotometer (Thermo Fisher Scientific, Waltham, Ma, USA) was utilized to assess RNA purity and amounts. RNA quality assessment utilized 1% agarose gel electrophoresis. High-quality RNA samples (OD 260/280 of 1.8–2.0, integrity > 7.0 and 28S:18S above 1.0) were sequenced on an Illumina NovaSeq 6000 system, generating 150-bp paired end reads.

2.4. Transcriptome Sequencing

Approximately 5 μg RNA/sample constituted the input material for RNA sample preparation. Index-coded samples were clustered with NEBNext® Ultra™ Directional RNA Library Prep Kit for Illumina® according to a protocol provided by the manufacturer. Upon clustering, the prepared libraries underwent sequencing on an Illumina NovaSeq 6000 (Illumina, San Francisco, CA, USA). The image data of the sequences yielded by the high-throughput sequencer were converted into sequence data (reads) by CASA V A base recognition to obtain FASTQ files. Raw RNA-Seq FASTQ data next underwent filtration with Fastp v [10], a quality control software that can quickly filter and correct FASTQ data to exclude adapter-containing, N-containing and low-quality (quality score below 20) reads, resulting in clean reads, and were mapped to the goat (*C. hircus*) (ARS1.2) reference genome [4] using HISAT2, a software uses a graph Ferragina Manzini index to align DNA and RNA sequences [11].

2.5. Quantification of Gene Expression

Reads mapped to a given gene were counted with featurerts for estimating the expression of various gene transcripts. Gene expression was determined from million mapped reads per kilobase (FPKM) values [12], the current commonest approach to estimate gene expression [13].

2.6. Differential Expression Analysis

The DESeq2 software (1.20.0) was used to analyze differential expression between the treatment and control groups. The Benjamini–Hochberg algorithm was utilized to adjust p values (p-adj) to control for false discovery rate. $|\log2$ (FoldChange)$| \geq 1$ and padj < 0.05 was set as the significance threshold for differential expression [14].

2.7. GO and KEGG Enrichment Analyses of DEGs

GO and KEGG analyses of DEGs were implemented with ClusterProfiler, correcting for gene length bias. KEGG is an information database based on molecular findings, particularly via genome sequencing and additional high-throughput techniques to produce large-scale molecular data sets, allowing a deep understanding of biological systems (http://www.genome.jp/kegg/, accessed on 21 July 2022) [15]. GO and KEGG terms with $|\log2$ (FoldChange)$| \geq 1$ and padj < 0.05 were deemed to be DEGs with significant enrichment [10].

2.8. Prediction of New Transcripts and Alternative Splicing Analysis

StringTie was utilized to build and identify previously reported and new transcripts from HISAT2 alignment data. StringTie utilizes a network-flow algorithm with optional de novo assembly to splice transcripts. Compared with cufflinks, StringTie has the following classifying AS events, which were assessed in various samples separately.

Advantages: (1) it yields more complete transcripts; (2) assembles more accurate transcripts; (3) better estimates the transcript's expression level and (4) has greater splicing speed [16,17]. rMATS (http://rnaseqmats.sourceforge.net/index.html, accessed on 22 July 2022) was utilized to classify events.

2.9. qRT-PCR for RNA-Seq Data Validation

By analyzing the results of RNA sequencing and combining with the results of other earlier studies [18–23], we screened *TGFBR1, TGFB1, EGFR, IGF1, MAPK3* and *SMAD4* associated with testicular development or spermatogenesis. These DEGs were significantly up-regulated or down-regulated in our RNA sequencing results. The six selected DEGs were examined by qRT-PCR to validate the correctness of our RNA-Seq findings. Total RNA (1000 ng) was utilized to produce complementary DNA (cDNA) with 2×SYBR Green qPCR Master Mix None ROX (Servicebio, Wuhan, China) at 25 °C (5 min), 42 °C (30 min) and 85 °C (5 s). The primers used for qRT-PCR are shown in Table 1. A CFX96 Real-Time PCR system (Bio-Rad, Hercules, CA, USA) was utilized for amplification in 15-μL reactions containing 2×qPCR Mix (7.5 μL), forward and reverse primers (10 pmol/μL, 0.75 μL each), cDNA (1000 ng/μL, 2 μL) and nuclease-free water (4 μL). The reaction conditions were: 1 cycle at 95 °C (30 s), followed by 40 cycles at 95 °C (15 s) and 60 °C (30 s), with fluorescence signals collected every 0.5 °C increase from 65 °C to 95 °C. Melting curves were utilized to assess primer specificity. Assays were carried out in triplicate and *GAPDH* was utilized for normalization in data analysis by the $2^{-\Delta\Delta Ct}$ method.

Table 1. Primers utilized in qRT-PCR.

Primer Name	Gene ID	Primer Sequence	Fragment Length	Annealing Temperature
GAPDH-F GAPDH-R	XM_005680968.3	ATGTTTGTGATGGGCGTGAA GGCGTGGACAGTGGTCATAAGT	153 bp	60 °C
TGFBR1-F TGFBR1-R	XM_018052233.1	TTCAAACGTGCTGACATCTATGC ACTGATGGATCGGAAGGTACAAG	128 bp	60 °C
SMAD4-F SMAD4-R	XM_018039535.1	CATAACAGCACTACCACCTGGACT GGATGATTAGAAATAGGAGGCTGG	173 bp	60 °C
TGFB1-F TGFB1-R	NM_001314142.1	CAACAATTCCTGGCGCTACCT ATGTCCACTTGAAGCGTGTTATCC	183 bp	60 °C
EGFR-F EGFR-R	XM_018067044.1	CCGTGCGATTCAGTAACAACC GGTCAATTTCTGGCAGTTCTCCTC	194 bp	60 °C
IGF1-F IGF1-R	NM_001285697.1	AATCAGCAGTCTTCCAACCCAA AGCAAGCACAGGGCCAGATA	114 bp	60 °C
MAPK3-F MAPK3-R	XM_018040780.1	CTGGACCGGATGTTGACCTTTA CTCCTTCAGTCGTTCCTTGGG	138 bp	60 °C

3. Results

3.1. Gene Expression Profiling during Testicular Development in Qianbei Ma Goats

To determine the genes associated with testicular development and sperm formation, 18 libraries, including 6 each from immature, sexually mature and physically mature testes were sequenced. Table 2 shows an overview of the information pertaining to raw and clean reads for all libraries. Error rates and GC contents for various libraries were determined for their quality control. All 18 libraries were of high quality.

Paired-end clean reads underwent alignment to the goat reference genome ARS1.2 that is commonly utilized for *C. hircus* (goats) [4] with HISAT2, and the expected number of fragments per kilobase of transcript sequence per million base pairs sequenced (FPKM) of each gene was calculated. Table 3 shows an overview of the information about uniquely mapped clean reads.

Additionally, an overview of the percentages of clean reads mapped to exon borders (junction reads) is shown in Table 4.

All RNA-Seq data in our study had been submitted to NCBI (accession number BioProject: PRJNA879963).

3.2. Alternative Splicing Data

In this work, AS events were categorized into five types with rMATS (http://rnaseq-mats.sourceforge.net/index.html, accessed on 21 July 2022). By determining the types and amounts of AS events, and analyzing each AS type, a large number of AS events were found in testicular development and sperm formation. Of all the DEGs detected between immature (I) and sexually mature (S) testes in our present work, 544 genes underwent AS events between sexually mature (S) and physically mature (P) testes, with no detected AS events (padj $\leq$ 0.05). Therefore, approximately 4% of genes showed AS between I and S, and no gene had AS between S and P in this study. The above findings indicated AS was very important in the complexity of gene expression during testis development, especially in the period from immaturity to sexual maturity.

Table 2. Statistics of RNA-Seq data quality.

Sample Name	Library Number	Raw Reads (n)	Clean Reads (n)	Error Rate	Q20	Q30	GC pct
I1	1	40,405,582	39,128,916	0.03	97.36	93.09	50.99
I2	2	43,841,646	42,366,502	0.03	97.41	93.2	51.16
I3	3	39,433,208	38,237,518	0.03	97.42	93.25	51.42
I4	4	42,396,474	40,926,952	0.03	97.43	93.23	51.26
I5	5	45,463,988	43,854,898	0.03	97.41	93.19	51.1
I6	6	39,351,212	38,000,364	0.03	97.48	93.36	50.91
S1	7	44,934,642	43,739,660	0.03	97.51	93.41	52.09
S2	8	43,025,142	41,894,994	0.03	97.39	93.17	52.36
S3	9	45,845,898	44,647,422	0.03	97.59	93.55	52.15
S4	10	41,237,336	40,020,062	0.03	97.37	93.11	52.29
S5	11	41,130,946	39,901,252	0.03	97.36	93.09	51.09
S6	12	41,242,240	40,019,490	0.03	97.51	93.4	51.9
P1	13	45,377,178	44,131,516	0.03	97.48	93.34	52.27
P2	14	44,664,918	43,577,078	0.03	97.49	93.36	52.06
P3	15	43,792,502	42,601,198	0.03	97.47	93.32	51.98
P4	16	43,921,138	42,702,250	0.03	97.5	93.39	52.17
P5	17	43,223,452	42,134,186	0.03	97.44	93.27	52.14
P6	18	45,571,906	44,622,618	0.03	97.3	92.91	50.37

Error rate: overall sequencing error rate for the data; Q20 and Q30: percentages of total bases with Phred values above 20 and 30, respectively; GC pct: percentage of C and G among the 4 bases in clean reads.

Table 3. Statistics of reads aligned with the reference genome.

Sample	Total Reads	Total Map	Unique Map	Multi Map	Positive Map	Negative Map
I1	39,128,916	37,563,635 (96.00%)	35,626,251 (91.05%)	1,937,384 (4.95%)	17,793,049 (45.47%)	17,833,202 (45.58%)
I2	42,366,502	40,625,554 (95.89%)	38,612,044 (91.14%)	2,013,510 (4.75%)	19,294,015 (45.54%)	19,318,029 (45.60%)
I3	38,237,518	36,735,876 (96.07%)	34,685,761 (90.71%)	2,050,115 (5.36%)	17,323,488 (45.30%)	17,362,273 (45.41%)
I4	40,926,952	39,282,398 (95.98%)	37,389,786 (91.36%)	1,892,612 (4.62%)	18,672,350 (45.62%)	18,717,436 (45.73%)
I5	43,854,898	42,092,675 (95.98%)	40,255,938 (91.79%)	1,836,737 (4.19%)	20,105,549 (45.85%)	20,150,389 (45.95%)
I6	38,000,364	36,500,392 (96.05%)	34,501,227 (90.79%)	1,999,165 (5.26%)	17,230,278 (45.34%)	17,270,949 (45.45%)
S1	43,739,660	42,076,928 (96.20%)	40,551,689 (92.71%)	1,525,239 (3.49%)	20,259,699 (46.32%)	20,291,990 (46.39%)
S2	41,894,994	40,318,684 (96.24%)	39,006,285 (93.10%)	1,312,399 (3.13%)	19,484,346 (46.51%)	19,521,939 (46.60%)
S3	44,647,422	43,061,457 (96.45%)	41,514,183 (92.98%)	1,547,274 (3.47%)	20,739,259 (46.45%)	20,774,924 (46.53%)
S4	40,020,062	38,465,949 (96.12%)	36,829,538 (92.03%)	1,636,411 (4.09%)	18,396,451 (45.97%)	18,433,087 (46.06%)
S5	39,901,252	38,334,062 (96.07%)	36,942,595 (92.59%)	1,391,467 (3.49%)	18,452,256 (46.24%)	18,490,339 (46.34%)
S6	40,019,490	38,522,907 (96.26%)	36,912,252 (92.24%)	1,610,655 (4.02%)	18,439,977 (46.08%)	18,472,275 (46.16%)
P1	44,131,516	42,514,601 (96.34%)	41,073,585 (93.07%)	1,441,016 (3.27%)	20,518,750 (46.49%)	20,554,835 (46.58%)
P2	43,577,078	41,944,011 (96.25%)	40,303,516 (92.49%)	1,640,495 (3.76%)	20,133,389 (46.20%)	20,170,127 (46.29%)
P3	42,601,198	41,016,519 (96.28%)	39,440,892 (92.58%)	1,575,627 (3.70%)	19,701,050 (46.25%)	19,739,842 (46.34%)
P4	42,702,250	41,098,071 (96.24%)	39,528,322 (92.57%)	1,569,749 (3.68%)	19,745,068 (46.24%)	19,783,254 (46.33%)
P5	42,134,186	40,510,371 (96.15%)	38,849,502 (92.20%)	1,660,869 (3.94%)	19,405,731 (46.06%)	19,443,771 (46.15%)
P6	44,622,618	38,550,615 (86.39%)	37,036,723 (83.00%)	1,513,892 (3.39%)	18,501,103 (41.46%)	18,535,620 (41.54%)

Sample: sample name; total reads: number of clean reads upon quality control; total map: number (percentage) of reads aligned to the reference genome; unique map: number (percentage) of reads aligned to a unique region of ARS1.2 (subsequently analyzed for quantitation); multi map: number (percentage) of reads with alignment to many locations of ARS1.2; positive and negative maps: numbers (percentages) of reads with alignment to the positive and negative strands of the reference genome, respectively.

Table 4. Statistics of reads aligned to genomic regions.

Sample Name	Exonic Region	Intronic Region	Intergenic Region
I1	4,466,151,533 (79.48%)	611,822,545 (10.89%)	541,085,383 (9.63%)
I2	4,779,622,398 (78.65%)	761,383,464 (12.53%)	535,946,498 (8.82%)
I3	4,432,093,997 (80.66%)	570,346,947 (10.38%)	492,580,965 (8.96%)
I4	4,614,052,978 (78.53%)	738,532,433 (12.57%)	523,236,881 (8.90%)
I5	5,160,159,063 (81.96%)	605,739,527 (9.62%)	530,252,297 (8.42%)
I6	4,298,336,210 (78.73%)	599,791,632 (10.99%)	561,800,986 (10.29%)
S1	5,004,805,293 (79.49%)	606,545,683 (9.63%)	684,973,145 (10.88%)
S2	4,891,821,345 (81.09%)	533,206,687 (8.84%)	607,543,338 (10.07%)
S3	5,173,907,638 (80.30%)	600,212,860 (9.32%)	668,983,341 (10.38%)
S4	4,579,030,919 (79.56%)	552,189,052 (9.59%)	623,968,138 (10.84%)
S5	4,488,498,885 (78.27%)	613,561,446 (10.70%)	632,649,316 (11.03%)
S6	4,595,667,229 (79.73%)	543,550,434 (9.43%)	624,459,696 (10.83%)
P1	5,058,268,703 (79.52%)	636,322,977 (10.00%)	666,220,762 (10.47%)
P2	4,921,081,484 (78.42%)	644,761,157 (10.27%)	709,616,381 (11.31%)
P3	4,871,901,450 (79.39%)	593,069,117 (9.66%)	671,905,236 (10.95%)
P4	4,918,914,052 (80.00%)	601,562,419 (9.78%)	628,332,353 (10.22%)
P5	4,886,718,972 (80.63%)	554,456,620 (9.15%)	619,529,425 (10.22%)
P6	4,658,554,613 (80.75%)	537,138,992 (9.31%)	573,327,628 (9.94%)

Exonic, intronic and intergenic regions: numbers (percentages) of reads with alignment to the exonic, intronic and intergenic regions, respectively.

To determine AS types associated with testicular genes, AS events were compared between the S group and the I group. The five known types of AS events include retained intron (RI), mutually exclusive exon (MXE), alternative 3′ splice site (A3SS), alternative 5′ splice site (A5SS) and skipped exon (SE). All five types of AS were found in the S vs. I group comparison. RI, MXE, A3SS and SE were found in the P vs. S group comparison. These findings showed SE as the most common AS event amounting to 368 in S vs. I. Other identified AS events were RI (48), MXE (110), A3SS (66) and A5SS (54) (Figure 1).

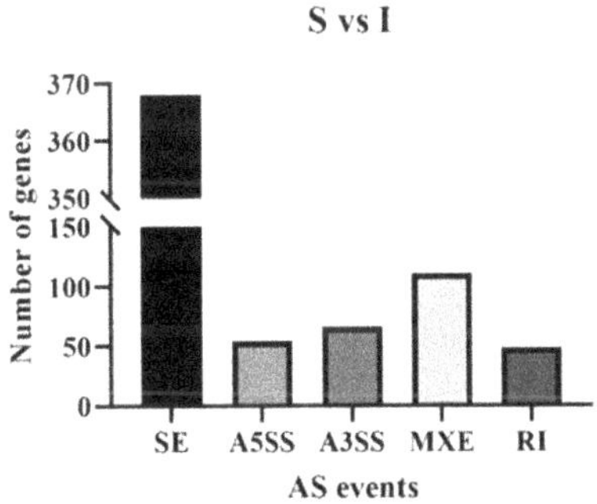

Figure 1. AS events among genes for immature and sexually mature group comparisons. X axis represents five known types of AS events, respectively, skipped exon (SE), alternative 5′ splice site (A5SS), alternative 3′ splice site (A3SS), mutually exclusive exon (MXE) and retained intron (RI).

3.3. Analysis of DEGs

In order to clarify the difference in gene expression in testis of Qianbei Ma goat at different developmental stages, we compared and analyzed the differentially expressed genes in three different developmental stages. DEGs were determined with $|\log2\ (\text{fold}-\text{change})| \geq 1$ and padj < 0.05. As a result, a total of 9,724 upregulated and 4,153 downregulated DEGs were detected between the I and S groups; 7 upregulated and 3 downregulated genes were detected between the S and P groups (Figure 2A1, Figure 2A2). Among all DEGs detected, 1112 genes were only expressed in the immature group, 379 genes were only expressed in the sexual mature group, 336 genes were only expressed in the physical mature group, and 13,299 genes were expressed in all three groups (Figure 2B). Figure 2C depicts hierarchical clustering, with DEGs for the eighteen libraries grouped into three clusters, the result suggested the I, S and P groups had differential expression patterns overall, but S and P had similar repetitive expression commonalities.

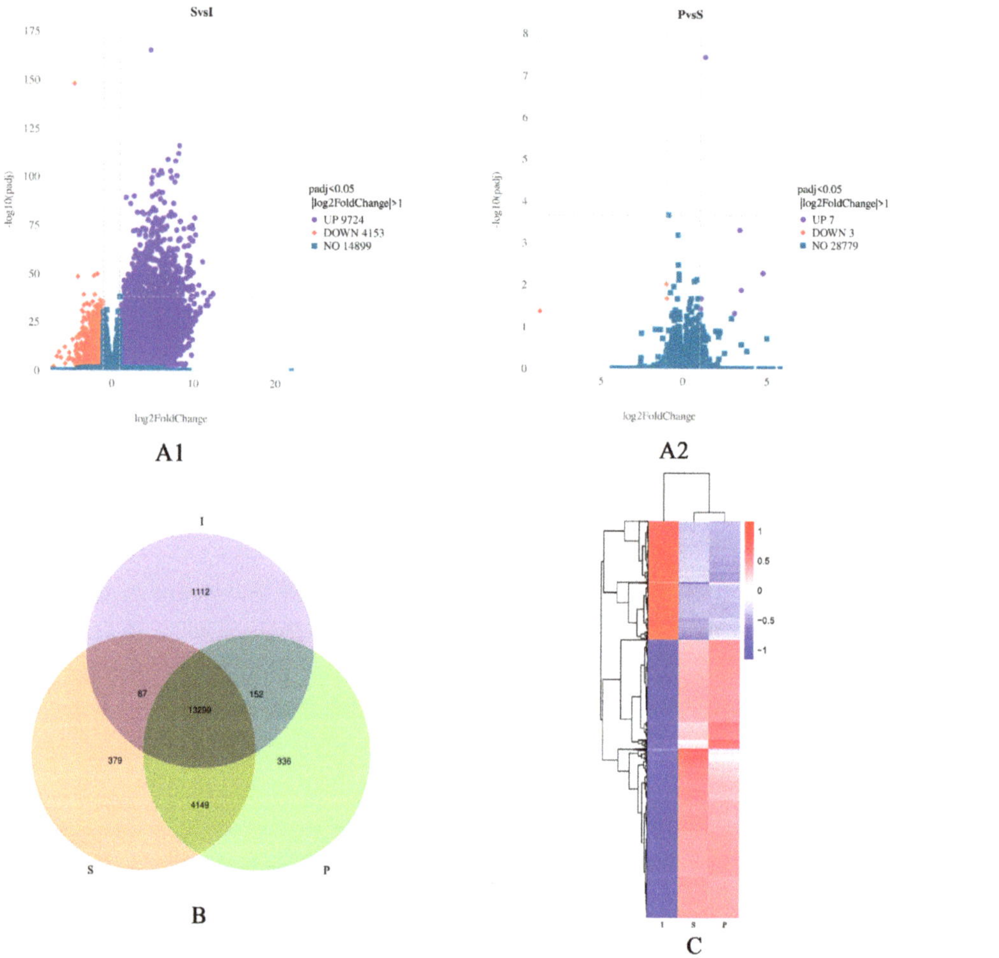

Figure 2. (**A1**): Volcanic plot of differentially expressed genes (DEGs) between the immature and sexually mature groups. (**A2**): Volcanic plot of DEGs between the sexually mature and physically mature groups. Highly significant differences in the expression of up− (purple) and down− (red) regulated genes were observed between immature (I) and sexually mature (S) testes, and between sexually mature (S) and physically mature (P) testes. Blue indicates no differential expression. (**B**): Venn diagram depicting gene expression patterns. The numbers of uniquely and commonly (FPKM > 1) expressed genes were shown. (**C**): Clustering of differentially expressed genes. The overall FPKM hierarchical clustering map was obtained with log10 (FPKM + 1) for normalization. Different columns represent groups, different rows represent genes, different colors represent expression levels (red: high, blue: low).

3.4. GO Analysis of DEGs

GO analysis was carried out to examine the functions of DEGs in testicular development. Totally, 106 GO terms associated with "biological processes", "molecular functions" and "cellular components" were markedly enriched between immature (I) and sexually mature (S) testes. We found 43 GO terms belonged to "biological process", 56 GO terms belonged to "molecular functions", and 7 GO terms belonged to "cellular components". Totally, 36 GO terms associated with "biological processes", "molecular functions" and "cellular components" had significant enrichment in the sexually mature (S) vs. physically mature (P) testes. We found 5 GO terms belonged to "biological processes", 21 GO terms belonged to "molecular functions", and 10 GO terms belonged to "cellular components" (Table 5, Figure 3A,B).

Table 5. Genes functions enriched through GO enrichment.

	S vs. I	P vs. S
BP	protein phosphorylation ion transport transmembrane transport reproduction reproductive process	sodium ion transport metal ion transport monovalent inorganic cation transport immune response immune system process
MF	ion channel activity channel activity passive transmembrane transporter activity substrate-specific channel activity scavenger receptor activity	voltage-gated sodium channel activity voltage-gated ion channel activity involved in regulation of postsynaptic membrane potential sodium channel activity chemokine activity chemokine receptor activity
CC	myosin complex extracellular matrix plasma membrane plasma membrane part cell periphery	voltage-gated sodium channel complex sodium channel complex ion channel complex cation channel complex integral component of plasma membrane

BP: biological processes; MF: molecular functions; CC: cellular components.

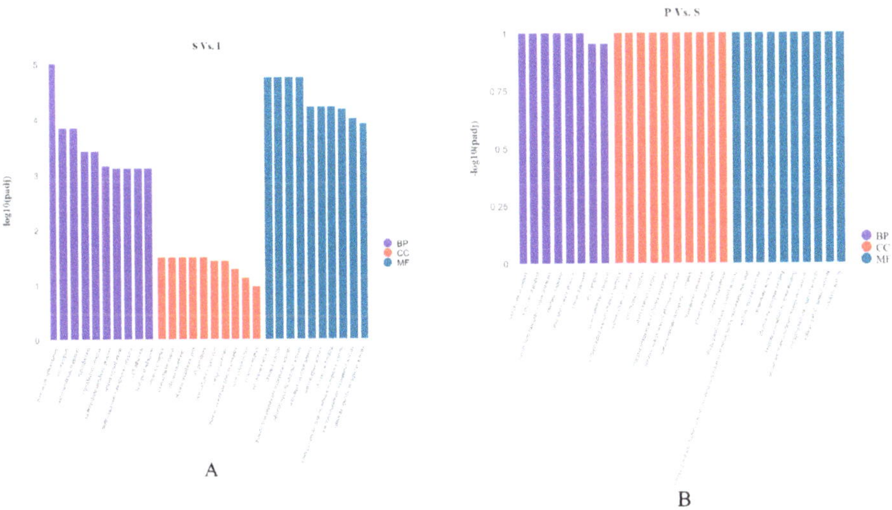

Figure 3. (**A**): Most enriched GO terms between immature (I) and sexually mature (S) testes. (**B**): Most enriched GO terms between sexually mature (S) and physically mature (P) testes. Abscissas and ordinates represent enriched GO terms and significance levels of GO enrichment, respectively. Purple, "biological processes"; red, "cellular components"; blue, "molecular functions".

3.5. KEGG Pathway Analysis of DEGs

KEGG pathway analysis of DEGs (|log2 (fold−change)| $\geq$ 1 and padj < 0.05) was carried out. For I vs. S, 8 KEGG pathways were upregulated, including "glycerolipid metabolism", "protein digestion and absorption" and "steroid hormone biosynthesis".

Totally, 90 pathways were downregulated, including +"lysosome", "MAPK signaling pathway" and "chemokine signaling pathway" (Figure 4A). For S vs. P, 7 KEGG pathways were upregulated, including "oxidative phosphorylation", "thermogenesis" and "steroid biosynthesis". However, no pathways were significantly downregulated (Figure 4B).

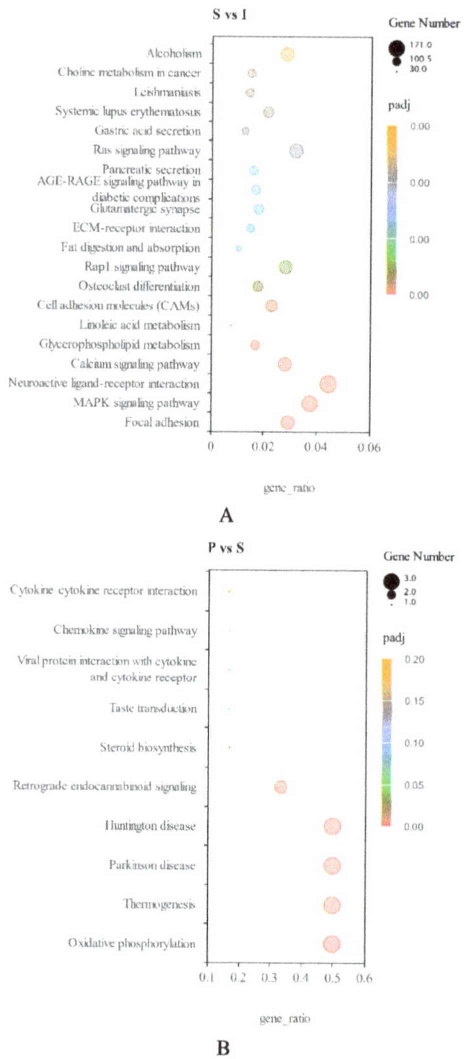

Figure 4. (**A**): Scatter plot of differentially expressed KEGG genes in immature (I) and sexually mature (S) testes. (**B**): Scatter plot of differentially expressed KEGG genes in sexually mature (S) and physically mature (P) testes. Ordinates and abscissas represent the names of KEGG pathways and gene ratios, respectively. Point sizes and colors represent the numbers of DEGs and the ranges of Q values, respectively.

3.6. *qRT-PCR Validation of DEGs*

To verify the DEGs in immature (I), sexually mature (S) and physically mature (P) testes, we selected 6 DEGs, including *TGFBR1*, *TGFB1*, *EGFR*, *IGF1*, *MAPK3* and *SMAD4*, to validate RNA-Seq data by qRT-PCR. qRT-PCR data corroborated RNA-Seq findings, suggesting the reliability of RNA-Seq data (Figure 5).

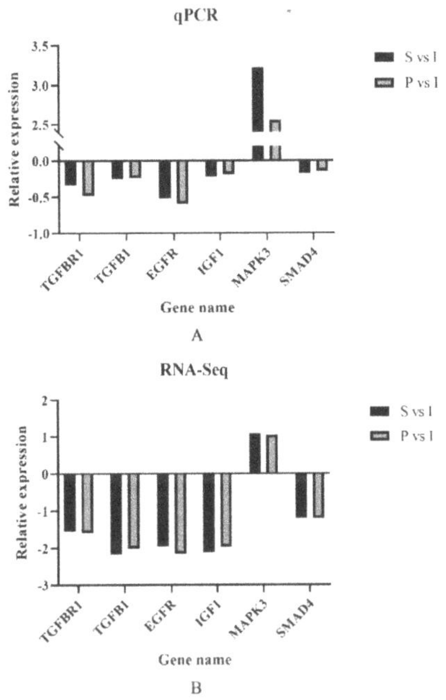

Figure 5. Relative expression of different genes in sexually mature (S) and physically mature (P) testes compared to immature (I) testes, respectively, using group I as a reference. (**A**): qPCR relative expression of S vs. I and P vs. I. (**B**): RNA−Seq data for S vs. I and P vs. I.

4. Discussion

With the current development of detection technology, more and more mRNAs on testis related to testis development and spermatogenesis have been reported. RNA-Seq has emerged as a tool for efficiently and inexpensively detecting new transcripts and genes. RNA-Seq methods have been broadly utilized to determine DEGs or gene expression patterns, new transcripts, AS events and SNPS, and have empowered studies examining porcine [24,25], cattle [26,27] and mouse [28,29] testicular development. In goats, the profiles of ovarian [30,31], uterine [32,33] and testicular [34,35] tissues under different conditions were recently compared by RNA-Seq. However, limited data on testicular development in goats were available. Breed and age represent major factors affecting testicular development. Here, RNA-Seq was performed to build a complete dataset that explains the spatiotemporal transcriptome of the testicular tissue in Qianbei Ma goats. Testicular growth and development constitute the key factors affecting goat reproduction. Therefore, identifying genes regulating testicular growth and development is critical. In this study, 13,887 genes were assessed by RNA-Seq in six immature, six sexually mature and six physically mature testes. Totally, 9724 genes were upregulated and 4153 were downregulated between immature and sexually mature testes; seven genes were upregulated and three were downregulated between sexually mature and physically mature testes which was similar to the hierarchical clustering analysis result in our study. The reason for this result is related to the development process of the testis of Qianbei Ma goat. We speculate that the testis develops from immature to sexual mature with various reproductive functions and also starts to produce sperm, which is a very complex process that requires the co-regulation of a large number of different genes. When the goats continue to develop from sexual maturity to physical maturity, the testes already had all the functions related to reproduction, so the differences in development were relatively small, the number of related DEGs was significantly reduced, which was similar to the results of Gong studies on testis of mice at three different developmental stages [36]. Using next-generation platforms, we determined most upregulated genes were associated with protein coding and might have functions in testicular development and sperm formation.

Alternative splicing (AS) represents a commonly encountered phenomenon in eukaryotes, which could lead to the production of various protein forms at different times under different circumstances, increasing species/body fitness. Although AS research is a known subfield of molecular biology, only in recent years has this subfield attracted sufficient attention. AS is critical for the complex proteomes and functions found in higher organisms. AS is an important mechanism in the regulation of gene expression and promotes proteome diversity [37]. It is estimated that about 95% of human multiple-exon gene expression is associated with AS events [38]. In metazoans, AS is critical for the production of various protein forms with functions in different cell events such as cell growth, differentiation and death [39]. Here, five AS events were observed, mostly involving ES. The effects of AS events on the functions of related genes can be predicted by a comprehensive analysis of AS events, and GO and KEGG analysis data [40,41]. In the current study, Sec insertion sequence binding protein 2 (*SECISBP2*) was the gene with the highest number of SE events, i.e., a total of ten SE events. *SECISBP2* underwent the most AS events during the development, indicating that SECISBP2 protein synthesis is complex and important to Qianbei Ma goats' testis development. Mutation of the gene altered thyroid hormone metabolism [42,43]. Thyroid hormones can modulate semen quality under physiological conditions by regulating testosterone and changing some semen indexes [44–46]. Thyroid hormone levels in the early stages of testicular development can also influence testicular development and spermatogenesis by regulating the length of time during which supporting cells proliferate [47]. Based on previous studies, we hypothesized that there is active synthesis of SECISBP2 during testicular development in Qianbei Ma goat, which is essential for testicular size and spermatogenesis.

Combining previous relevant reports, and KEGG and GO data in the current study, the genes involved in the regulation of testis development and sperm formation through protein phosphorylation were mainly *TGFB1*, *EGFR* and *IGF1*, which have critical functions in testis growth, hormone secretion, spermatogenesis and Leydig cell differentiation.

Transforming growth factor β-1 (*TGFB1*) plays multiple biological roles, including the control of proliferative and differentiation potentials of cells in rams [48]. In male lambs, *TGFB1* regulates tight junctions in Sertoli cells and controls spermatogenesis. It modifies the blood-testicular barrier (BTB) by downregulating tight junction proteins [49]. *TGFB1* might play an important role in testicular development because of its high expression in the immature testis and markedly reduced expression in sexual maturity, as spermatogenesis begins. A comparable expression pattern was found for TGFB receptor type 1 (*TGFBR1*) [50]. In addition, loss of *TGFB1* resulted in lower testosterone levels in the testis and serum, and decreased the ability to mate with females [51]. Based on previous studies, we speculated that *TGFB1* may not only directly regulate goat testicular development and sperm formation, but also ensure the normal development of male external genitalia and affect fertility.

Epidermal growth factor receptor (*EGFR*) represents a receptor gene for EGF and controls testicular function in the human, mouse, rat and livestock species as well as in alpacas [20]. In yak and bovine, EGF and *EGFR* were critical paracrine and/or autocrine modulators of testicular development and sperm formation and regulate testosterone production by testicular interstitial cells [52,53]. We speculate that EGF and *EGFR* may also be expressed in various goat testicular cells, stimulate testosterone secretion, and regulate testis development and spermatogenesis.

Insulin-like growth factor I (*IGF1*) contributes to the regulation of testicular function [54]. Pitetti et al. indicated growth factors of the insulin family play essential roles by controlling SC number, testis size and daily sperm production [55,56]. Both *IGF1* and its receptor *IGF1R* were expressed in testicles, and their hormones act directly on male gonads [57,58]. In immature testes, *IGF1* promotes the development of sustentacular cells, Leydig cells and gonocytes. In mature testis, the *IGF1* gene induces spermatogenesis and regulates Leydig cell function [59,60]. *IGF1* may act as an autocrine/paracrine or endocrine signal to regulate testicular steroid production as well as germ cell and Sertoli cell functions [61]. *IGF1* plays different roles in testicular function at different stages of testicular development [54,62]. We speculate that high IGF1 and IGF1R protein amounts in the immature testis may suggest they highly promote the development and differentiation of sustentacular cells, Leydig cells and gonocytes in goat testis during sexual maturity.

Transcriptome data revealed the MAPK pathway is implicated in goat testicular development, while *TGFBR1*, *TGFB1*, *EGFR* and *IGF1* were enriched in this pathway and downregulated during sexual maturation, as key genes that regulate testis development and spermatogenesis [63,64]. Multiple reports suggest MAPK signaling is a critical regulator of testis growth and development, testis cell proliferation, differentiation and apoptosis and testosterone secretion, thus affecting male fertility [65–67]. One of the key downstream target genes of MAPK signaling is *IGF1*, which together with other genes in the pathway controls testis cell proliferation, testis volume development, hormone secretion and spermatogenesis, and is often reported to be associated with male fertility [55,68,69]. We speculate that MAPK signaling is a critical regulatory pathway in goat testis development and spermatogenesis. As an essential male fecundity-related gene in the MAPK signaling pathway, *IGF1* modulates goat testis growth and development and can affect the functions of various cells of goat testis by regulating other downstream genes in the signaling pathway. During male goat sexual maturation, *IGF1* regulates the development of the testis, spermatogenesis and hormone synthesis through the MAPK signaling pathway and other cooperative genes.

5. Conclusions

This study first used RNA-Seq to profile the expression of genes during testicular development in Qianbei Ma goats. We identified 544 genes with AS events between I and S, which suggests that AS of differential genes might be critical in the regulation of testicular development in goats. Totally, eight KEGG pathways were upregulated and 90 were downregulated between I and S. Totally, seven KEGG pathways were upregulated between S and P. Among the six screened DEGs (*TGFBR1*, *TGFB1*, *EGFR*, *IGF1*, *MAPK3* and *SMAD4*) *TGFBR1*, *TGFB1*, *EGFR*, *IGF1* and *MAPK3* belonged to the "MAPK signaling pathway", corresponding to the GO term "protein phosphorylation". However, the specific function and pathway mechanism of the selected genes still need to be further studied at the cellular level. The findings should further our understanding of gene regulation during testicular development and sperm formation. Our study also suggests that RNA-seq has great potential in the study of the transcriptome during the development of goat testis.

Author Contributions: Methodology, Y.Z. and X.C.; software, W.T.; investigation, Y.R. and T.J.; resources, X.C.; validation, data curation, X.C.; writing—manuscript drafting, Y.Z.; writing—manuscript revision, X.T., W.G. and K.F. All authors have read and agreed to the published version of the manuscript.

Funding: The current study was supported by the National Natural Science Foundation of China (No. 32260835), the Science and Technology Project of Guizhou Province (Qian Kehe Foundation-ZK [(2021)] General 151), and the Guizhou High Level Innovative Talents Project (Qian Kehe Platform Talents [(2022)] 021-1).

Institutional Review Board Statement: The animals used in our study were consented by their owners. All procedures involving animals were approved and authorized by Guizhou University. The laboratory animal castration protocol for this study was approved by the Laboratory Animal Ethics of Guizhou University (No. EAE-GZU-2021-P024, Guiyang, China; 30 March 2021).

Informed Consent Statement: Not applicable.

Data Availability Statement: The raw data from sequencing are available at NCBI under BioProject ID: PRJNA879963.

Acknowledgments: The authors are thankful to Fuxing Husbandry Co., Ltd. (Guizhou, China) for providing the experimental Qianbei Ma goats.

Conflicts of Interest: The authors declare no conflict of interest.

References

1. Hecht, N.B. Molecular mechanisms of male germ cell differentiation. *BioEssays* **1998**, *20*, 555–561. [CrossRef]
2. Eddy, E.M.; O'Brien, D.A. Gene expression during mammalian meiosis. *Curr. Top. Dev. Biol.* **1998**, *37*, 141–200. [PubMed]
3. Hess, R.A.; De Franca, L.R. Spermatogenesis and cycle of the seminiferous epithelium. *Adv. Exp. Med. Biol.* **2008**, *636*, 1–15. [PubMed]
4. Mortazavi, A.; Williams, B.A.; McCue, K.; Schaeffer, L.; Wold, B. Mapping and quantifying mammalian transcriptomes by RNA-Seq. *Nat. Methods* **2008**, *5*, 621–628. [CrossRef]
5. Marioni, J.C.; Mason, C.E.; Mane, S.M.; Stephens, M.; Gilad, Y. RNA-seq: An assessment of technical reproducibility and comparison with gene expression arrays. *Genome Res.* **2008**, *18*, 1509–1517. [CrossRef]
6. Wang, Z.; Gerstein, M.; Snyder, M. RNA-Seq: A revolutionary tool for transcriptomics. *Nat. Reviews. Genet.* **2010**, *10*, 57–63. [CrossRef]
7. Ramsköld, D.; Wang, E.T.; Burge, C.B.; Sandberg, R. An abundance of ubiquitously expressed genes revealed by tissue transcriptome sequence data. *PLoS Comput. Biol.* **2009**, *5*, e1000598. [CrossRef]
8. Djureinovic, D.; Fagerberg, L.; Hallstrom, B.; Danielsson, A.; Lindskog, C.; Uhlen, M.; Ponten, F. The human testis-specific proteome defined by transcriptomics and antibody-based profiling. *Mol. Hum. Reprod.* **2014**, *20*, 476–488. [CrossRef]
9. Anand, M.; Prasad, B.V. The computational analysis of human testis transcriptome reveals closer ties to pluripotency. *J. Hum. Reprod. Sci.* **2012**, *5*, 266–273.
10. Fu, K.; Chen, X.; Guo, W.; Zhou, Z.; Zhang, Y.; Ji, T.; Yang, P.; Tian, X.; Wang, W.; Zou, Y. Effects of N Acetylcysteine on the Expression of Genes Associated with Reproductive Performance in the Goat Uterus during Early Gestation. *Animals* **2022**, *12*, 2431. [CrossRef]
11. Kim, D.; Paggi, J.M.; Park, C.; Bennett, C.; Salzberg, S.L. Graph-based Genome Alignment and Genotyping with HISAT2 and HISAT-genotype. *Nat. Biotechnol.* **2019**, *37*, 907–915. [CrossRef] [PubMed]
12. Bray, N.L.; Pimentel, H.; Melsted, P.; Pachter, L. Erratum: Near-optimal probabilistic RNA-seq quantification. *Nat. Biotechnol.* **2016**, *34*, 888. [CrossRef]
13. Trapnell, C.; Williams, B.A.; Pertea, G.; Mortazavi, A.; Kwan, G.; van Baren, M.J.; Salzberg, S.L.; Wold, B.J.; Pachter, L. Transcript assembly and quantification by RNA-Seq reveals unannotated transcripts and isoform switching during cell differentiation. *Nat. Biotechnol.* **2010**, *28*, 511–515. [CrossRef] [PubMed]
14. Anders, S.; Huber, W. Differential expression analysis for sequence count data. *Genome Biol.* **2010**, *11*, R106. [CrossRef]
15. Kanehisa, M.; Araki, M.; Goto, S.; Hattori, M.; Hirakawa, M.; Itoh, M.; Katayama, T.; Kawashima, S.; Okuda, S.; Tokimatsu, T.; et al. KEGG for linking genomes to life and the environment. *Nucleic Acids Res.* **2008**, *36*, D480–D484. [CrossRef]
16. Shumate, A.; Wong, B.; Pertea, G.; Pertea, M. Improved Transcriptome Assembly Using a Hybrid of Long and Short Reads with StringTie. *PLoS Comput. Biol.* **2022**, *18*, e1009730. [CrossRef] [PubMed]
17. Pertea, M. StringTie Enables Improved Reconstruction of a Transcriptome from RNA-seq Reads. *Nat. Biotechnol.* **2015**, *33*, 290–296. [CrossRef]
18. Moraveji, S.F.; Esfandiari, F.; Taleahmad, S.; Nikeghbalian, S.; Sayahpour, F.A.; Masoudi, N.S.; Shahverdi, A.; Baharvand, H. Suppression of transforming growth factor-beta signaling enhances spermatogonial proliferation and spermatogenesis recovery following chemotherapy. *Hum. Reprod.* **2019**, *34*, 2430–2442. [CrossRef]
19. Memon, M.A.; Anway, M.D.; Covert, T.R.; Uzumcu, M.; Skinner, M.K. Transforming growth factor beta (TGF beta 1, TGF beta 2 and TGF beta 3) null-mutant phenotypes in embryonic gonadal development. *Mol. Cell. Endocrinol.* **2008**, *294*, 70–80. [CrossRef]
20. He, J.; Dong, C.; You, R.; Zhu, Z.; Lv, L.; Smith, G.W. Localization of epidermal growth factor (EGF) and its receptor (EGFR) during postnatal testis development in the alpaca (*Lama pacos*). *Anim. Reprod. Sci.* **2009**, *116*, 155–161. [CrossRef]
21. Baker, J.; Hardy, M.P. Effects of an Igf1 gene null mutation on mouse reproduction. *Mol. Endocrinol.* **1996**, *10*, 903–918.
22. Manesh, K.P.S.; Saradha, B.; Suresh, C.S. Telomere Signaling and Maintenance Pathways in Spermatozoa of Infertile Men Treated with Antioxidants: An in silico Approach Using Bioinformatic Analysis. *Front. Cell Dev. Biol.* **2021**, *9*, 768510.
23. He, J.; Zhang, X.; Wen, X.; Zhao, L. Expression and localization of Smad2 and Smad4 in domestic cat testis during testicular development and spermatogenesis. *Acta Anat. Sin.* **2010**, *41*, 598–602.
24. Zhou, C.; Zhang, L.; Xu, D.; Ding, H.; Zheng, S.; Liu, M. MeDIP-seq and RNA-seq Analysis during Porcine Testis Development Reveals Functional DMR at the Promoter of LDHC. *Genomics* **2022**, *114*, 110467. [CrossRef]
25. Zhang, L.; Li, F.; Lei, P.; Guo, M.; Liu, R.; Wang, L.; Yu, T.; Lv, Y.; Zhang, T.; Zeng, W.; et al. Single-cell RNA-sequencing Reveals the Dynamic Process and Novel Markers in Porcine Spermatogenesis. *J. Anim. Sci. Biotechnol.* **2021**, *12*, 122. [CrossRef] [PubMed]
26. Chang, T.; Yang, Y.; Retzel, E.F.; Liu, W. Male-specific Region of the Bovine Y Chromosome Is Gene Rich with a High Transcriptomic Activity in Testis Development. *Proc. Natl. Acad. Sci. USA* **2013**, *110*, 12373–12378. [CrossRef]
27. Gao, Y.; Wu, M.; Fan, Y.; Li, S.; Lai, Z.; Huang, Y.; Lan, X.; Lei, C.; Chen, H.; Dang, R. Identification and Characterization of Circular RNAs in Qinchuan Cattle Testis. *R. Soc. Open Sci.* **2018**, *5*, 180413. [CrossRef]

28. Fu, Y.; Wu, P.; Beane, T.; Zamore, P.D.; Weng, Z. Elimination of PCR Duplicates in RNA-seq and Small RNA-seq Using Unique Molecular Identifiers. *BMC Genom.* **2018**, *19*, 531. [CrossRef] [PubMed]
29. Green, C.D.; Ma, Q.; Manske, G.L.; Shami, A.N.; Zheng, X.; Marini, S.; Moritz, L.; Sultan, C.; Gurczynski, S.J.; Moore, B.B.; et al. A Comprehensive Roadmap of Murine Spermatogenesis Defined by Single-Cell RNA-Seq. *Dev. Cell* **2018**, *46*, 651–667. [CrossRef]
30. Zou, X.; Lu, T.; Zhao, Z.; Liu, G.; Lian, Z.; Guo, Y.; Sun, B.; Liu, D.; Li, Y. Comprehensive Analysis of MRNAs and MiRNAs in the Ovarian Follicles of Uniparous and Multiple Goats at Estrus Phase. *BMC Genom.* **2020**, *21*, 267. [CrossRef]
31. Zi, X.; Lu, J.; Zhou, H.; Ma, L.; Xia, W.; Xiong, X.; Lan, D.; Wu, X. Comparative Analysis of Ovarian Transcriptomes between Prolific and Non-prolific Goat Breeds via High-throughput Sequencing. *Reprod. Domest. Anim.* **2018**, *53*, 344–351. [CrossRef] [PubMed]
32. Hong, L.; Hu, Q.; Zang, X.; Xie, Y.; Zhou, C.; Zou, X.; Li, Y.; Deng, M.; Guo, Y.; Liu, G.; et al. Analysis and Screening of Reproductive Long Non-coding RNAs Through Genome-Wide Analyses of Goat Endometrium During the Pre-attachment Phase. *Front. Genet.* **2020**, *11*, 568017. [CrossRef]
33. Liu, H.; Wang, C.; Li, Z.; Shang, C.; Zhang, X.; Zhang, R.; Wang, A.; Jin, Y.; Lin, P. Transcriptomic Analysis of STAT1/3 in the Goat Endometrium During Embryo Implantation. *Front. Vet. Sci.* **2021**, *8*, 757759. [CrossRef]
34. Bo, D.; Jiang, X.; Liu, G.; Hu, R.; Chong, Y. RNA-Seq Implies Divergent Regulation Patterns of LincRNA on Spermatogenesis and Testis Growth in Goats. *Animals* **2021**, *11*, 625. [CrossRef] [PubMed]
35. Zhao, J.; Yang, P.; Yang, H.; Wang, Z.; El-Samahy, M.A.; Wang, F.; Zhang, Y. Dietary Supplementation with Metformin Improves Testis Function and Semen Quality and Increases Antioxidants and Autophagy Capacity in Goats. *Theriogenology* **2022**, *188*, 79–89. [CrossRef]
36. Gong, W.; Pan, L.; Lin, Q.; Zhou, Y.; Xin, C.; Yu, X.; Cui, P.; Hu, S.; Yu, J. Transcriptome Profiling of the Developing Postnatal Mouse Testis Using Next-generation Sequencing. *Sci. China* **2013**, *56*, 1–12. [CrossRef]
37. Shen, F.; Geng, Y.; Zhang, L.; Luo, L.; Yan, G.; Hou, R.; Yue, B.; Zhang, X. Transcriptome Analysis Reveals the Alternative Splicing Changes in the Immune-Related Genes of the Giant Panda (*Ailuropoda melanoleuca*), in Response to the Canine Distemper Vaccine. *Zool. Sci.* **2022**, *39*, 275. [CrossRef]
38. Pan, Q.; Shai, O.; Lee, L.J.; Frey, B.J.; Blencowe, B.J. Deep surveying of alternative splicing complexity in the human transcriptome by high-throughput sequencing. *Nat. Genet.* **2008**, *40*, 1413–1415. [CrossRef] [PubMed]
39. Hagiwara, M.; Kuroyanagi, H. 2SE01 Regulatory Mechanisms of Alternative Splicing in Metazoans. *Seibutsu Butsuri* **2005**, *45*, S23. [CrossRef]
40. Liu, X.; He, P.; Zhang, Z.; Gong, P.; Niu, Y.; Bao, Z.; Yang, Y.; Gan, L.; Muhuyati. Candidate Genes and Their Alternative Splicing May Be Potential Biomarkers of Acute Myocardial Infarction: A Study of Mouse Model. *BMC Cardiovasc. Disord.* **2022**, *22*, 505. [CrossRef]
41. Zhu, Z.; Ni, S.; Zhang, J.; Yuan, Y.; Bai, Y.; Yin, X.; Zhu, Z. Genome-wide Analysis of Dysregulated RNA-binding Proteins and Alternative Splicing Genes in Keloid. *Front. Genet.* **2023**, *14*, 1118999. [CrossRef]
42. Schomburg, L.; Liao, X.; Majed, F.A.; Abdullah, M.S.Y.; Refetoff, S.; Boran, G.; Dumitrescu, A.M.; Lado-Abeal, J.; Moeller, L.C.; Weiss, R.E. Mutations in SECISBP2 Result in Abnormal Thyroid Hormone Metabolism. *Nat. Genet.* **2005**, *37*, 1247–1252.
43. Schoenmakers, E.; Chatterjee, K. Identification of Genetic Disorders Causing Disruption of Selenoprotein Biosynthesis. *Methods Mol. Biol.* **2018**, *1661*, 325–335.
44. Mohammadabadi, T.; Tabatbaei, S.; Ghezi, Z.; Swelum, A.A. Effect of Dietary Palm Kernel on Semen Quality, Reproductive and Thyroid Hormones and Blood Chemistry Parameters of Arabi Rams. *Anim. Nutr. Feed Technol.* **2020**, *20*, 93. [CrossRef]
45. Mogheiseh, A.; Vara, N.; Ayaseh, M.; Jalali, P. Effects of Cabergoline on Thyroid Hormones and Semen Quality of Dog. *Top. Companion Anim. Med.* **2017**, *32*, 13–15. [CrossRef]
46. Quartuccio, M.; Fazio, E.; Medica, P.; Cristarella, S.; Emmanuele, G.; Sinagra, L.; Liotta, L. Correlation between Sperm Parameters and Circulating Thyroid Hormones and Testosterone Concentrations in Labrador Retriever Dog. *Ital. J. Anim. Sci.* **2021**, *20*, 947–954. [CrossRef]
47. Majdic, G.; Snoj, T.; Horvat, A.; Mrkun, J.; Kosec, M.; Cestnik, V. Higher Thyroid Hormone Levels in Neonatal Life Result in Reduced Testis Volume in Postpubertal Bulls. *Int. J. Androl.* **1998**, *21*, 352–357. [CrossRef] [PubMed]
48. Ghasemzadeh-Hasankolaei, M.; Sedighi-Gilani, M.A.; Eslaminejad, M.B. Induction of Ram Bone Marrow Mesenchymal Stem Cells into Germ Cell Lineage Using Transforming Growth Factor-β Superfamily Growth Factors. *Reprod. Domest. Anim.* **2014**, *49*, 588–598. [CrossRef] [PubMed]
49. Rojas-García, P.P.; Recabarren, M.P.; Sir-Petermann, T.; Rey, R.; Palma, S.; Carrasco, A.; Perez-Marin, C.C.; Padmanabhan, V.; Recabarren, S.E. Altered Testicular Development as a Consequence of Increase Number of Sertoli Cell in Male Lambs Exposed Prenatally to Excess Testosterone. *Endocrine* **2013**, *43*, 705–713. [CrossRef]
50. Ingman, W.V.; Robertson, S.A. The Essential Roles of TGFB1 in Reproduction. *Cytokine Growth Factor Rev.* **2009**, *20*, 233–239. [CrossRef]
51. McGrath, L.J.; Ingman, W.V.; Robker, R.L.; Robertson, S.A. Exogenous Transforming Growth Factor Beta1 Replacement and Fertility in Male Tgfb1 Null Mutant Mice. *Reprod. Fertil. Dev.* **2009**, *21*, 561–570. [CrossRef] [PubMed]
52. Pan, Y.; Cui, Y.; Yu, S.; Zhang, Q.; Fan, J.; Abdul, R.B.; Yang, K. The Expression of Epidermal Growth Factor (EGF) and Its Receptor (EGFR) During Post-Natal Testes Development in the Yak. *Reprod. Domest. Anim.* **2014**, *49*, 970–977. [CrossRef] [PubMed]

53. Kassab, M.; Abd-Elmaksoud, A.; Ali, M.A. Localization of the Epidermal Growth Factor (EGF) and Epidermal Growth Factor Receptor (EGFR) in the Bovine Testis. *J. Mol. Histol.* **2007**, *38*, 207–214. [CrossRef]
54. Cannarella, R.; Condorelli, R.A.; La Vignera, S.; Calogero, A.E. Effects of the Insulin-like Growth Factor System on Testicular Differentiation and Function: A Review of the Literature. *Andrology* **2018**, *6*, 3–9. [CrossRef] [PubMed]
55. Pitetti, J.; Calvel, P.; Zimmermann, C.; Conne, B.; Papaioannou, M.D.; Aubry, F.; Cederroth, C.R.; Urner, F.; Fumel, B.; Crausaz, M.; et al. An Essential Role for Insulin and IGF1 Receptors in Regulating Sertoli Cell Proliferation, Testis Size, and FSH Action in Mice. *Mol. Endocrinol.* **2013**, *27*, 814–827. [CrossRef]
56. Cannarella, R.; La Vignera, S.; Condorelli, R.A.; Calogero, A.E. The IGF1/FSH Ratio Correlates with Sperm Count and Testicular Volume. *Endocrines* **2022**, *3*, 624–632. [CrossRef]
57. Müller, L.; Kowalewski, M.P.; Reichler, I.M.; Kollár, E.; Balogh, O. Different Expression of Leptin and IGF1 in the Adult and Prepubertal Testis in Dogs. *Reprod. Domest. Anim.* **2017**, *52*, 187–192. [CrossRef]
58. Radovic, S.M.; Starovlah, I.M.; Capo, I.; Miljkovic, D.; Nef, S.; Kostic, T.S.; Andric, S.A. Insulin/IGF1 Signaling Regulates the Mitochondrial Biogenesis Markers in Steroidogenic Cells of Prepubertal Testis, but Not Ovary. *Biol. Reprod.* **2019**, *100*, 253–267. [CrossRef] [PubMed]
59. Neirijnck, Y.; Calvel, P.; Kilcoyne, K.R.; Kühne, F.; Stévant, I.; Griffeth, R.J.; Pitetti, J.; Andric, S.A.; Hu, M.; Pralong, F.; et al. Insulin and IGF1 Receptors Are Essential for the Development and Steroidogenic Function of Adult Leydig Cells. *FASEB J.* **2018**, *32*, 3321–3335. [CrossRef]
60. Rojas, M.; Conei, D.; Bustos-Obregon, E. Epithelial-Mesenchymal Transitions in the Development of Testis/Interacciones Epitelio-Mesenquimaticas En El Desarrollo Testicular. *Int. J. Morphol.* **2017**, *35*, 1444. [CrossRef]
61. Stratikopoulos, E.; Szabolcs, M.; Dragatsis, I.; Klinakis, A.; Efstratiadis, A. Hormonal Action of IGF1 in Postnatal Mouse Growth. *Proc. Natl. Acad. Sci. USA* **2008**, *105*, 19378–19383. [CrossRef]
62. Pitetti, J.; Calvel, P.; Romero, Y.; Conne, B.; Vy, T.; Papaioannou, M.D.; Schaad, O.; Docquier, M.; Herrera, P.L.; Wilhelm, D.; et al. Insulin and IGF1 Receptors Are Essential for XX and XY Gonadal Differentiation and Adrenal Development in Mice. *PLoS Genet.* **2013**, *9*, E1003160. [CrossRef] [PubMed]
63. Warr, N.; Carre, G.; Siggers, P.; Faleato, J.V.; Brixey, R.; Pope, M.; Bogani, D.; Childers, M.; Wells, S.; Scudamore, C.L.; et al. Gadd45γ and Map3k4 Interactions Regulate Mouse Testis Determination via P38 MAPK-Mediated Control of Sry Expression. *Dev. Cell* **2012**, *23*, 1020–1031. [CrossRef] [PubMed]
64. Liu, J.; Ren, L.; Wei, J.; Zhang, J.; Zhu, Y.; Li, X.; Jing, L.; Duan, J.; Zhou, X.; Sun, Z. Fine Particle Matter Disrupts the Blood–testis Barrier by Activating TGF-β3/p38 MAPK Pathway and Decreasing Testosterone Secretion in Rat. *Environ. Toxicol.* **2018**, *33*, 711–719. [CrossRef]
65. Shen, L.; Tang, X.; Wei, Y.; Long, C.; Tan, B.; Wu, S.; Sun, M.; Zhou, Y.; Cao, X.; Wei, G. Vitamin E and Vitamin C Attenuate Di-(2-ethylhexyl) Phthalate-induced Blood-testis Barrier Disruption by P38 MAPK in Immature SD Rats. *Reprod. Toxicol.* **2018**, *81*, 17–27. [CrossRef]
66. Luo, D.; He, Z.; Yu, C.; Guan, Q. Role of P38 MAPK Signalling in Testis Development and Male Fertility. *Oxidative Med. Cell. Longev.* **2022**, *2022*, 6891897. [CrossRef]
67. Wang, X.; Zhang, X.; Sun, K.; Wang, S.; Gong, D. Polystyrene Microplastics Induce Apoptosis and Necroptosis in Swine Testis Cells via ROS/MAPK/HIF1α Pathway. *Environ. Toxicol.* **2022**, *37*, 2483–2492. [CrossRef] [PubMed]
68. Gonzalez, C.R.; Dorfman, V.B.; Vitullo, A.D. IGF1 Regulation of BOULE and CDC25A Transcripts via a Testosterone-independent Pathway in Spermatogenesis of Adult Mice. *Reprod. Biol.* **2015**, *15*, 48–55. [CrossRef] [PubMed]
69. Wang, H.; Shan, B.; Duan, Y.; Zhu, J.; Jiang, L.; Liu, Y.; Zhang, Y.; Qi, F.; Niu, S. Effects of Heshouwuyin on Gene Expression of the Insulin/IGF Signalling Pathway in Rat Testis and Spermatogenic Cells. *Pharm. Biol.* **2020**, *58*, 1208–1219. [CrossRef]

genes

MDPI

Article

Salidroside Supplementation Affects In Vitro Maturation and Preimplantation Embryonic Development by Promoting Meiotic Resumption

Shuming Shi [1,†], Zhaojun Geng [1,†], Xianfeng Yu [1], Bing Hu [2], Liying Liu [1], Zhichao Chi [1], Linyi Qu [1], Mingjun Zhang [2,*] and Yongxun Jin [1,*]

[1] Jilin Provincial Key Laboratory of Animal Model, College of Animal Science, Jilin University, Changchun 130062, China; shism20@mails.jlu.edu.cn (S.S.); gengzj20@mails.jlu.edu.cn (Z.G.); xianfeng79@jlu.edu.cn (X.Y.); liuliying22@mails.jlu.edu.cn (L.L.); chizc9919@mails.jlu.edu.cn (Z.C.); quly9919@mails.jlu.edu.cn (L.Q.)

[2] Animal Genome Editing Technology Innovation Center, College of Animal Science, Jilin University, Changchun 130062, China; hubing21@mails.jlu.edu.cn

* Correspondence: mjzhang@jlu.edu.cn (M.Z.); jyx0429@126.com (Y.J.)

† These authors contributed equally to this work.

Simple Summary: The in vitro maturation (IVM) of oocytes is often delayed due to oxidative stress and mitochondrial function, resulting in a decrease in oocyte quality. We used a porcine oocyte model to check the effects of salidroside supplementation during IVM on the quality and developmental competence of oocytes. Using a variety of experimental methods, the efficiency of salidroside in protecting mature oocytes, apoptosis, and mitochondrial function was observed.

Abstract: Salidroside (Sal) possesses several pharmacological activities, such as antiaging, and anti-inflammatory, antioxidant, anticancer activities, and proliferation-promoting activities, but the effects of Sal on oocytes have rarely been reported. In the present study, we evaluated the beneficial effects of Sal, which is mainly found in the roots of Rhodiola. Porcine cumulus oocyte complexes were cultured in IVM medium supplemented (with 250 μmol/L) with Sal or not supplemented with Sal. The maturation rate in the Sal group increased from 88.34 ± 4.32% to 94.12 ± 2.29%, and the blastocyst rate in the Sal group increased from 30.35 ± 3.20% to 52.14 ± 7.32% compared with that in the control group. The experimental groups showed significant improvements in the cumulus expansion area. Sal reduced oocyte levels of reactive oxygen species (ROS) and enhanced intracellular GSH levels. Sal supplementation enhanced the mitochondrial membrane potential (MMP), ATP level, and mtDNA copy number, which shows that Sal enhances the cytoplasmic maturation of oocytes. Oocytes in the Sal group exhibited slowed apoptosis and reduced DNA breakage. Cell cycle signals and oocyte meiosis play important roles in oocyte maturation. The mRNA expressions of the MAPK pathway and MAPK phosphorylation increased significantly in the Sal group. The mRNA expression of the oocyte meiosis gene also increased significantly. These results show that Sal enhances the nuclear maturation of oocytes. Moreover, Sal increased the number of blastocyst cells, the proliferation of blastocysts, and the expressions of pluripotency genes. Sal down-regulated apoptosis-related genes and the apoptotic cell rate of blastocysts. In summary, our results demonstrate that Sal is helpful to improving the quality of porcine oocytes in vitro, and their subsequent embryonic development.

Keywords: embryos; MAPK; maturation; mitochondria; oocytes; Sal; ROS

Citation: Shi, S.; Geng, Z.; Yu, X.; Hu, B.; Liu, L.; Chi, Z.; Qu, L.; Zhang, M.; Jin, Y. Salidroside Supplementation Affects In Vitro Maturation and Preimplantation Embryonic Development by Promoting Meiotic Resumption. *Genes* **2023**, *14*, 1729. https://doi.org/10.3390/genes14091729

Academic Editor: Bao Yuan

Received: 9 August 2023
Revised: 24 August 2023
Accepted: 27 August 2023
Published: 30 August 2023

1. Introduction

In vitro porcine embryo production (IVP) is now being used in the development of gamete/embryo biology and agriculture, as well as in the production of cloned and transgenic pigs. Therefore, it has certain advantages for research and commercial use [1,2].

Even though the porcine in vitro maturation system has been established, the maturation rate is not ideal. Due to its sensitivity to the in vitro maturation environment, the quality of oocytes plays a decisive role in embryo quality, and it is necessary to further improve the maturation rate [3,4], lay a foundation for the improvement of human embryo reproduction technology, and strengthen its application in agriculture and medical research [5].

In vitro maturation (IVM) is a complex process controlled by multiple intrinsic and extrinsic factors, studies in mice have shown that perturbations of mitochondrial function affect the maturation of oocytes [6]. Oxidative stress has become a damaging factor of oocytes [7]. Currently, to promote the production of in vitro embryos, supplements are used during the in vitro maturation of oocytes such as antioxidants, which can counteract some of the negative effects of oxidative stress [8]. Mitogen-activated protein kinases (MAPKs) are serine/threonine protein kinases that widely exist in mammalian cells [9]. The MAPK signaling pathway plays an important role in cell proliferation, differentiation, cell survival, and apoptosis [10]. During the process of oocyte maturation, MAPK maintains the stability of the spindle structure and promotes oocyte maturation [5].

Sal is mainly found in the roots of Rhodiola. It has been extensively used for medicinal purposes, because it has many biological and pharmacological properties, and has many active properties [11], such as anticancer [12], antioxidant [13], antiaging, antidiabetic, antidepression, antilipid, anti-inflammatory immune regulation [14], promoting multiplication, and so on [15]. Sal influences the PI3K/Akt, AMPK and MEK/ERK pathways, which are the main reasons for its promotion of proliferation [15,16]. Sal has well-documented beneficial biological and physiological functions such as anticancer, anti-oxidative, and slowing down apoptosis. However, there is little information available in the literature on the mode of action of Sal during oocyte IVM.

Therefore, we hypothesized that Sal would enhance the developmental potential of oocytes because it has a protective effect. More importantly, the molecular mechanisms by which Sal affect oocyte maturation are still unknown. In this study, we used pigs as a model for evaluating whether adding Sal to the culture medium can improve IVM oocyte quality and subsequent embryonic development. We first examined the first polar body excretion and parthenogenetic activation of embryonic development, and then assessed oocytes and preimplantation embryos' molecular changes, factors affecting cytoplasmic maturation including ROS levels, mitochondrial membrane potential (MMP), and mtDNA copy number, factors affecting nuclear maturation including proliferation and apoptosis levels, and MAPK level changes.

2. Materials and Methods

All reagents except those specially marked were purchased from Sigma Aldrich (St. Louis, MO, USA).

2.1. Oocyte Collection and Culture

Porcine ovaries were collected from the local slaughterhouse, and then transported to the laboratory in 37 °C 0.9% saline in a thermos. Follicular fluid was extracted from these 3–6 mm follicles using a syringe with a 1.2 mm needle, and collected under a stereomicroscope (SMZ800N, Nikon, Tokyo, Japan). This experiment used cumulus oocyte complexes (COCs), which are oocytes with three or more layers of cumulus cells that are evenly distributed on the surface. The subsequent COCs were washed three times in medium 199 containing HEPES (GIBCO, Waltham, MA, USA, 12350039), 0.1% PVA, and cultured in IVM (tissue culture medium 199 supplemented with 0.6 mM L-cysteine, 0.91 mM Na pyruvate, 10 ng/mL EGF, 10% porcine follicular fluid, 10 IU/mL follicle-stimulating hormone, 10 IU/mL luteinizing hormone, and 0.1% penicillin streptomycin), and in order to explore the optimal dosing concentration of the treatment group, 50 oocytes were placed in 500 μL of IVM supplemented with 50 μM, 250 μM, and 400 μM Sal. They were equally divided into treated and untreated groups, and were added to the treatment group of 250 μM Sal covered with mineral oil incubate in 38.5 °C under 5% CO_2 and 100% moisture for 42 h.

A picture for each COC was taken before and at the end of the in vitro culture, and the cumulus expansion was calculated as the ratio between the final cumulus area and the initial cumulus area. The cumulus area was measured by analysis via NIH ImageJ software 1.52a (National Institutes of Health, Bethesda, MD, USA).

2.2. Parthenogenetic Activation

The cumulus cells were removed with 0.1% hyaluronidase, and the oocytes without cumulus cells were treated in 5 mm ionomycin (pzm-5 configuration) for 5 min, then washed in pzm-5 three times. Then, the embryos were placed in bicarbonate-buffered porcine zygote medium 5 (PZM-5) containing 0.4 mg/mL bovine serum albumin (BSA) and 7.5 μg/mL CB for 4 h to inhibit the excretion of the pseudo-second polar body. The blastocyst rates and diameters were analyzed on day 6, and the blastocyst diameters were analyzed via NIH ImageJ software 1.52a (National Institutes of Health, Bethesda, MD, USA).

2.3. Count the Number of Cells in the Blastocyst

To quantify the total number of cells in the blastocysts, the blastocysts of the treatment group and untreated group on the seventh day were selected and fixed with 4% paraformaldehyde with pH = 7.4. Then, the fixed blastocysts were washed with PBS–PVA (0.1% PVA) three times, and placed into the permeabilization solution containing 0.5% TritonX-100 for 20 min at room temperature. The blastocysts were again washed with PBS–PVA three times; the 5 ng/mL Hoechst 33,342 stained nuclei were incubated at room temperature for 10 min, then gently placed on slides and sealed with an anti-fluorescence quencher. Next, the blastocysts were gently mounted onto glass slides and examined under a fluorescence microscope (IX73, Olympus, Tokyo, Japan). NIH ImageJ software was used to analyze the total number of nuclei.

2.4. Evaluation of ROS and GSH Levels in Oocytes

To measure the ROS and GSH levels in the M2 oocytes of the treatment group and untreated group, the M2 oocytes in each group were divided into four groups and cultured in PBS–BSA medium containing 10 mM 20, 70-dichlorodihydrofluorescein diacetate (Thermo Fisher Scientific, Waltham, MA, USA, C400), and 10 mM 4-chloromethyl-6,8-difluoro-7-hydroxycoumarin (Thermo Fisher Scientific, C12881) for 15 and 30 min. Then, the oocytes were washed with 0.1% PBS–BSA for 3 times. A digital camera (E179168; Nikon) was connected to a fluorescence microscope to take pictures, with the fluorescence photos saved as TIFF files, and the fluorescence intensity was analyzed using NIH ImageJ software.

2.5. Mitochondrial Membrane Potential Assay

To measure the mitochondrial membrane potential (MMP), M2-stage oocytes were washed three times with PBS–PVA, and then washed with 2 mM 5,50,6,60-tetrachloro-1,10,3,30-tetraethyl-imidacarbocyanine iodide (JC-1, Beyotime, Haimen, China) for 30 min. The oocytes were again washed three times with PBS–PVA, and a fluorescence microscope was used to connect to the camera and capture the red/green fluorescence signal. The MMP was calculated as the ratio of red fluorescence and analyzed according to the manufacturer's instructions. The fluorescence intensity of the images was processed and measured using NIH ImageJ software.

2.6. Determination of ATP Content

The oocyte ATP content was determined using an Enhanced ATP Assay Kit (Beyotime, China) according to the manufacturer's instructions, and the results are shown in the microplate reader.

2.7. Mitochondrial Copy Number Detection

The total DNA was extracted from 200 MII-stage oocytes using a DNA isolation kit (QIAGEN, Hilden, Germany, 80284) according to the instructions. Through fluorescence quantification to detect the mitochondrial copy number, its principle is to extract mitochondrial *ATP6*, *ND1*, and *COX1* genes, and the nuclear single-copy gene *GCG* was used to detect the relative concentrations of the mitochondrial and nuclear genomes. Since the genes in mitochondria are single-copy genes, the mitochondrial gene and single-copy nuclear gene relative concentration ratio can be used to determine a single tissue cell. Quantitative fluorescence PCR was performed using the total DNA as the template, and each sample was quantified using specific primers for mitochondrial genes and nuclear single-copy genes. The statistics using the formula for the PCR primers used to amplify each gene are shown in Table 1.

Table 1. Primer sequences of Mitochondrial DNA copy number in porcine oocyte for RT-PCR.

Gene Name	Sequence	Amplicon Size (bp)	Accession Number
ATF6	F: TATTTGCCTCTTTCATTGCCC R: GGATCGAGATTGTGCGGTTAT	123	NC_000845.1
ND1	F: GCCACATCCTCAATCTCCAT R: GATTAGAGGGTAGGGTATTGGTAG	99	NC_000845.1
COX1	F: TCCAATGGACATTATGGCTC R: GAAGACATCTCGGCTGAACT	220	NC_010443.5
GCG	F: GAATCAACACCATCGGTCAAAT R: CTCCACCCATAGAATGCCCAGT	198	NC_010457.5
GADPH	F: CCCAGAATATCATCCCTGCT R: CTGCTTCACCACCTTCTTGA	185	NC_010447.5

F, forward; R, reverse.

For the quantification of the mitochondrial DNA copy number, using the single-copy nuclear gene *GCG* as the reference gene, the multiple of the mitochondrial gene expression relative to *GCG* is the mitochondrial copy number of cells in the tissue. The calculation formula is as follows:

$$N = 2^{-Ctmt}/2^{-Ctn}$$

N: mtDNA copy number; Ctmt: Ct value for mitochondrial genes; Ctn: Ct value for nuclear single-copy genes.

2.8. Quantitative Real-Time PCR (qRT-PCR)

The total mRNA was extracted from 200 M2 stage oocytes and 50 blastocysts using an mRNA isolation kit (QIAGEN, Germany, 80284) according to the manufacturer's instructions. The gene expressions were quantified using the CFX96 real-time system (Bio-Rad, Harkles, CA, USA). The internal control group was the expression level of GAPDH. The statistics for the formula for the PCR primers used to amplify each gene are shown in Table 2.

Table 2. The names of the tested genes, sequences of primers, sizes of PCR products, and accession numbers for RT-qPCR experiments.

Gene Name	Sequence	Amplicon Size (bp)	Accession Number
Caspase3	F: GTGGGATTGAGACGGACAGTGG R: TTCGCCAGGAATAGTAACCAGGTG	114	NM_214131.1
BCL-2	F: TGGATGACCGAGTACCTGAA R: CAGCCAGGAGAAATCAAACA	120	NM_001166486.1

Table 2. *Cont.*

Gene Name	Sequence	Amplicon Size (bp)	Accession Number
BAX	F: CACCAAGAAGCTGAGCGAGTGT R: TCGGAAAAAGACCTCTCGGGGA	118	NM_173894
COX2	F: GGCTGCGGGAACATAATAGA R: GCAGCTCTGGGTCAAACTTC	183	NM_214321.1
MEK	F: TCATCGACTCCATGGCCAAC R: AGATGTCCGACTGCACGGAGTA	96	NM_001244550.1
ERK1	F: AGCCCTTTTGAGCATCAGACC R: AATGACGTTCTCGTGGCGG	84	XM_021088019.1
ERK2	F: CAAACCTTCCAACCTGCTGC R: TACTCCGTCAGGAACCCTGT	111	XM_021071922.1
C-MOS	F: GGTGGTGGCCTACAATCTCC R: TCAGCTTGTAGAGCGCGAAG	165	NM_001113219.1
GDF9	F: GTCTCCAACAAGAGAGAGATTC R: CTGCCAGAAGAGTCATGTTAC	109	NM_001001909.1
CDK1	F: TAATAAGCTGGGATCTACCACATC R: TGGCTACCACTTGACCTGTA	130	NM_001159304.2
Cyclin B	F: AGCTAGTGGTGGCTTCAAGG R: GCGCCATGACTTCCTGTA	101	NM_001170768.1
NANOG	F: AGGACAGCCCTGATTCTTCCACAA R: AAAGTTCTTGCATCTGCTGGAGGC	198	XM_021092390.1
CDX2	F: AGCCAAGTGAAAACCAGGAC R: TGCGGTTCTGAAACCAGATT	178	NM_001278769.1
SOX2	F: GCGGAGTGGAAACTTTTGTCC R: CGGGAAGCGTGTACTTATCCTT	157	NM_001123197.1
OCT4	F: GGCTTCAGACTTCGCCTCC R: AACCTGAGGTCCACAGTATGC	226	XM_021097869.1
Caspase9	F: GGCCACTGCCTCATCATCAA R: GGAGGTGGCTGGCCTTG	163	XM_013998997.2
GADPH	F: CCCAGAATATCATCCCTGCT R: CTGCTTCACCACCTTCTTGA	185	NM_001034034.2

F, forward; R, reverse.

2.9. Western Blotting

A total of 200 M2 stage oocytes were collected, and the protein was extracted by RIPA (Beyotime, P0013 K); a 10–15% sodium dodecyl sulfate (SDS) separation gel and a concentration gel were prepared. The diluted primary antibodies H_2AX (1:2000, 10856-1-AP, Proteintech, Rosemont, IL, USA) and p44/42 MAPK (1:1000, 9101, Cell Signaling) were added to the membrane and incubated overnight at 4 °C. The membranes were washed five times with TBST (10 min each time), supplemented with horseradish peroxidase-marked secondary antibody (1:5000; Proteintech, SA00001–1), oscillated, and then incubated at room temperature for 1 h. After incubation, each membrane was washed five times with TBST (10 min each time) and reacted with enhanced chemiluminescence solution (Tanon, 180–5001) at room temperature for 5 s, and the relative expressions of target proteins were measured using GAPDH (1:5000, 10494-1-AP, Proteintech) as the reference.

2.10. Assessment of Blastocyst Apoptosis

For the TUNEL (Beyotime, C1086S) assays, the blastocysts were incubated with 50 μL of TUNEL reaction mixture for 1 h at 37 °C in the dark. A fluorescence microscope connected to a digital camera was used to take pictures to capture the fluorescence signal. The images were processed and measured using NIH ImageJ software.

2.11. Statistics of Blastocyst Cell Proliferation

The propagation of blastocyst cells was cytochemically detected according to the manufacturer's instructions (C0071L, Beyotime, China). Briefly, the blastocysts were incubated with the EdU staining buffer for 1 h, fixed with 4% polyformaldehyde, and stained for nuclei with Hoechst. The stained blastocysts were scanned and photographed under a microscope, and the numbers of positive cells were counted.

2.12. Statistical Analysis

All of the data were analyzed using SPSS software version 11.0 (IBM, Armonk, NY, USA). All of the experiments were performed more than 3 times. Comparisons of data between groups were performed with Student's *t*-test. The values are expressed as the mean ± standard deviation, and significant differences are represented with * ($p < 0.05$), ** ($p < 0.01$), and *** ($p < 0.001$).

3. Results

3.1. Protective Effects of Sal at Different Concentrations on Porcine Oocytes

To determine the optimal concentration of Sal, different concentrations of Sal (0 μM, 50 μM, 250 μM, 400 μM) were added to the IVM culture medium. After 44 h of culture, the results showed that compared with the control group, Sal had the highest maturity rate at the concentration of 250 μM (Figure 1A,C 88.34 ± 4.32% vs. 88.83 ± 2.36% vs. 94.12 ± 2.29% vs. 89.71 ± 4.25%). The cleavage rate after 30 h showed statistically significant parthenogenetic activation, but there was no significant difference in the cleavage rate (Figure 1D 81.78 ± 5.18% vs. 86.40 ± 5.16% vs. 86.22 ± 6.65% vs. 84.76 ± 6.53%). The blastocyst rate was counted on the sixth day of parthenogenetic activation. The blastocyst rate was also the highest among oocytes under 250 μM Sal culture (Figure 1B,E 30.35 ± 3.20% vs. 38.89 ± 5.38% vs. 52.14 ± 7.32% vs. 28.58 ± 3.21%), and had the best protective effect on oocytes.

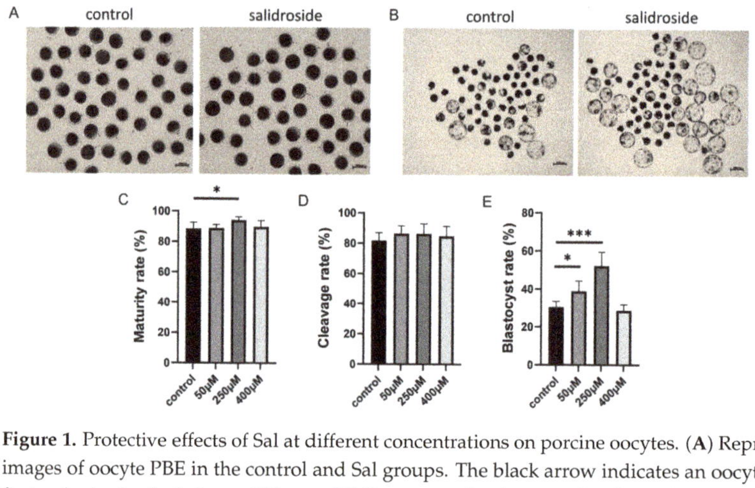

Figure 1. Protective effects of Sal at different concentrations on porcine oocytes. (**A**) Representative images of oocyte PBE in the control and Sal groups. The black arrow indicates an oocyte with the first polar body. Scale bar = 100 μm. (**B**) Representative images of embryo development on day 6 in the control and Sal groups. Scale bar = 200 μm. (**C**) Maturation rate of oocytes in the control (n = 292) and Sal groups (50 μM n = 301; 250 μM n = 296; 400 μM n = 294). (**D**) Cleavage rates of porcine parthenogenetic embryos at 30 h after parthenogenetic activation in the control (n = 281) and Sal groups (50 μM n = 286; 250 μM n = 272; 400 μM n = 278). (**E**) Blastocyst formation rates of porcine parthenogenetic embryos on day 6 in the control (n = 281) and Sal groups (50 μM n = 286; 250 μM n = 272; 400 μM n = 278). The data were obtained from three separate experiments. Values are expressed as the mean ± standard deviation, and significant differences are represented with * ($p < 0.05$) and *** ($p < 0.001$).

Based on the above results, a concentration of 250 μM was used in the follow-up experiment. Photographs of each COC were taken before and after in vitro culture, and the ratio between the final cumulus area and the initial cumulus area was calculated. The Sal treatment group significantly increased the area of cumulus expansion (Figure 2A,B). To investigate the reasons for the increase in cumulus expansion area, we collected digested cumulus cells for flow cytometry reactive oxygen species detection and mitochondrial DNA copy number detection, and found that the data in the treatment group were higher than those in the control group (Figure S1).

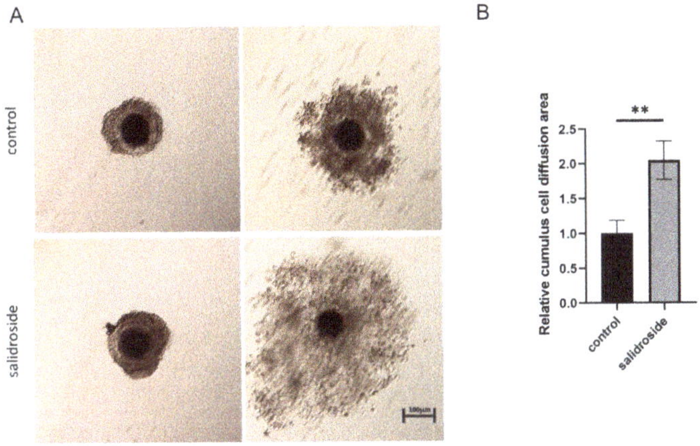

Figure 2. Protective effects of Sal at different concentrations on porcine oocytes. (**A**) Representative images of cumulus cell expansion status before and after IVM in the control and Sal groups. Scale bar = 100 μm. (**B**) Relative cumulus cell expansion areas in the control (n = 31) and Sal groups (n = 33). The data were obtained from three separate experiments. Significant differences are represented with ** ($p < 0.01$).

3.2. Effects of Sal Supplementation on the Levels of Oxidative Stress in Porcine Oocytes

Since Sal has free radical scavenging properties, we further evaluated whether Sal supplementation during IVM can improve the oxidative stress of porcine oocytes. As shown in Figure 3A,B, the ROS level in the porcine oocytes in the Sal group was significantly lower than that in the control group during the M2 stage, and Sal was not added during IVM. Based on these results, we further evaluated the level of GSH in porcine M2 oocytes, as shown in Figure 3C,D. Compared with the control group, the GSH level in the Sal-supplemented group was significantly higher.

3.3. Effects of Sal Supplementation on the Mitochondrial Function of Porcine Oocytes

Mitochondria are the energy factories of oocytes, one of the indicators of oocyte quality, and play an important role in cell metabolism, growth, and proliferation [17]. We used the JC-1 fluorescence reaction to evaluate the mitochondrial membrane potential (MMP) of the oocytes. Representative images of JC-1 staining are shown in Figure 4A. The quantitative results showed that the average relative value of fluorescence intensity (red/green) in the Sal-supplemented group was significantly higher than that in the control group (Figure 4B). We measured the concentration of ATP in oocytes between different treatment groups, and found that the ATP concentration in the treatment group treatment was significantly higher than that in the control group (Figure 4C). Some studies have suggested that the mitochondrial DNA (mtDNA) copy number could be used as a biomarker of human oocyte viability [18]. We measured that the mtDNA copy number in oocytes between different treatment groups was significantly higher than that in the control group (Figure 4D).

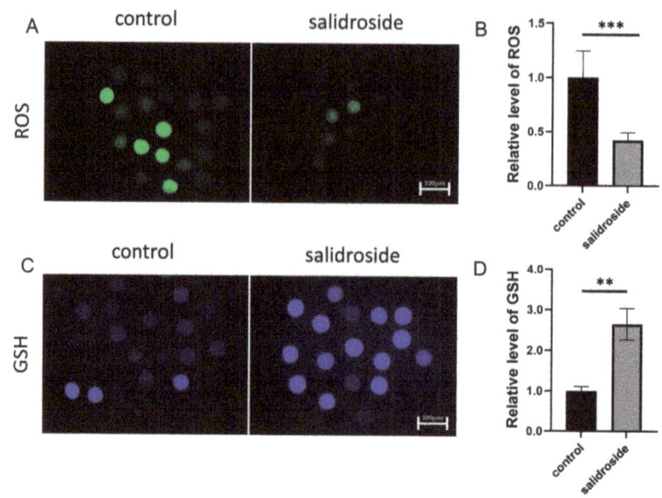

Figure 3. Effects of Sal supplementation on the levels of oxidative stress in porcine oocytes. (**A**) Representative fluorescence images of intracellular H_2DCFDA (ROS) levels in the control and Sal groups. Scale bar = 100 μm. (**B**) Representative fluorescence images of intracellular glutathione (ROS) levels in the control (n = 62) and Sal groups (n = 59). Scale bar = 100 μm. (**C**) Relative intracellular GSH levels in the control and Sal groups. (**D**) Relative intracellular GSH levels in the control (n = 53) and Sal groups (n = 54). Significant differences are represented with ** ($p < 0.01$) and *** ($p < 0.001$). The data were obtained from three separate experiments.

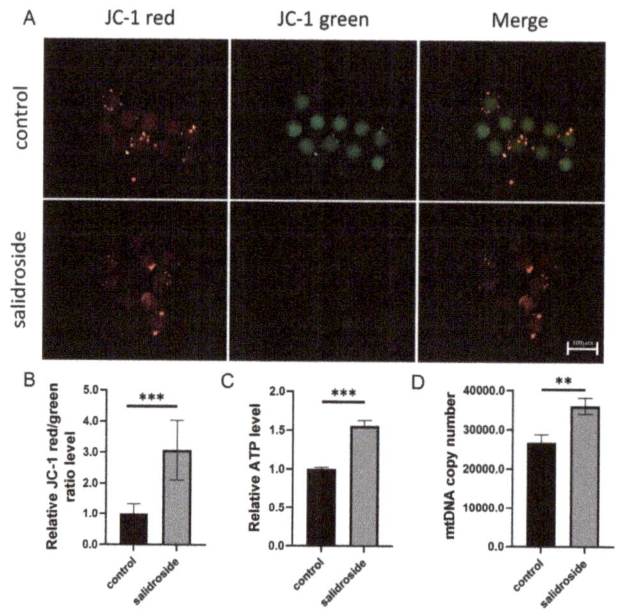

Figure 4. Effects of Sal supplementation on the mitochondrial function of porcine oocytes. (**A**) Representative fluorescence images of JC-1 fluorescence staining in the control and Sal groups. Scale bar = 100 μm. (**B**) Relative intracellular MMP levels in the control (n = 40) and Sal groups (n = 36). (**C**) Relative ATP levels in the control and Sal groups. (**D**) Mitochondrial DNA copy numbers in the control and Sal groups. The data were obtained from three separate experiments. Significant differences are represented with ** ($p < 0.01$) and *** ($p < 0.001$).

3.4. Effects of Sal Supplementation on Apoptosis and DNA Damage in Porcine Oocytes

Based on the mRNA expression, Sal significantly down-regulated the apoptosis related gene *Caspase-3* (Figure 5A) and the mRNA expression of *BAX/BCL-2* (Figure 5B), and significantly down-regulated the DNA breakage marker protein γH2AX (Figure 5C,D).

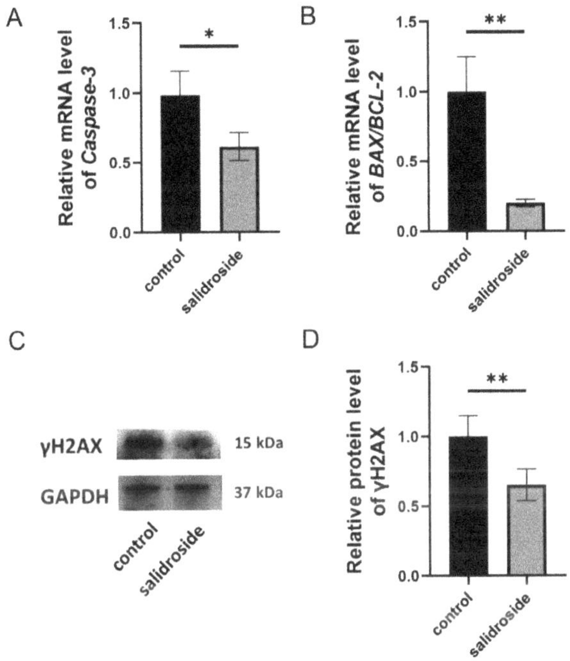

Figure 5. Effects of Sal supplementation on apoptosis and DNA damage in porcine oocytes. (**A**) Relative gene expressions of *Caspase-3* in the control and Sal groups. (**B**) Relative gene expressions of *BAX/BCL-2* in the control and Sal groups. (**C**) The protein expression levels of γH2AX in the control and Sal groups. (**D**) Relative levels of γH2AX in the control and Sal groups. The data were obtained from three separate experiments. Significant differences are represented with * ($p < 0.05$), and ** ($p < 0.01$).

3.5. Effects of Sal Supplementation on MAPK Activation and Meiosis in Porcine Oocytes

The porcine COCs were cultured for 44 h in IVM media containing 250 μM Sal, after cumulus cells were removed from the COCs, the *C-MOS*, *MEK*, *ERK1*, and *ERK2* mRNA expression levels in the oocytes were determined. We measured that the mRNA expressions in oocytes between different treatment groups were significantly higher than those in the control group (Figure 6C–F), and significantly increased the phosphorylated MAPK protein (Figure 6A,B). The mRNA expression of the oocyte maturation genes (*CDK1*, *Cyclin B*, and *GFP-1*) also increased significantly (Figure 6G).

3.6. The Addition of Sal Can Reduce the Apoptosis and Increase the Proliferation of PA Blastocysts

We analyzed the total cell numbers and the ability of blastocysts to proliferate on day 6. The analysis of the total blastocyst cell number showed that it was significantly increased in the Sal-supplemented group compared with the untreated group (Figure 7). The number of blastocyst proliferations in the treatment group was significantly higher than that in the untreated group (Figure 8A,B), and Sal significantly increased the expression of *COX-2*, which is involved in blastocyst formation, and the expression of the pluripotency genes *NANOG*, *CDX2*, *SOX2*, and *OCT-4* (Figure 8C).

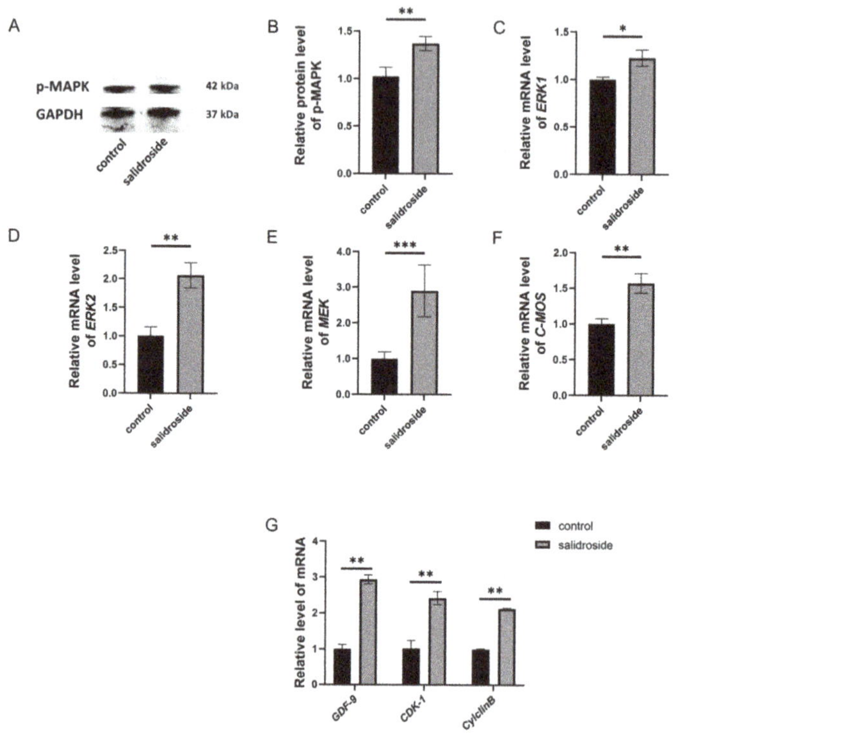

Figure 6. Effects of Sal supplementation on MAPK activation and meiosis of porcine oocytes. (**A**) The protein expression levels of p-MAPK in the control and Sal groups. (**B**) Relative levels of p-MAPK in the control and Sal groups. (**C**) Relative gene expressions of *ERK1* in the control and Sal groups. (**D**) Relative gene expressions of *ERK2* in the control and Sal groups. (**E**) Relative gene expressions of *MEK* in the control and Sal groups. (**F**) Relative gene expressions of *C-MOS* in the control and Sal groups. (**G**) Relative gene expressions of *GDF-9*, *CDK-1*, and *Cyclin B* in the control and Sal groups. The data were obtained from three separate experiments. Significant differences are represented with * ($p < 0.05$), ** ($p < 0.01$), and *** ($p < 0.001$).

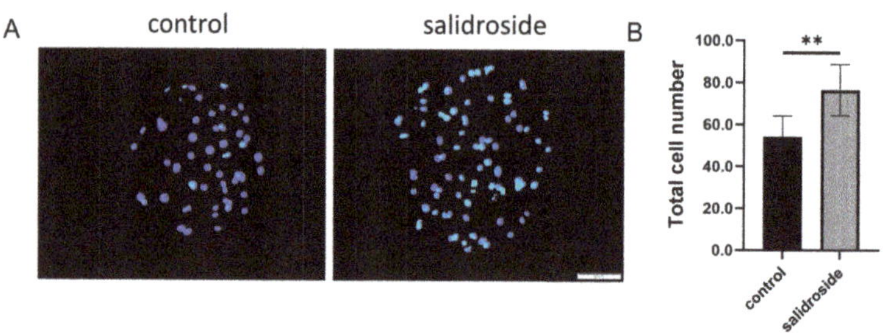

Figure 7. The addition of Sal can reduce the apoptosis of PA blastocysts and increase the proliferation of PA blastocysts. (**A**) Hoechst 33,342 staining of blastocysts on day 7 in the control and Sal groups. Scale bar = 50 μm. (**B**) Blastocyst total cell numbers in the control (n = 43) and Sal groups (n = 40). The data were obtained from three separate experiments. Significant differences are represented with ** ($p < 0.01$).

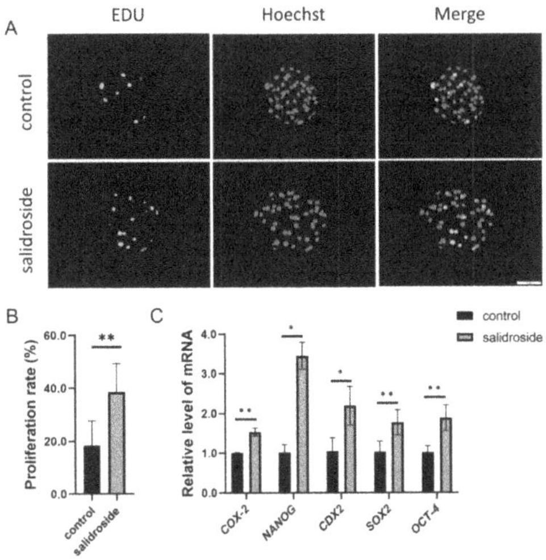

Figure 8. The addition of Sal can reduce the apoptosis of PA blastocysts and increase the proliferation of PA blastocysts. (**A**) Staining of EDU in blastocysts. Scale bar = 50 μm. (**B**) Rates of cell proliferation in blastocysts developing in the control (n = 61) and Sal groups (n = 67). (**C**) Relative gene expressions of *COX-2, NANOG, CDX2, SOX2,* and *OCT-4* at blastocyst stage in the control and Sal groups. The data were obtained from three separate experiments. Significant differences are represented with * ($p < 0.05$) and ** ($p < 0.01$).

Using TUNEL staining for the blastocysts obtained on the sixth day after parthenogenetic activation of oocytes in the treatment group and the untreated group, the number of apoptotic cells was significantly lower than that in the untreated group (Figure 9A,B), and down-regulated the apoptosis-related gene *Caspase-9* during the blastocyst stage (Figure 9C).

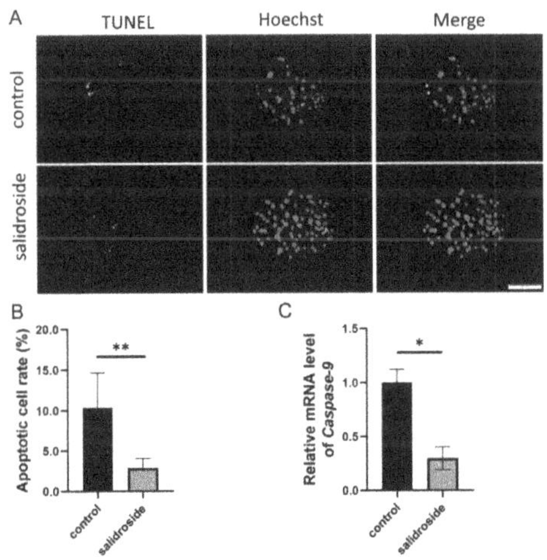

Figure 9. The addition of Sal can reduce the apoptosis of PA blastocysts and increase the proliferation of PA blastocysts. (**A**) Staining of TUNEL in blastocysts. Scale bar = 50 μm. (**B**) Rates of cell apoptosis in developing blastocysts in the control (n = 23) and Sal groups (n = 20). (**C**) Relative gene expressions of *Caspase-9* at the blastocyst stage in the control and Sal groups. The data were obtained from three separate experiments. Significant differences are represented with * ($p < 0.05$) and ** ($p < 0.01$).

4. Discussion

Sal is a tyrosol-like glycoside that mainly exists in the roots of Rhodiola rosea plants, and has various biological and pharmacological properties. Numerous studies have been conducted over the past decade to investigate the medicinal properties of Sal, and anticancer, antioxidative, antiaging, antidiabetic, antidepressant, antihyperlipidemic, anti-inflammatory, and immunomodulatory effects were reported [19]. Although there have been studies indicating that Sal improves the lipid content in oocytes [20], we found that Sal promotes oocyte maturation while increasing mitochondrial activity, while reducing ROS and GSH levels in oocytes. This is reflected in aspects such as maturation, metabolism, cell communication, apoptosis, and oxidative stress. Our results provide new evidence for the improvement of oocyte quality and the subsequent effects of Sal.

To explore the impact of Sal on oocyte quality at a deeper level, considering the mechanism of oocytes and the in vivo physiology of oocyte maturation, the maturation of oocytes is closely related to the expansion of cumulus cells [21], cumulus cells are required for the transfer of energy or nutrients to support oocyte maturation. The highly active mitochondria in cumulus cells provide energy substrates for the maturation of oocytes [22], and have a critical role in protecting oocytes against oxidative stress [23]. The proliferative ability of cumulus cells is closely related to the nuclear and qualitative maturation of oocytes [24]. Sal promotes proliferation of vascular smooth muscle and PASMC by protecting mitochondria against oxidative stress [25,26]. Sal can reduce oxidative stress in human granulosa cell lines and reduce the occurrence of polycystic ovary syndrome [27]. Our results reveal the ability of Sal to promote the expansion of cumulus cells in terms of the energy supply to oocytes, which may be due to Sal reducing oxidative stress levels in cumulus cells and increasing mitochondrial DNA copy numbers.

Cumulus cell expansion is considered to be an important marker of oocyte maturation [28]. MPF consists of p-cdk1 and Cyclin B1, and is the main promoter of meiosis, which is essential for IVM in mammalian oocytes [29]. The activation of *MPF* and *ERK1/2* requires the activation of maternal *Ccnb1* and *MOS* mRNA translation [30], respectively. Taken together, the cooperation and positive feedback active action of *ERK1/2* and *CDK1* lead to the fine-tuning of mRNA translation and cell cycle progression during mouse oocyte maturation. MAPK plays a key role in promoting and stabilizing the spindle pole during oocyte meiosis [31], and the formation of the spindle structure is essential for chromosome segregation and cytokinesis [32]. Related studies have shown that in porcine oocytes, MAPK is mainly located at the poles of the metaphase spindle, and treatment with U0126 inhibits MAPK activity, resulting in the inhibition of chromosome segregation, PB1 discharge, and MII spindle formation [33]. The activation of MAPK plays an important role in oocyte maturation. Additionally, in most mammalian oocytes, GFP plays an important role in granulosa cell development and oocyte fertilization [34]. Sal can protect human bone marrow-derived endothelial progenitor cells [35], PC-12 cells [36], and alveolar epithelial cells through the activation of the ERK1/2 and MAPK signaling pathways [37]. The results indicate that Sal can increase the mRNA expression of the MAPK pathway (*C-MOS*, *MEK*, and *ERK1/2*) and that MAPK phosphorylation significantly increases, enhancing oocyte maturation by increasing the expressions of genes (*CDK1*, *CyclinB*, and *GFP-1*) related to oocyte maturation. This also suggests the potential mechanism by which Sal promotes oocyte maturation, which may be through activating the MAPK signaling pathway and maintaining spindle stability; further research is needed. This also explains the significantly higher exclusion rate of the first polar body in the treatment group.

Nevertheless, more studies are needed to further explore the mechanisms affecting maturity. There are many factors that affect the development of oocytes in vitro, among which oxidative stress is an important factor leading to oocyte damage [38]. Pig oocytes have a higher lipid content than other animals, and are particularly susceptible to oxidative stress [39], and GSH is an antioxidant that can scavenge ROS [40]. Particularly important in maintaining the redox state of the oocyte, a loss of GSH can lead to oxidative stress and the apoptosis of germ cells [41]. In related studies, Sal can increase the level of GSH and reduce the level of ROS in human umbilical vein endothelial cells (HUVECs) [42] and human RPE cells [43], and then reduce cell apoptosis to protect cells. In the present study, the addition of Sal during IVM reduced the accumulation of ROS in oocytes and increased the accumulation of GSH. We speculate that Sal may reduce oxidative stress by removing intracellular ROS and increasing GSH synthesis, thereby increasing the quality of porcine oocytes.

Mitochondria play an important role in the function of mammalian oocytes [44,45]. Mitochondria regulate the mitochondrial membrane potential through ATP [46]. Mitochondrial dysfunction and reduced mitochondrial membrane potential are the causes of poor oocyte quality and early embryo dysfunction [47]. That is, an abnormality of MMP is closely related to fertilization, germ cell development, and apoptosis [48]. Mitochondria synthesize adenosine triphosphate (ATP) via β-oxidation, a process that involves the electron transport chain [49]. During the meiotic maturation of oocytes, mitochondrial integrity is considered to be one of the indicators of cytoplasmic maturation [50]. Due to the complete maternal inheritance of mitochondria, mtDNA in oocytes is the only source of mtDNA in embryos [51], and numerous studies have shown a positive correlation between the mtDNA copy number and oocyte quality [18]. Numerous studies have shown that Sal can protect various cells, such as PC12 cells [52], HT22 cells [53], and 2BS cells [54], by regulating mitochondrial function and ROS levels. After treatment of diabetic mice with Sal, mitochondrial DNA copies and electron transfer chain proteins in the kidneys of mice significantly increased [55]. In our study, it was further confirmed that the MMP and mtDNA copy numbers of the treatment group significantly increased, and the ATP content significantly increased, which was also verified by previous experiments. The effect of Sal on promoting the maturation of oocytes may be due to the protection of mitochondria by Sal.

BCL2 family proteins, BCL2 and BAX, modulate mitochondrial membrane potential and activation of caspases [56]. In the mitochondrial apoptosis pathway, BCL2 has an anti-apoptotic role, whereas BAX promotes apoptosis following mitochondrial damage [57]. The protective effect of Sal on mitochondrial membrane potential and ATP production has been described above, and Sal inhibited cancer cell activation by inhibiting BCL2 in nasopharyngeal carcinoma cells [58] and human hepatocellular carcinoma [59]. Therefore, the expression levels of mitochondrial genes *BAX* and *BCL2* were analyzed. In our experiments we found via Q-PCR that Sal inhibited *BAX/BCL-2*. In further experiments, the expression of the *Caspase-3* gene and the expression of γH_2AX protein were significantly reduced. This further illustrates that Sal protects against embryonic apoptosis and DNA damage based on mitochondria.

Oocyte quality is a key factor in determining embryo development [60]. Therefore, our study shows the treatment group after PA increases the developmental potential of the blastocysts. *COX2* affects the embryonic implantation ability during embryonic development [61]. In our results, SAL increased the expression of *COX-2* in blastocysts and the expression of totipotency-related genes (*OCT-4*, *CDX2*, *SOX2*, and *NANOG*) [62]. After EdU staining and TUNEL staining, the proliferation of PA blastocyst cells in the treatment group increased significantly, decreased apoptotic cells significantly, and the expression of the apoptosis gene *Caspase-9* slowed down the occurrence of apoptosis during embryonic development. This also further verified the positive effects of Sal on the quality and maturation of oocytes.

5. Conclusions

These results indicate that Sal modifies the expressions of key genes involved in oocyte cytoplasmic development and maturation, enhances their developmental ability, and is involved in modifying oocyte maturation, mitochondrial metabolism, apoptosis, and oxidative stress levels; these effects improve embryonic development and quality, and increase embryo yield, and support an optimized plan for efficient IVP embryo production.

Supplementary Materials: The following supporting information can be downloaded at: https://www.mdpi.com/article/10.3390/genes14091729/s1, Figure S1: Analysis of ROS levels and mitochondrial DNA copy number in cumulus cells.

Author Contributions: Conceptualization, M.Z. and Y.J.; validation, S.S. and Z.G.; formal analysis, B.H. and L.L; investigation, Z.C. and L.Q.; resources, M.Z., Y.J. and X.Y.; data curation, S.S.; writing—original draft preparation, S.S., Z.G. and X.Y.; writing—review and editing, X.Y.; visualization, L.L.; supervision, M.Z. and Y.J.; project administration, M.Z. and Y.J.; funding acquisition, M.Z. and X.Y. All authors have read and agreed to the published version of the manuscript.

Funding: This research was funded by the earmarked fund for the Jilin Provincial Natural Science Foundation of China (20210101014JC), and an international cooperation project (3D5205987203).

Data Availability Statement: The data presented in this study are available on request from the corresponding author.

Conflicts of Interest: The authors declare no conflict of interest.

References

1. Yoshioka, K. Development and application of a chemically defined medium for the in vitro production of porcine embryos. *J. Reprod. Dev.* **2011**, *57*, 9–16. [CrossRef]
2. Sekhar, R.V.; Patel, S.G.; Guthikonda, A.P.; Reid, M.; Balasubramanyam, A.; Taffet, G.E.; Jahoor, F. Deficient synthesis of glutathione underlies oxidative stress in aging and can be corrected by dietary cysteine and glycine supplementation. *Am. J. Clin. Nutr.* **2011**, *94*, 847–853. [CrossRef] [PubMed]
3. Brussow, K.P.; Torner, H.; Kanitz, W.; Ratky, J. In vitro technologies related to pig embryo transfer. *Reprod. Nutr. Dev.* **2000**, *40*, 469–480. [CrossRef] [PubMed]
4. Day, B.N. Reproductive biotechnologies: Current status in porcine reproduction. *Anim. Reprod. Sci.* **2000**, *60*, 161–172. [CrossRef] [PubMed]
5. Tatemoto, H.; Muto, N. Mitogen-activated protein kinase regulates normal transition from metaphase to interphase following parthenogenetic activation in porcine oocytes. *Zygote* **2001**, *9*, 15–23. [CrossRef] [PubMed]
6. Hu, W.; Zhang, Y.; Wang, D.; Yang, T.; Qi, J.; Zhang, Y.; Jiang, H.; Zhang, J.; Sun, B.; Liang, S. Iron Overload-Induced Ferroptosis Impairs Porcine Oocyte Maturation and Subsequent Embryonic Developmental Competence in vitro. *Front. Cell Dev. Biol.* **2021**, *9*, 673291. [CrossRef]
7. Chatzispyrou, I.A.; Alders, M.; Guerrero-Castillo, S.; Zapata Perez, R.; Haagmans, M.A.; Mouchiroud, L.; Koster, J.; Ofman, R.; Baas, F.; Waterham, H.R.; et al. A homozygous missense mutation in ERAL1, encoding a mitochondrial rRNA chaperone, causes Perrault syndrome. *Hum. Mol. Genet.* **2017**, *26*, 2541–2550. [CrossRef] [PubMed]
8. van der Reest, J.; Nardini Cecchino, G.; Haigis, M.C.; Kordowitzki, P. Mitochondria: Their relevance during oocyte ageing. *Ageing Res. Rev.* **2021**, *70*, 101378. [CrossRef] [PubMed]
9. Guo, B.; Dong, W.; Huo, J.; Sun, G.; Qin, Z.; Liu, X.; Zhang, B.; Wang, W. Integrated Metabolomics and Network Pharmacology Analysis Immunomodulatory Mechanisms of Qifenggubiao Granules. *Front. Pharmacol.* **2022**, *13*, 828175. [CrossRef]
10. Yu, H.; Lin, Y.; Zhong, Y.; Guo, X.; Lin, Y.; Yang, S.; Liu, J.; Xie, X.; Sun, Y.; Wang, D.; et al. Impaired AT2 to AT1 cell transition in PM2.5-induced mouse model of chronic obstructive pulmonary disease. *Respir. Res.* **2022**, *23*, 70. [CrossRef]
11. Chiang, H.M.; Chen, H.C.; Wu, C.S.; Wu, P.Y.; Wen, K.C. Rhodiola plants: Chemistry and biological activity. *J. Food Drug Anal.* **2015**, *23*, 359–369. [CrossRef] [PubMed]
12. Wang, J.; Li, J.Z.; Lu, A.X.; Zhang, K.F.; Li, B.J. Anticancer effect of Sal on A549 lung cancer cells through inhibition of oxidative stress and phospho-p38 expression. *Oncol. Lett.* **2014**, *7*, 1159–1164. [CrossRef] [PubMed]
13. Li, F.; Tang, H.; Xiao, F.; Gong, J.; Peng, Y.; Meng, X. Protective effect of Sal from Rhodiolae Radix on diabetes-induced oxidative stress in mice. *Molecules* **2011**, *16*, 9912–9924. [CrossRef] [PubMed]
14. Recio, M.C.; Giner, R.M.; Manez, S. Immunmodulatory and Antiproliferative Properties of Rhodiola Species. *Planta Med.* **2016**, *82*, 952–960. [CrossRef] [PubMed]
15. Fu, S.; Yan, M.; Fan, Q.; Xu, J. Sal promotes osteoblast proliferation and differentiation via the activation of AMPK to inhibit bone resorption of knee osteoarthritis mice. *Tissue Cell* **2022**, *79*, 101917. [CrossRef] [PubMed]
16. Ying, Y.; Luo, J. Sal promotes human periodontal ligament cell proliferation and osteocalcin secretion via ERK1/2 and PI3K/Akt signaling pathways. *Exp. Ther. Med.* **2018**, *15*, 5041–5045. [CrossRef] [PubMed]
17. Wang, N.; Zhang, C.; Chen, M.; Shi, Z.; Zhou, Y.; Shi, X.; Zhou, W.; Zhu, Z. Characterization of MicroRNAs Associated with Reproduction in the Brown Planthopper, Nilaparvata lugens. *Int. J. Mol. Sci.* **2022**, *23*, 7808. [CrossRef]
18. Cecchino, G.N.; Garcia-Velasco, J.A. Mitochondrial DNA copy number as a predictor of embryo viability. *Fertil. Steril.* **2019**, *111*, 205–211. [CrossRef]
19. Zhang, X.; Xie, L.; Long, J.; Xie, Q.; Zheng, Y.; Liu, K.; Li, X. Sal: A review of its recent advances in synthetic pathways and pharmacological properties. *Chem. Biol. Interact.* **2021**, *339*, 109268. [CrossRef]

20. Li, J.; Wang, R.; Chen, Q.; Tian, Y.; Gao, L.; Lei, A. Sal improves porcine oocyte maturation and subsequent embryonic development by promoting lipid metabolism. *Theriogenology* **2022**, *192*, 89–96. [CrossRef]
21. Lee, S.; Kang, H.G.; Jeong, P.S.; Nanjidsuren, T.; Song, B.S.; Jin, Y.B.; Lee, S.R.; Kim, S.U.; Sim, B.W. Effect of Oocyte Quality Assessed by Brilliant Cresyl Blue (BCB) Staining on Cumulus Cell Expansion and Sonic Hedgehog Signaling in Porcine during In Vitro Maturation. *Int. J. Mol. Sci.* **2020**, *21*, 4423. [CrossRef] [PubMed]
22. Sreerangaraja Urs, D.B.; Wu, W.H.; Komrskova, K.; Postlerova, P.; Lin, Y.F.; Tzeng, C.R.; Kao, S.H. Mitochondrial Function in Modulating Human Granulosa Cell Steroidogenesis and Female Fertility. *Int. J. Mol. Sci.* **2020**, *21*, 3592. [CrossRef] [PubMed]
23. Tatemoto, H.; Sakurai, N.; Muto, N. Protection of porcine oocytes against apoptotic cell death caused by oxidative stress during In vitro maturation: Role of cumulus cells. *Biol. Reprod.* **2000**, *63*, 805–810. [CrossRef] [PubMed]
24. Okazaki, T.; Nishibori, M.; Yamashita, Y.; Shimada, M. LH reduces proliferative activity of cumulus cells and accelerates GVBD of porcine oocytes. *Mol. Cell Endocrinol.* **2003**, *209*, 43–50. [CrossRef]
25. Zhuang, X.; Maimaitijiang, A.; Li, Y.; Shi, H.; Jiang, X. Sal inhibits high-glucose induced proliferation of vascular smooth muscle cells via inhibiting mitochondrial fission and oxidative stress. *Exp. Ther. Med.* **2017**, *14*, 515–524. [CrossRef]
26. Huang, X.; Zou, L.; Yu, X.; Chen, M.; Guo, R.; Cai, H.; Yao, D.; Xu, X.; Chen, Y.; Ding, C.; et al. Sal attenuates chronic hypoxia-induced pulmonary hypertension via adenosine A2a receptor related mitochondria-dependent apoptosis pathway. *J. Mol. Cell Cardiol.* **2015**, *82*, 153–166. [CrossRef]
27. Ji, R.; Jia, F.Y.; Chen, X.; Wang, Z.H.; Jin, W.Y.; Yang, J. Sal alleviates oxidative stress and apoptosis via AMPK/Nrf2 pathway in DHT-induced human granulosa cell line KGN. *Arch. Biochem. Biophys.* **2022**, *715*, 109094. [CrossRef]
28. Chen, L.; Wert, S.E.; Hendrix, E.M.; Russell, P.T.; Cannon, M.; Larsen, W.J. Hyaluronic acid synthesis and gap junction endocytosis are necessary for normal expansion of the cumulus mass. *Mol. Reprod. Dev.* **1990**, *26*, 236–247. [CrossRef]
29. Guo, M.; Cao, Y.; Wang, T.; Song, X.; Liu, Z.; Zhou, E.; Deng, X.; Zhang, N.; Yang, Z. Baicalin inhibits Staphylococcus aureus-induced apoptosis by regulating TLR2 and TLR2-related apoptotic factors in the mouse mammary glands. *Eur. J. Pharmacol.* **2014**, *723*, 481–488. [CrossRef]
30. Kalous, J.; Tetkova, A.; Kubelka, M.; Susor, A. Importance of ERK1/2 in Regulation of Protein Translation during Oocyte Meiosis. *Int. J. Mol. Sci.* **2018**, *19*, 698. [CrossRef]
31. Wang, F.; Zhang, L.; Duan, X.; Zhang, G.L.; Wang, Z.B.; Wang, Q.; Xiong, B.; Sun, S.C. RhoA-mediated FMNL1 regulates GM130 for actin assembly and phosphorylates MAPK for spindle formation in mouse oocyte meiosis. *Cell Cycle* **2015**, *14*, 2835–2843. [CrossRef] [PubMed]
32. Li, J.; Ha, S.; Li, Z.; Huang, Y.; Lin, E.; Xiao, W. Aurora B prevents aneuploidy via MAD2 during the first mitotic cleavage in oxidatively damaged embryos. *Cell Prolif.* **2019**, *52*, e12657. [CrossRef] [PubMed]
33. Huo, L.J.; Fan, H.Y.; Liang, C.G.; Yu, L.Z.; Zhong, Z.S.; Chen, D.Y.; Sun, Q.Y. Regulation of ubiquitin-proteasome pathway on pig oocyte meiotic maturation and fertilization. *Biol. Reprod.* **2004**, *71*, 853–862. [CrossRef] [PubMed]
34. Hsieh, M.; Lee, D.; Panigone, S.; Horner, K.; Chen, R.; Theologis, A.; Lee, D.C.; Threadgill, D.W.; Conti, M. Luteinizing hormone-dependent activation of the epidermal growth factor network is essential for ovulation. *Mol. Cell Biol.* **2007**, *27*, 1914–1924. [CrossRef]
35. Tang, Y.; Vater, C.; Jacobi, A.; Liebers, C.; Zou, X.; Stiehler, M. Sal exerts angiogenic and cytoprotective effects on human bone marrow-derived endothelial progenitor cells via Akt/mTOR/p70S6K and MAPK signalling pathways. *Br. J. Pharmacol.* **2014**, *171*, 2440–2456. [CrossRef]
36. Liao, Z.L.; Su, H.; Tan, Y.F.; Qiu, Y.J.; Zhu, J.P.; Chen, Y.; Lin, S.S.; Wu, M.H.; Mao, Y.P.; Hu, J.J.; et al. Sal protects PC-12 cells against amyloid beta-induced apoptosis by activation of the ERK1/2 and AKT signaling pathways. *Int. J. Mol. Med.* **2019**, *43*, 1769–1777. [CrossRef]
37. Guo, B.; Zuo, Z.; Di, X.; Huang, Y.; Gong, G.; Xu, B.; Wang, L.; Zhang, X.; Liang, Z.; Hou, Y.; et al. Sal attenuates HALI via IL-17A-mediated ferroptosis of alveolar epithelial cells by regulating Act1-TRAF6-p38 MAPK pathway. *Cell Commun. Signal* **2022**, *20*, 183. [CrossRef]
38. Orsi, N.M.; Leese, H.J. Protection against reactive oxygen species during mouse preimplantation embryo development: Role of EDTA, oxygen tension, catalase, superoxide dismutase and pyruvate. *Mol. Reprod. Dev.* **2001**, *59*, 44–53. [CrossRef]
39. Dunning, K.R.; Russell, D.L.; Robker, R.L. Lipids and oocyte developmental competence: The role of fatty acids and beta-oxidation. *Reproduction* **2014**, *148*, R15–R27. [CrossRef]
40. Gasparrini, B.; Boccia, L.; Marchandise, J.; Di Palo, R.; George, F.; Donnay, I.; Zicarelli, L. Enrichment of in vitro maturation medium for buffalo (*Bubalus bubalis*) oocytes with thiol compounds: Effects of cystine on glutathione synthesis and embryo development. *Theriogenology* **2006**, *65*, 275–287. [CrossRef]
41. Lim, J.; Luderer, U. Glutathione deficiency sensitizes cultured embryonic mouse ovaries to benzo[a]pyrene-induced germ cell apoptosis. *Toxicol. Appl. Pharmacol.* **2018**, *352*, 38–45. [CrossRef] [PubMed]
42. Hu, R.; Wang, M.Q.; Ni, S.H.; Wang, M.; Liu, L.Y.; You, H.Y.; Wu, X.H.; Wang, Y.J.; Lu, L.; Wei, L.B. Sal ameliorates endothelial inflammation and oxidative stress by regulating the AMPK/NF-kappaB/NLRP3 signaling pathway in AGEs-induced HUVECs. *Eur. J. Pharmacol.* **2020**, *867*, 172797. [CrossRef] [PubMed]
43. Yin, Y.; Liu, D.; Tian, D. Sal prevents hydroperoxide-induced oxidative stress and apoptosis in retinal pigment epithelium cells. *Exp. Ther. Med.* **2018**, *16*, 2363–2368. [CrossRef] [PubMed]

44. Ramalho-Santos, J.; Varum, S.; Amaral, S.; Mota, P.C.; Sousa, A.P.; Amaral, A. Mitochondrial functionality in reproduction: From gonads and gametes to embryos and embryonic stem cells. *Hum. Reprod. Update* **2009**, *15*, 553–572. [CrossRef]
45. May-Panloup, P.; Chretien, M.F.; Malthiery, Y.; Reynier, P. Mitochondrial DNA in the oocyte and the developing embryo. *Curr. Top. Dev. Biol.* **2007**, *77*, 51–83. [CrossRef]
46. Yao, X.; Jiang, H.; Liang, S.; Shen, X.; Gao, Q.; Xu, Y.N.; Kim, N.H. Laminarin enhances the quality of aged pig oocytes by reducing oxidative stress. *J. Reprod. Dev.* **2018**, *64*, 489–494. [CrossRef]
47. Liang, S.; Jin, Y.X.; Yuan, B.; Zhang, J.B.; Kim, N.H. Melatonin enhances the developmental competence of porcine somatic cell nuclear transfer embryos by preventing DNA damage induced by oxidative stress. *Sci. Rep.* **2017**, *7*, 11114. [CrossRef]
48. Bisht, S.; Dada, R. Oxidative stress: Major executioner in disease pathology, role in sperm DNA damage and preventive strategies. *Front. Biosci. Sch.* **2017**, *9*, 420–447. [CrossRef]
49. Chen, W.; Sun, Y.; Sun, Q.; Zhang, J.; Jiang, M.; Chang, C.; Huang, X.; Wang, C.; Wang, P.; Zhang, Z.; et al. MFN2 Plays a Distinct Role from MFN1 in Regulating Spermatogonial Differentiation. *Stem Cell Rep.* **2020**, *14*, 803–817. [CrossRef]
50. Zhang, Y.; ShiYang, X.; Zhang, Y.; Li, Y.; Shi, X.; Xiong, B. Exposure to aristolochic acid I compromises the maturational competency of porcine oocytes via oxidative stress-induced DNA damage. *Aging* **2019**, *11*, 2241–2252. [CrossRef]
51. Singh, I.; Parte, P. Heterogeneity in the Epigenetic Landscape of Murine Testis-Specific Histone Variants TH2A and TH2B Sharing the Same Bi-Directional Promoter. *Front. Cell Dev. Biol.* **2021**, *9*, 755751. [CrossRef]
52. Tang, Y.; Hou, Y.; Zeng, Y.; Hu, Y.; Zhang, Y.; Wang, X.; Meng, X. Sal attenuates $CoCl_2$-simulated hypoxia injury in PC12 cells partly by mitochondrial protection. *Eur. J. Pharmacol.* **2021**, *912*, 174617. [CrossRef]
53. Wang, X.; Tang, Y.; Xie, N.; Bai, J.; Jiang, S.; Zhang, Y.; Hou, Y.; Meng, X. Sal, a phenyl ethanol glycoside from Rhodiola crenulata, orchestrates hypoxic mitochondrial dynamics homeostasis by stimulating Sirt1/p53/Drp1 signaling. *J. Ethnopharmacol.* **2022**, *293*, 115278. [CrossRef]
54. Mao, G.X.; Xu, X.G.; Wang, S.Y.; Li, H.F.; Zhang, J.; Zhang, Z.S.; Su, H.L.; Chen, S.S.; Xing, W.M.; Wang, Y.Z.; et al. Sal Delays Cellular Senescence by Stimulating Mitochondrial Biogenesis Partly through a miR-22/SIRT-1 Pathway. *Oxid. Med. Cell Longev.* **2019**, *2019*, 5276096. [CrossRef]
55. Xue, H.; Li, P.; Luo, Y.; Wu, C.; Liu, Y.; Qin, X.; Huang, X.; Sun, C. Sal stimulates the Sirt1/PGC-1alpha axis and ameliorates diabetic nephropathy in mice. *Phytomedicine* **2019**, *54*, 240–247. [CrossRef]
56. Daniel, L.L.; Daniels, C.R.; Harirforoosh, S.; Foster, C.R.; Singh, M.; Singh, K. Deficiency of ataxia telangiectasia mutated kinase delays inflammatory response in the heart following myocardial infarction. *J. Am. Heart Assoc.* **2014**, *3*, e001286. [CrossRef]
57. Miranda, M.A.; Marcato, P.D.; Mondal, A.; Chowdhury, N.; Gebeyehu, A.; Surapaneni, S.K.; Bentley, M.; Amaral, R.; Pan, C.X.; Singh, M. Cytotoxic and chemosensitizing effects of glycoalkaloidic extract on 2D and 3D models using RT4 and patient derived xenografts bladder cancer cells. *Mater. Sci. Eng. C Mater. Biol. Appl.* **2021**, *119*, 111460. [CrossRef]
58. Liu, S.; Li, Y.; Li, Z. Sal suppresses the activation of nasopharyngeal carcinoma cells via targeting miR-4262/GRP78 axis. *Cell Cycle* **2022**, *21*, 720–729. [CrossRef]
59. Ding, S.Y.; Wang, M.T.; Dai, D.F.; Peng, J.L.; Wu, W.L. Sal induces apoptosis and triggers endoplasmic reticulum stress in human hepatocellular carcinoma. *Biochem. Biophys. Res. Commun.* **2020**, *527*, 1057–1063. [CrossRef]
60. Marteil, G.; Richard-Parpaillon, L.; Kubiak, J.Z. Role of oocyte quality in meiotic maturation and embryonic development. *Reprod. Biol.* **2009**, *9*, 203–224. [CrossRef]
61. Pakrasi, P.L.; Jain, A.K. Cyclooxygenase-2-derived endogenous prostacyclin reduces apoptosis and enhances embryo viability in mouse. *Prostaglandins Leukot. Essent. Fatty Acids* **2008**, *79*, 27–33. [CrossRef] [PubMed]
62. Rizzino, A.; Wuebben, E.L. Sox2/Oct4: A delicately balanced partnership in pluripotent stem cells and embryogenesis. *Biochim. Biophys. Acta* **2016**, *1859*, 780–791. [CrossRef] [PubMed]

genes

Article

Transcriptome and Metabolome Analyses Reveal the Mechanism of Corpus Luteum Cyst Formation in Pigs

Jiage Dai [1,2,†], Jiabao Cai [1,3,†], Taipeng Zhang [3], Mingyue Pang [1,4], Xiaoling Xu [1], Jiahua Bai [1], Yan Liu [1] and Yusheng Qin [1,*]

1 Institute of Animal Husbandry and Veterinary Medicine, Beijing Academy of Agriculture and Forestry Sciences, Beijing 100097, China; djg730@126.com (J.D.); c18833075975@163.com (J.C.); jasmine769@163.com (M.P.); xu_xiaoling1980@163.com (X.X.); bai_jiahua@126.com (J.B.); liuyanxms@163.com (Y.L.)
2 College of Animal Sciences and Technology, China Agricultural University, Beijing 100193, China
3 College of Life Sciences and Food Engineering, Hebei University of Engineering, Handan 056038, China; ztp5331365@163.com
4 Animal Science and Technology College, Beijing University of Agriculture, Beijing 102206, China
* Correspondence: blackberrysheng@163.com
† These authors contributed equally to this work.

Abstract: Corpus luteum cysts are a serious reproductive disorder that affects the reproductive performance of sows. In this study, transcriptome and metabolome datasets of porcine normal and cyst luteal granulosa cells were generated to explore the molecular mechanism of luteal cyst formation. We obtained 28.9 Gb of high−quality transcriptome data from luteum tissue samples and identified 1048 significantly differentially expressed genes between the cyst and normal corpus luteum samples. Most of the differentially expressed genes were involved in cancer and immune signaling pathways. Furthermore, 22,622 information-containing positive and negative ions were obtained through gas chromatography−mass spectrometry, and 1106 metabolites were successfully annotated. Important differentially abundant metabolites and pathways were identified, among which abnormal lipid and choline metabolism were involved in the formation of luteal cysts. The relationships between granulosa cells of luteal cysts and cancer, immune-related signaling pathways, and abnormalities of lipid and choline metabolism were elaborated, providing new entry points for studying the pathogenesis of porcine luteal cysts.

Keywords: luteal cyst; choline; Pi3k-Akt; MAPK; ALDH

Citation: Dai, J.; Cai, J.; Zhang, T.; Pang, M.; Xu, X.; Bai, J.; Liu, Y.; Qin, Y. Transcriptome and Metabolome Analyses Reveal the Mechanism of Corpus Luteum Cyst Formation in Pigs. *Genes* **2023**, *14*, 1848. https://doi.org/10.3390/genes14101848

Academic Editor: Bao Yuan

Received: 4 August 2023
Revised: 16 September 2023
Accepted: 21 September 2023
Published: 23 September 2023

1. Introduction

The normal structure and function of ovaries are very important for the estrus cycle and ovulation in sows. Ovarian cysts are a serious reproductive disorder in sows, and because the clinical symptoms are mild, affected sows are easily overlooked during production, resulting in decreased productivity [1].

The factors involved and the mechanism of luteal cyst formation are complicated [2–4]. In the normal estrous cycle of sows, the luteinizing hormone peak causes ovulation, and the ovulating follicles form the luteum and secrete progesterone to maintain the pregnancy. If there is no pregnancy, the corpus luteum dissolves around the sixteenth day of the estrus cycle and the next estrus cycle begins. Disruption of the hypothalamus–pituitary–gonadal axis and disturbance of steroid and gonadotropin receptor expression in the endocrine signaling pathway can lead to cyst formation [5–8]. Stress, metabolic disorders, and proliferation/apoptosis can also lead to cyst formation [9].

The pathogenesis of luteal cysts is a complicated process that has not been thoroughly investigated, and the specific mechanism is still unclear. The protein and gene expression patterns of follicular granulosa cells are different at different stages [10]. We speculated that the transcriptome and metabolome levels may be different between granulosa cells in cystic luteum and normal luteum. This study may provide a new entry point for studying the pathogenesis of porcine luteal cysts through transcriptome and metabolome analyses, as well as theoretical guidance for the prevention of luteal cysts.

2. Materials and Methods

2.1. Sample Collection

Cystic corpus luteum and normal corpus luteum were obtained from large-scale commercial pig farms in Beijing (Figure 1). Long−term non−estrus sows were tested through B−ultrasound, identified by professionals as having corpus luteum cysts, and slaughtered. The collected corpus luteum was placed in a 37 °C saline solution containing penicillin−streptomycin solution and transported to the laboratory. The outer membrane of the luteum was peeled off with sterile tooth tweezers. An appropriate amount of luteum tissue was biopsied using eye tweezers, placed in a sterile centrifuge tube, quickly frozen with liquid nitrogen, and transferred to a −80 °C refrigerator. Normal corpus luteum was collected, processed, and stored in the same way.

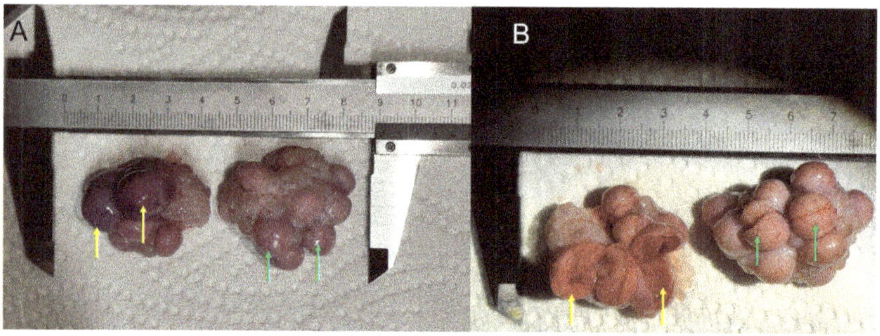

Figure 1. Corpus luteum. Cystic corpus luteum (**A**) and normal corpus luteum (**B**). Corpus luteum cysts ((**A**) yellow arrows) and normal corpus luteum ((**A**) green arrows); Vertical section through corpus luteum cysts ((**B**) yellow arrows) and through normal corpus luteum ((**B**) green arrows).

2.2. RNA Extraction

RNA was extracted from the tissue samples using TRIzol reagent according to the manufacturer's instructions (Invitrogen, Carlsbad, CA, USA). The RNA integrity was assessed using 1% agarose gel assays. The RNA concentration was measured using a Nanodrop microspectrophotometer, and the sample quality was determined using an Agilent 2100 electrophoretogram (Agilent Technologies, Palo Alto, CA, USA).

2.3. Library Preparation and Sequencing

We used 3 μL mRNA from each sample and enriched it for poly(A) mRNAs using oligo-dT magnetic beads (Epicentre, Madison, WI, USA). The mRNAs were disrupted through ultrasound, and first-strand cDNA was synthesized using an M-MuLV reverse transcriptase system (NEB# M0253L; New England Biolabs, Ipswich, MA, USA) with six-base random primers. Second-strand cDNA was synthesized using endonuclease (RNaseH) and a DNA polymerase I system. After purification, end repair, and ligation of the sequencing adapters, cDNA that was approximately 200-bp long was screened for using AMPure XP beads (Agencourt, Beckman Coulter, Brea, CA, USA). After amplification, the cDNA was purified with AMPure XP beads, and PCR products were obtained to complete the library. The library concentration and size of the inserted fragment were tested using Qubit 2.0 (Invitrogen, USA) and an Agilent 2100 Bioanalyzer (Agilent Technologies, CA, USA). Agarose gel electrophoresis and a nanophotometer spectrophotometer were used to measure the RNA integrity and purity, after passing the library inspection. To obtain high-quality data, we removed reads that contained adapters, reads with >10% N bases (all of them were A bases), and/or reads with $Q \leq 20$ bases, which made up >50% of the read sequence. RNA sequencing of the resultant clean reads was performed on an Illumina-HiSeq X sequencing platform with a read length of 150 bp.

2.4. Identification of Single Nucleotide Polymorphisms (SNPs)

After quality control, the sequencing data were compared with the pig reference genome (*Sus scrofa* isolate TJ Tabasco breed Duroc, GCF_000003025.6_Sscrofa11.1). The reads were mapped to chromosome coordinates, duplicate reads were removed, and reads with mass values < 30 were filtered out. The Genome Analysis Toolkit (GATK) was used to detect SNP loci [11]. SNPs related to differentially expressed genes (DEGs) were analyzed to identify genes that were potentially related to the occurrence of luteal cysts.

2.5. Analysis of Differentially Expressed Genes

The DESeq2 software [12] was used to analyze the DEGs, and |log2 (fold change)| > 1 and FDR < 0.05 were used as the screening criteria. The selected DEGs were annotated using the Gene Ontology (GO; http://geneontology.org/, accessed on 5 April 2022) and Kyoto Encyclopedia of Genes and Genomes (KEGG) databases (http://www.genome.jp/kegg/, accessed on 5 April 2022).

2.6. Screening and Quantitative Fluorescence Analysis of Luteal Cyst Genes

The KEGG pathways were ranked based on q−values; low q−values indicated significant differences between the corresponding pathways. The selected DEGs were verified through quantitative PCR (qPCR), and the expression patterns of genes that were differentially expressed between normal and cyst luteal granulosa cells were analyzed. The primer sequences were synthesized by Shanghai Bioengineering Co., Ltd. (Shanghai, China) and are presented in Supplementary Table S1. Quantitative amplification of the cDNA was performed using an iScript kit (Bio−Rad Laboratories, Hercules, CA, USA). The Bio−Rad Chrome real-time qPCR system was used to perform qPCR, and the $2^{-\Delta\Delta CT}$ method was used to calculate the relative expression level of each gene. Each group had three technical replicates and three biological replicates.

2.7. Metabolite Identification and Differential Metabolite Analysis

Vanquish ultra-high performance liquid chromatography (UHPLC) (Thermo Fisher Scientific, Waltham, MA, USA) and an Orbitrap Q Exactive TM HF−X mass spectrometer (Thermo Fisher Scientific, Sunnyvale, CA, USA) were used for liquid chromatography tandem mass spectrometry (LC−MS/MS) analysis [13]. The acquired LC-MS raw data were processed using Compound Discover 3.1 (CD3.1, Thermo Fisher Scientific), and retention time, mass-charge ratio, and other parameters were screened. Peak alignment of the different samples was performed using a retention time deviation of 0.2 min and mass deviation of 5 ppm, to increase the accuracy of the identification. Peaks were extracted according to the set mass deviation of 5 ppm, signal strength deviation of 30%, signal-to-noise ratio of 3, minimum signal strength of 100,000, as well as ion and other information. The peak area was quantified, and the target ion was integrated. The molecular formula was predicted using molecular ion peak and fragment ion, and compared with the mzCloud (https://www.mzcloud.org/, accessed on 5 April 2022), mzVault, and Masslist databases [14]. Background ions were removed using blank samples, and the quantitative results were normalized to obtain the identification and quantitative results. The hypothesis test value (p-value) of potential differentially abundant metabolites was obtained using Student's t-test, and the fold change values of metabolites were calculated by comparing the average value of each peak area.

2.8. Analysis of Differentially Abundant Metabolite Pathways

Differentially abundant metabolite pathways were annotated through enrichment and topological analyses. The Chemical Entities of Biological Interest (http://www.ebi.ac.uk/chebi/init.do/, accessed on 5 April 2022), National Institute of Standards and Technology (http://www.nist.gov/index.html/, accessed on 3 April 2022), KEGG (http://www.genome.jp/kegg/, accessed on 6 April 2022), and other databases were used to identify metabolites, explain the biological functions, and build pathways. MetaboAnalyst 5.0 (http://www.metaboanalyst.ca/MetaboAnalyst/, accessed on 3 April 2022) was used to analyze the different metabolites in the KEGG pathways. The KEGG and PubChem databases were used to determine whether there were significant changes in the pathways that involved cyst and normal corpus luteum genes, and to evaluate the potential effects of metabolite concentrations on specific pathways based on the location of the metabolites in the pathways. The aim was to identify key enzymes and rate-limiting enzymes that play important roles in the significant differentially abundant metabolic pathways.

2.9. Combined Metabolomic and Transcriptomic Analyses

Based on the gene expression and metabolite content data, KEGG pathway maps were annotated for differentially abundant metabolites and differentially expressed genes to obtain common synthesis pathway information.

3. Results

3.1. Transcriptome Sequencing, Statistical Assessment of Sequence Quality, and Expression Analysis

Transcriptome analyses were performed on normal and cystic luteal granulosa cells to identify the genes related to the development of luteal cysts. We obtained 57,229,092, 39,761,374, and 40,244,178 raw reads from the corpus luteum samples, and 49,227,446, 46,621,552, and 57,313,272 raw reads from the corpus luteum cysts (Figure 2A). The raw data were quality tested using FastQC [15]. For all of the samples, the Q20 values were >95%, indicating that the sequencing data were of reliable quality with a low error rate. After obtaining high-quality clean reads (Supplementary Table S2), HISAT2 was used to map the sequencing data to the reference genome [16]. The comparison rate of all the processed data was >95% (Supplementary Table S3), indicating that the sequence quality was sufficient for subsequent gene expression analysis (Figure 2B). More than 85% of the sequencing data mapped to the exon regions of the genes and other reads mapped to the intron or intergenic regions, further indicating the high quality of the sequencing data. The principal component analysis (PCA) results of the normal and cystic luteal cells are shown in Figure 2C; PC1 was 83.7% and PC2 was 12.8%. The degree of separation between PC1 and PC2 in the scatter plot indicated significant differences between the samples. The distribution trend of the gene expression was consistent among the cystic and normal corpus luteum samples, and the log10 of the Fragments Per Kilobase of the transcript per Million mapped reads (FPKM) values were between −2 and 2 (Figure 2D). All of the raw sequence data have been deposited in the China National Gene Bank (CNGB) Nucleotide Sequence Archive (CNSA) (https://db.cngb.org/cnsa/, accessed on 3 April 2022) under accession number CNP0004552.

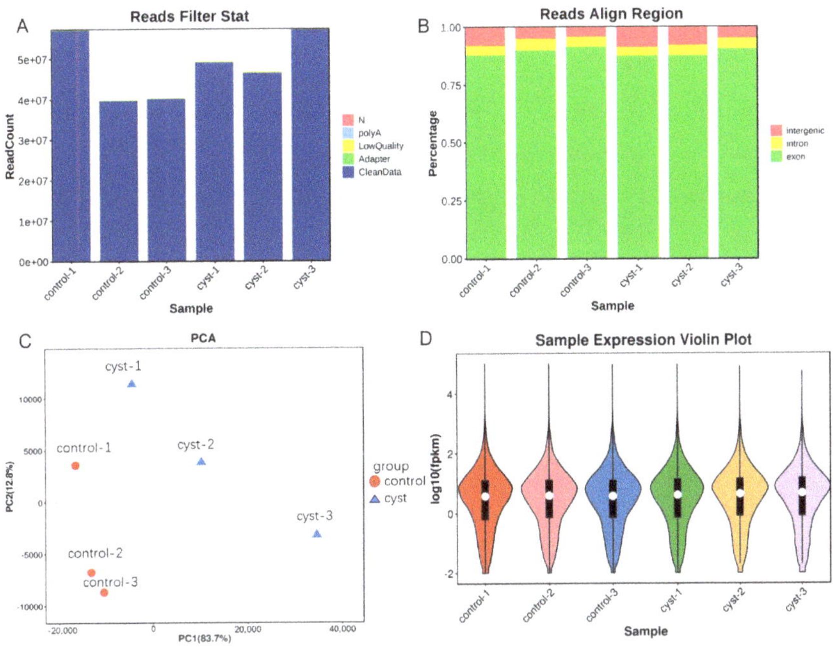

Figure 2. Characteristics of the sequencing data, quality control, and gene expression analysis. (**A**) Preprocessing of raw reads. (**B**) Regions of the reference genome to which the clean reads mapped, pink is intergenic, yellow is intron, green is exon. (**C**) Representative principal component analysis (PCA). (**D**) Distribution of gene expression in the samples, red is control-1, pink is control−2, blue is control−3, green is cyst−1, yellow is cyst−2, purple is cyst−3.

3.2. *Identification and Functional Annotation of Differentially Expressed Genes*

We identified 1048 DEGs between the normal corpus luteum and cystic corpus luteum granulosa cells; 892 genes were significantly up−regulated, and 156 genes were significantly down-regulated. To predict the functions of the DEGs, we used GO and KEGG enrichment. The GO analysis showed that 558 terms were significantly enriched ($p < 10^{-5}$); 527, 23, and 8 terms were under the biological process, cellular component, and molecular function categories, respectively. The top 20 most significantly enriched GO terms are shown in Figure 3. Many of these terms are associated with immune processes, including the immune system process, regulation of immune system process, immune response, immune effector process, regulation of immune system process, leukocyte activation, innate immune response, positive regulation of immune system process, and lymphocyte activation.

The top 10 most significantly enriched GO terms were selected as the main nodes of a directed acyclic graph (Figure 4), and the hierarchical relationship of each node was analyzed to determine the overlap of genes in different GO categories. Analysis of the directed acyclic graph showed that the DEGs affected the occurrence of corpus luteum cysts mainly by activating the immune system or by regulating the biological adhesion process. The activation of the immune system further promoted the leukocyte activation process and activated the immune response. These two biological processes were regulated through cell surface receptor signaling.

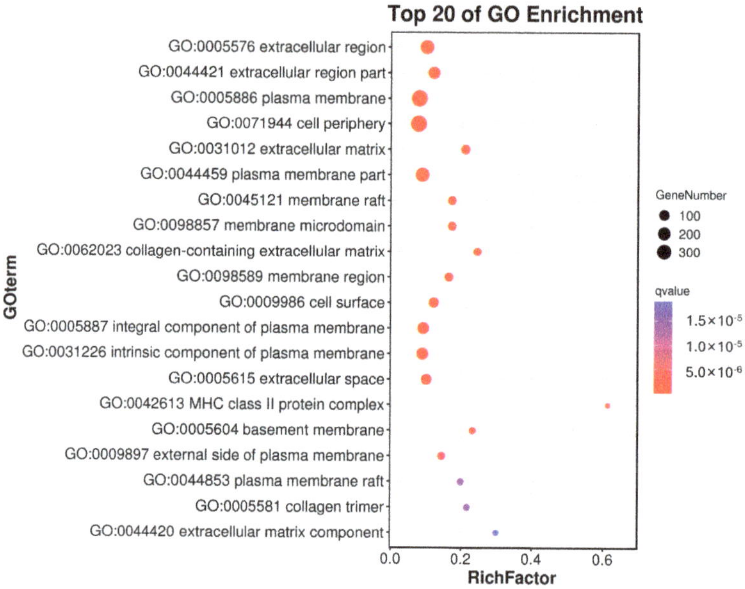

Figure 3. The top 20 most significantly enriched Gene Ontology (GO) terms. GO functional annotation of the differentially expressed genes. Distribution of the significantly enriched GO terms in the cellular component and molecular function categories.

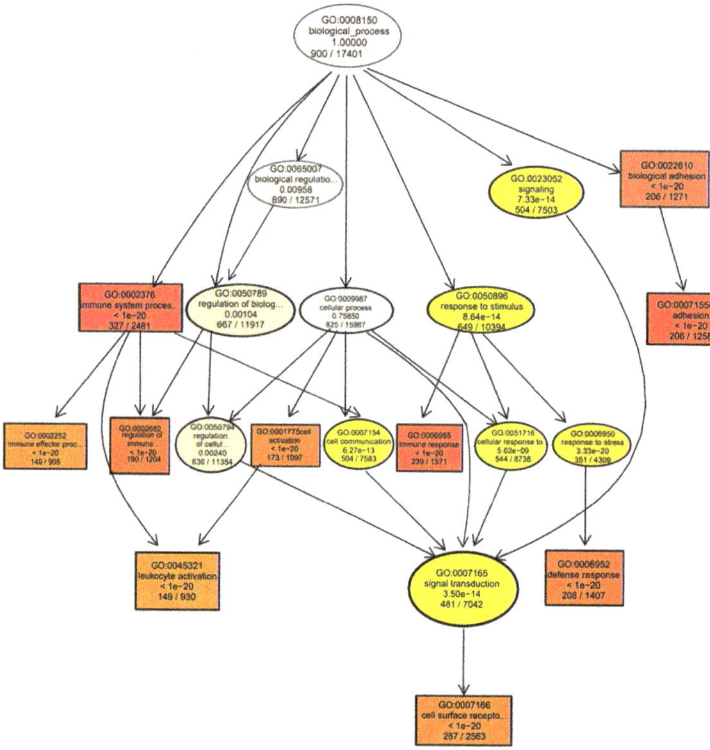

Figure 4. Directed acyclic graph of highly enriched Gene Ontology (GO) terms showing the hierarchical relationship of each node.

The KEGG analysis showed that 79 pathways were significantly enriched. The top 20 significantly enriched pathways were classified into five main categories: cellular processes, metabolism, environmental information processing, human diseases, and organismal systems (Figure 5). In total, 10 of the top 20 pathways were related to diseases that mostly involve tumor formation, including cell adhesion molecules, Th1 and Th2 cell differentiation, the chemokine signaling pathway, the B cell receptor signaling pathway, the tumor necrosis factor signaling pathway, the NF−kappa B signaling pathway, and the retinol signaling pathway.

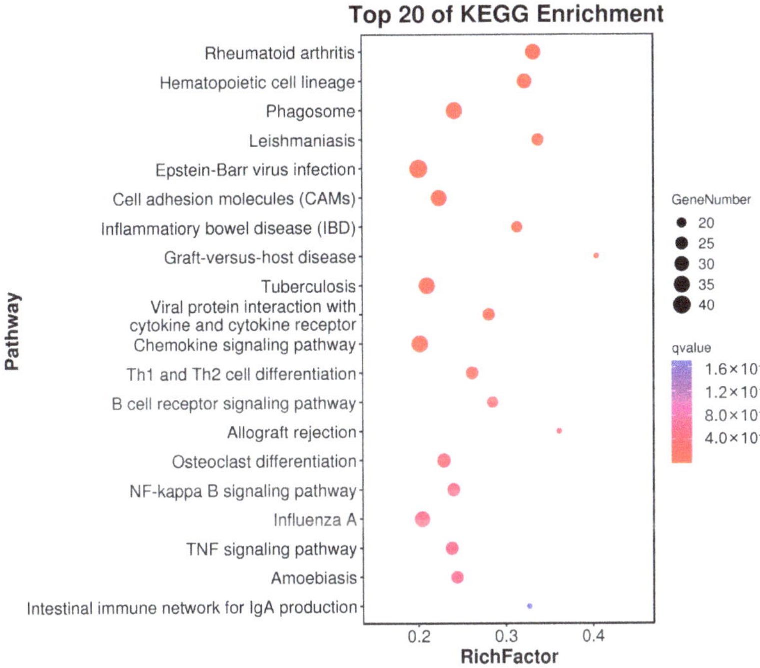

Figure 5. Kyoto Encyclopedia of Genes and Genomes (KEGG) pathway enrichment analysis of the differentially expressed genes.

3.3. Genes Related to Luteal Cysts Identified by Transcriptome Data Analysis

Fifty of the significantly enriched KEGG pathways were selected for further analysis, and the genes involved in these pathways were sorted according to their fold change (Supplementary Table S4). The selected genes were mainly involved in pathways in cancer, cell adhesion molecules, cytokine-cytokine receptor interaction, transcriptional mis-regulation in cancers, the MAPK signaling pathway, the PI3K-Akt signaling pathway, the cAMP signaling pathway, and neuroactive ligand-receptor interaction. Among them, cell adhesion molecules are closely related to the GO terms that relate to the biological adhesion process. Further analysis of the DEGs (fold change > 2) related to cell adhesion molecules identified *ICOS*, *CD8B*, *PDCD1*, *CD2*, *SIGLEC1*, *CD22*, and *CD80*. These genes play important roles in stimulating the immune response processes and are candidate marker genes for the early detection of luteal cysts.

3.4. Metabolome Analysis and Principal Component Analysis

Metabolites in the cystic and normal corpus luteum were detected through gas chromatography-mass spectrometry in the positive and negative ion modes to ensure that the metabolite coverage rate was high and the detection was effective. We identified 22,622 original data peaks. After bias filtering, missing value recoding, and data standardization, 22,622 peaks were retained (Table 1), and 1106 metabolites were annotated.

Table 1. Metabolite detection in luteum tissues by gas chromatography-mass spectrometry.

Type	All	Keep	Known	Unknown
POS [1]	10,079	10,079	626	9453
NEG [2]	12,543	12,543	480	12,063
Total	22,622	22,622	1106	21,516

[1] POS means positive ion mode. [2] NEG means negative ion mode.

The PCA of the metabolic profiles of the differentially treated samples (Figure 6) showed obvious quadrantal differences in the metabolic patterns on the PC1 axis. After 200 permutations of the cystic and normal corpus luteum, different random R2 and Q2 values were obtained. R2Y and Q2Y were used to evaluate the validity of the model, and orthogonal projections to latent structures discriminant analysis (OPLS−DA) of PC1 and PC2 were completed. The scoring charts showed that each stage was clearly divided (Figure 6), indicating that cystic and normal corpus luteum have different metabolic patterns, and may therefore be physiologically different.

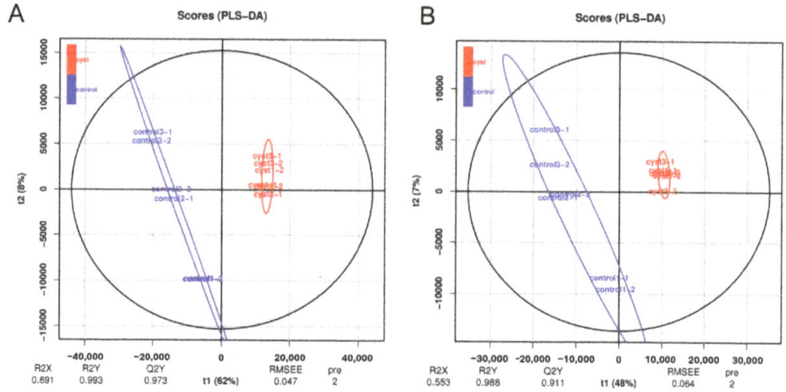

Figure 6. Principal component analyses of metabolic profiles were obtained using the positive ion mode, POS (**A**) and negative ion mode, NEG (**B**) models.

3.5. KEGG Annotation of Differentially Abundant Metabolites in the Metabolome

The analysis of the differential metabolism showed that the number of down-regulated metabolites was significantly higher than the number of up−regulated metabolites in the cystic corpus luteum compared with the normal corpus luteum (Figure 7A,B). The genes in the pathways that were mapped and enriched by the differentially abundant metabolites were annotated with KEGG pathways (Figure 7C). The top 20 enriched pathways were mainly related to metabolic and genetic information processing diseases. Among them, seven significantly enriched pathways were identified, namely glycerophospholipid metabolism, cysteine and methionine metabolism, ascorbate and aldarate metabolism, glyoxylate and dicarboxylate metabolism, phenylalanine metabolism, glucagon signaling pathway, and choline metabolism in cancer.

3.6. Genes Related to Luteal Cyst Occurrence Based on the Metabolome

We focused on choline metabolism in the cancer pathway based on the location of the differentially abundant metabolites in the metabolic pathways. The cancer pathway is associated with the PI3K-Akt and MAPK signaling pathways, which is consistent with the results from the transcriptome analysis. These findings indicate that choline metabolism in the luteal cyst tissues is related to regulation of the MAPK and PI3K−Akt signaling pathways. *CSF3*, *COL1A1*, *PIK3R5*, *COL9A2*, and *COL1A2* had high differential expression levels in the PI3K-Akt signaling pathway, whereas, in the MAPK signaling pathway, the expression of *MAP4K1* and *FLNC* were significantly increased (by at least four times).

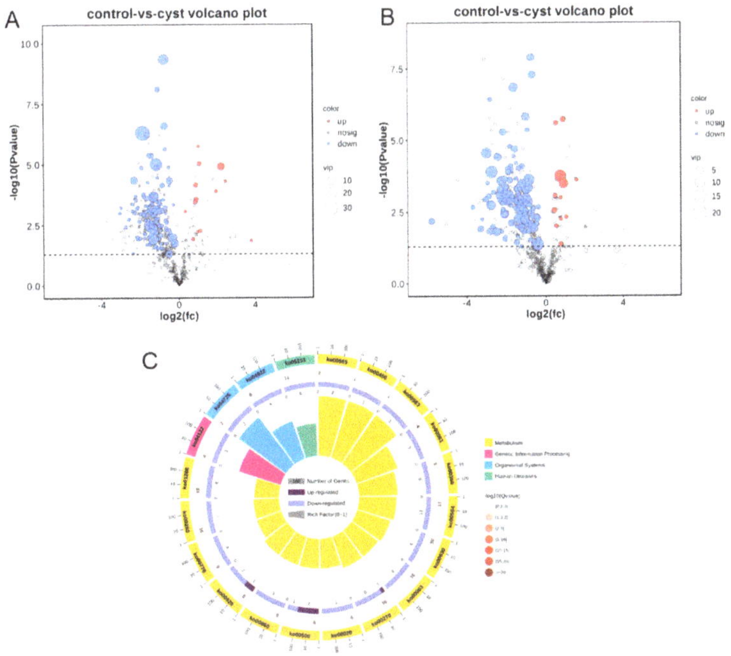

Figure 7. Volcano plots of differentially abundant metabolites between cyst and normal corpus luteum in positive ion mode, POS (**A**) and in negative ion mode, NEG (**B**) modes. Circular plot of KEGG orthology enrichment of differentially abundant metabolites (**C**).

3.7. Combined Metabolomic and Transcriptomic Analysis

To better understand the interaction between granulosa cell metabolites and genes in the formation of luteal cysts, we conducted KEGG common pathway enrichment analysis for the differentially abundant metabolites and DEGs. A total of 26 pathways with significant differences were enriched in the KEGG pathway map, such as pathways in cancer, metabolic pathways, ovarian steroidogenesis, sphingolipid metabolism, fatty acid metabolism, and cAMP signaling. Pathways in cancer are at the top of the list, and we speculate that this pathway plays a key role in the formation of luteal cysts.

3.8. Verification of the Expression of Selected Genes

We selected the DEGs related to luteal cyst occurrence and verified their expression through RT−qPCR. The expression levels of the DEGs with specific functions were consistent with the RNA sequencing data (Figure 8).

3.9. SNPs in DEGs and Their Effects on Gene Expression

We detected 433,755 SNPs, 15,206 deletions, and 31,041 insertions in the DEGs (Supplementary File S1: snp. annot). These variants were located in 38 non-synonymous sites in the DEG sequences (Supplementary File S2: non-synonymous.snp.txt), including in five adhesion−related genes (*CD22*, *CD8B*, *CD80*, *PDCD1*, and *SIGLEC1*), two PI3K−Akt pathway genes (*PIK3R5*, *COL9A2*), and *FLNC* in the MAPK pathway. Combining these findings with the SNP variant results (Table 2), we identified these eight genes to be key in luteal cysts occurrence. We consider these genes to be candidate markers for the detection of luteal cysts because the pathways that these genes are involved in were identified as key pathways for the formation of luteal cysts.

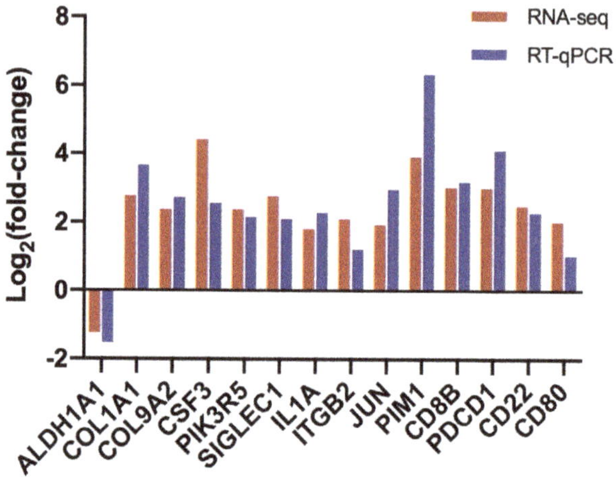

Figure 8. Verification of differentially expressed genes (DEG) expression by RT−qPCR. Aldehyde dehydrogenases 1 family member A1(ALDH1A1) gene; Collagen type I alpha 1 (COL1A1); Colony Stimulating Factor 3 (CSF3); Sialic Acid Binding Ig Like Lectin 1 (SIGLEC1); Interleukin 1 Alpha (IL1A); Integrin Subunit Beta 2(ITGB2); (Jun Proto−Oncogene) JUN; Pim1 Proto−Oncogene, Se−ine/Threonine Kinase (PIM1); CD8 Subunit Beta (CD8B); Programmed Cell Death 1 (PDCD1); CD22 Molecule (CD22); CD80 Molecule (CD80).

Table 2. Number of non-synonymous variants in eight candidate marker genes.

Gene	Number
FLNC	4
PDCD1	4
COL9A2	3
CD22	4
CD8B	3
PIK3R5	4
CD80	1
SIGLEC1	15

4. Discussion

Most of the sequencing reads had Q20s > 95%, and the qPCR verification showed that the sequencing results were reliable. The DEGs were annotated, with functionally enriched GO terms and KEGG pathways, and a large number of pathways and GO terms related to luteal cysts were identified. The predicted biological processes and pathways will help to further the understanding of the regulatory mechanisms underlying luteal cysts.

Luteal cysts can affect normal estrus in sows [17]. In this study, many of the top 20 functionally enriched GO terms assigned to the DEGs in luteal cysts were genes involved in immune-related biological processes, including *SIGLEC1*, *PDCD1*, *ICOS*, and *MHCII*, and to significantly up-regulated genes enriched in cell adhesion molecule signaling pathways. The increased expression of *SIGLEC1* is closely related to tumor formation and metastasis [18–20], and the increased expression of *ICOS* leads to poor prognosis and significantly increased *PDCD1* expression in ovarian tumors [21–23]. The increased expression of these immune-related genes may promote the proliferation and differentiation of granulosa cells and the formation of cysts. The increased expression of *MHCII* enhances the recognition of tumors by the immune system and enhances anti-tumor immunity [24,25], thereby promoting a good prognosis [26,27]. These results indicate that substantial inflammatory and anti-inflammatory reactions occur in the luteal cells of cysts.

The KEGG enrichment showed that the DEGs of luteal granulosa cells are related to tumor occurrence involving the PI3K/Akt, JAK/STAT, and MAPK signaling pathways. PIK3C promotes the transmission of PI3K signaling and stimulates the formation of tumors [28–30], while *PIK3R5* is highly expressed in tumors [31,32]. *FGF* and *FGFR* are up-regulated in various tumor tissues [33,34], and FGFR4 activates a downstream oncogenic signaling pathway to promote cancer development [35–37]. Interleukins play important roles in the immune response [38,39]. *IL6* expression is significantly increased in tumor cells and *IL6* is used as a marker of advanced colorectal cancer [40–43]. The expression of *IL4R* is also up-regulated in various tumors [44–46]. *CSF1* expression is significantly increased in hepatocellular carcinoma and is positively correlated with tumor severity [47–50]. The expression of CSF1R in colorectal cancer tissues is significantly higher than that in paracancerous tissues [51]. These enriched tumor-related signaling pathways may play a key role in the regulation of luteal cyst development. Because immunity can both inhibit and promote the occurrence of tumors [52], it has been speculated that cysts may belong to a special state of tumor pathology. These findings indicate that exploring changes in the microenvironment of cystic luteal cells from the perspective of immunology and tumor occurrence can guide the analysis of luteal cyst occurrence and prognosis.

Aldehyde dehydrogenase (ALDH) oxidizes retinol, the main form of vitamin A, to retinal, and then irreversibly oxidizes retinal to retinoic acid. This metabolic process avoids the cytotoxicity of retinal [53]. Moreover, retinoic acid and cellular retinoic acid-binding protein 1 (CRABP1) are involved in the regulation of gene expression, which can inhibit tumor formation [54,55]. Significantly reduced ALDH expression decreases retinoic acid synthesis and impairs the retinoic acid signaling pathway, leading to glioblastoma. Moreover, in *Helicobacter pylori*-induced gastric cancer, the expression of ALDH1 and RALDH2 in the retinol metabolic pathway are decreased [56], and the intestinal retinoic acid level in a colorectal cancer mouse model was significantly decreased [57]. Granulosa cells are the main cells responsible for retinol uptake and metabolism to retinoic acid in the ovary [58,59]. In this study, the retinol metabolism pathway was one of the most significantly enriched pathways in granulosa cells of luteum cysts, and the expression of genes that encode key enzymes in this pathway, such as retinol dehydrogenase and ALDH1A, was significantly reduced. Therefore, we hypothesized that retinoic acid production was blocked in the granulosa cells of the corpus luteum, which disabled the retinol metabolism pathway, resulting in the abnormal degeneration of the corpus luteum and the formation of corpus luteum cysts.

Abnormal lipid uptake, synthesis, and anabolism are closely related to the malignant transformation of tumor cells [60,61]. Glycerophospholipids, the basic skeleton of biofilm systems, participate in transcription, energy metabolism, and signal transduction [62]. The follicular fluid of polycystic follicle syndrome has a decreased abundance of glycerol phospholipids [63], which affects follicular development and reduces the rate of insemination [64,65]. Phosphatidylcholine metabolism is also abnormal in polycystic follicle syndrome, which affects cell proliferation and differentiation [66,67]. These lipids are closely related to follicle development and the ability of the ovaries to respond to gonadotropins [68]. In this study, the metabolomics profiling of the luteal granulosa cells showed that the levels of glycerophospholipids and phosphocholine were significantly reduced, which may inhibit the response of luteal granulosa cells to gonadotropin signals [69]. We speculate that changes in the microenvironment that affect lipid metabolism may lead to the occurrence of luteum cysts.

5. Conclusions

The formation of luteal cysts is closely related to cancer- and immune-related signaling pathways. Cell adhesion molecules such as CD22, CD8B, CD80, PDCD1, SIGLEC1, PIK3R5, COL9A2, and FLNC can be used as markers for the initial identification of luteal cysts. Abnormal lipid metabolism is also involved in the formation of luteal cysts. Our results provide a new entry point for exploring the pathogenesis of porcine corpus luteum cysts.

Supplementary Materials: The following supporting information can be downloaded at: https://www.mdpi.com/article/10.3390/genes14101848/s1, Table S1: The names and sequences of 15 genes; Table S2: Data filtering statistics table; Table S3: Comparison of numbers with reference statistics; Table S4: control-vs-Cyst Pathway Enrichment; Supplementary File S1: snp. annot; Supplementary File S2: non-synonymous.snp.txt.

Author Contributions: Writing—original draft preparation, J.D. and J.C.; data curation, T.Z. and M.P.; supervision, X.X.; conceptualization, J.B.; funding acquisition and supervision, Y.L.; funding acquisition and project administration, Y.Q. All authors have read and agreed to the published version of the manuscript.

Funding: This study was supported by the National Key Research and Development Program, China (2022YFD1300302), the Beijing Innovation Consortium of Livestock Research System (BAIC05-2023), the Scientific and Technological Innovation Ability Construction Project of the Beijing Academy of Agriculture and Forestry Sciences (KJCX20230218), the Beijing Academy of Agriculture and Forestry Sciences Animal Husbandry and Veterinary Graduate Reform and Development (XMS202321), and the Youth Research Fund of the Beijing Academy of Agriculture and Forestry Sciences (QNJJ202029).

Institutional Review Board Statement: Not applicable.

Informed Consent Statement: Not applicable.

Data Availability Statement: Not applicable.

Acknowledgments: We thank FC Jiang for their contribution to this project.

Conflicts of Interest: The authors declare no conflict of interest.

References

1. Tummaruk, P.; Kesdangsakonwut, S.; Kunavongkrit, A. Relationships among specific reasons for culling, reproductive data, and gross morphology of the genital tracts in gilts culled due to reproductive failure in Thailand. *Theriogenology* **2009**, *71*, 369–375. [CrossRef]
2. Szulańczyk-Mencel, K.; Rzasa, A.; Bielas, W. Relationships between ovarian cysts and morphological and hormonal state of ovarian cortex in sows. *Anim. Reprod. Sci.* **2010**, *121*, 273–278. [CrossRef]
3. Yücel-Tenekeci, G.; Sepici-Dinçel, A.; Özkul, İ.A. Pathomorphological Lesions in the Ovaries of Water Buffaloes. *Acta Sci. Vet.* **2022**, *50*.
4. Tummaruk, P.; Kesdangsakonwut, S. Factors affecting the incidence of cystic ovaries in replacement gilts. *Comp. Clin. Pathol.* **2012**, *21*, 1–7. [CrossRef]
5. Amweg, A.N.; Salvetti, N.R.; Stangaferro, M.L.; Paredes, A.H.; Lara, H.H.; Rodríguez, F.M.; Ortega, H.H. Ovarian localization of 11β-hydroxysteroid dehydrogenase (11βHSD): Effects of ACTH stimulation and its relationship with bovine cystic ovarian disease. *Domest. Anim. Endocrinol.* **2013**, *45*, 126–140. [PubMed]
6. Vanholder, T.; Opsomer, G.; de Kruif, A. Aetiology and pathogenesis of cystic ovarian follicles in dairy cattle: A review. *Reprod. Nutr. Dev.* **2006**, *46*, 105–119. [CrossRef]
7. Marelli, B.E.; Diaz, P.U.; Salvetti, N.R.; Rey, F.; Ortega, H.H. mRNA expression pattern of gonadotropin receptors in bovine follicular cysts. *Reprod. Biol.* **2014**, *14*, 276–281. [CrossRef]
8. Salvetti, N.R.; Alfaro, N.S.; Velázquez, M.M.; Amweg, A.N.; Matiller, V.; Díaz, P.U.; Ortega, H.H. Alteration in localization of steroid hormone receptors and coregulatory proteins in follicles from cows with induced ovarian follicular cysts. *Reproduction* **2012**, *144*, 723–735. [CrossRef] [PubMed]
9. Zulu, V.C.; Sawamukai, Y.; Nakada, K.; Moriyoshi, M. Relationship among insulin-like growth factor-I, blood metabolites and postpartum ovarian function in dairy cows. *J. Vet. Med. Sci.* **2002**, *64*, 879–885. [CrossRef]
10. Fortin, C.S.; Leader, A.; Mahutte, N.; Hamilton, S.; Léveillé, M.C.; Villeneuve, M.; Sirard, M.A. Gene expression analysis of follicular cells revealed inflammation as a potential IVF failure cause. *J. Assist. Reprod. Gen.* **2019**, *36*, 1195–1210. [CrossRef] [PubMed]
11. Yao, Z.; You, F.M.; N'Diaye, A.; Knox, R.E.; McCartney, C.; Hiebert, C.W.; Pozniak, C.; Xu, W. Evaluation of variant calling tools for large plant genome re-sequencing. *BMC Bioinform.* **2020**, *21*, 360. [CrossRef]
12. Hu, G.; Yue, X.; Song, J.; Xing, G.; Chen, J.; Wang, H.; Su, N.; Cui, J. Calcium Positively Mediates Blue Light-Induced Anthocyanin Accumulation in Hypocotyl of Soybean Sprouts. *Front. Plant Sci.* **2021**, *12*, 983. [CrossRef]
13. Want, E.J.; Masson, P.; Michopoulos, F.; Wilson, I.D.; Theodoridis, G.; Plumb, R.S.; Shockcor, J.; Loftus, N.; Holmes, E.; Nicholson, J. Global metabolic profiling of animal and human tissues via UPLC-MS. *Nat. Protoc.* **2013**, *8*, 17–32. [CrossRef]
14. Fu, Y.P.; Li, C.Y.; Peng, X.; Wangensteen, H.; Inngjerdingen, K.T.; Zou, Y.F. Pectic polysaccharides from Aconitum carmichaelii leaves protects against DSS-induced ulcerative colitis in mice through modulations of metabolism and microbiota composition. *Biomed. Pharmacother* **2022**, *155*, 113767. [CrossRef] [PubMed]

15. Agarwal, D.; Kuhns, R.; Dimitriou, C.N.; Barlow, E.; Wahlin, K.J.; Enke, R.A. Bulk RNA sequencing analysis of developing human induced pluripotent cell-derived retinal organoids. *Sci. Data* **2022**, *9*, 759. [CrossRef]
16. Ding, N.; Yuan, Z.; Zhang, X.; Chen, J.; Zhou, S.; Deng, Y. Programmable cross-ribosome-binding sites to fine-tune the dynamic range of transcription factor-based biosensor. *Nucleic Acids Res.* **2020**, *48*, 10602–10613. [CrossRef] [PubMed]
17. Toriumi, H.; Tsumagari, S.; Kuwahara, Y.; Ichikawa, Y.; Takeishi, M.; Sakai, T. Development of a method of diagnosing ovarian disorders in sows and gilts using uterine ultrasonography. *J. Vet. Med. Sci.* **2003**, *65*, 243–247. [CrossRef]
18. Briem, O.; Källberg, E.; Kimbung, S.; Veerla, S.; Stenström, J.; Hatschek, T.; Hagerling, C.; Hedenfalk, I.; Leandersson, K. $CD169^+$ Macrophages in Primary Breast Tumors Associate with Tertiary Lymphoid Structures, T_{regs} and a Worse Prognosis for Patients with Advanced Breast Cancer. *Cancers* **2023**, *15*, 1262. [CrossRef]
19. Cassetta, L.; Fragkogianni, S.; Sims, A.H.; Swierczak, A.; Forrester, L.M.; Zhang, H.; Soong, D.Y.H.; Cotechini, T.; Anur, P.; Lin, E.Y.; et al. Human tumor-associated macrophage and monocyte transcriptional landscapes reveal cancer-specific reprogramming, biomarkers, and therapeutic targets. *Cancer Cell* **2019**, *35*, 588–602.e10. [CrossRef]
20. Yu, Y.; Peng, W. Recent progress in targeting the sialylated glycan-SIGLEC axis in cancer immunotherapy. *Cancer Biol. Med.* **2023**, *20*, 369. [CrossRef] [PubMed]
21. Faget, J.; Sisirak, V.; Blay, J.Y.; Caux, C.; Bendriss-Vermare, N.; Ménétrier-Caux, C. ICOS is associated with poor prognosis in breast cancer as it promotes the amplification of immunosuppressive $CD4^+$ T cells by plasmacytoid dendritic cells. *Onocimmunocogy* **2013**, *2*, e23185. [CrossRef]
22. Amatore, F.; Gorvel, L.; Olive, D. Role of Inducible Co-Stimulator (ICOS) in cancer immunotherapy. *Expert. Opin. Biol. Ther.* **2020**, *20*, 141–150. [CrossRef] [PubMed]
23. Zhao, X.; Wang, Y.; Jiang, X.; Mo, B.; Wang, C.; Tang, M.; Rong, Y.; Zhang, G.; Hu, M.; Cai, H. Comprehensive analysis of the role of ICOS (CD278) in pan-cancer prognosis and immunotherapy. *BMC Cancer* **2023**, *23*, 194. [CrossRef] [PubMed]
24. Accolla, R.S.; Ramia, E.; Tedeschi, A.; Forlani, G. CIITA-driven MHC class II expressing tumor cells as antigen presenting cell performers: Toward the construction of an optimal anti-tumor vaccine. *Front. Immunol.* **2019**, *10*, 1806. [CrossRef] [PubMed]
25. Chatterjee, F.; Spranger, S. MHC-dressing on dendritic cells: Boosting anti-tumor immunity via unconventional tumor antigen presentation. In *Seminars in Immunology*; Elsevier: Amsterdam, The Netherlands, 2023; Volume 66, p. 101710.
26. Thomas, P.; Srivastava, S. MHC-II molecules present RhoC-derived peptides on the surface of tumour cells. *bioRxiv* **2022**, *15*, 492002.
27. Jiang, X.; Ying, Q.; Xia, W.; Li, J.; Shi, N.; Feng, Q.; Tang, A.; Yi, X. An immune-lncRNA risk model to predict prognosis for patients with head and neck squamous cell carcinoma. *bioRxiv* **2022**, 483771.
28. Samuels, Y.; Diaz, L.A.; Schmidt-Kittler, O.; Cummins, J.M.; Delong, L.; Cheong, I.; Rago, C.; Huso, D.L.; Lengauer, C.; Kinzler, K.W. Mutant PIK3CA promotes cell growth and invasion of human cancer cells. *Cancer Cell* **2005**, *7*, 561–573. [CrossRef]
29. Hoxhaj, G.; Manning, B.D. The PI3K–AKT network at the interface of oncogenic signalling and cancer metabolism. *Nat. Rev. Cancer* **2020**, *20*, 74–88. [CrossRef]
30. Ren, A.A.; Snellings, D.A.; Su, Y.S.; Hong, C.C.; Castro, M.; Tang, A.T.; Detter, M.R.; Hobson, N.; Girard, R.; Romanos, S.; et al. PIK3CA and CCM mutations fuel cavernomas through a cancer-like mechanism. *Nature* **2021**, *594*, 271–276. [CrossRef]
31. Paula, L.M.; Moraes, L.H.F.; Canto, A.L.; Dos Santos, L.; Martin, A.A.; Rogatto, S.R.; De Azevedo Canevari, R. Analysis of molecular markers as predictive factors of lymph node involvement in breast carcinoma. *Oncol. Lett.* **2017**, *13*, 488–496. [CrossRef]
32. Li, L.Y.; Kim, H.J.; Park, S.A.; Lee, S.H.; Kim, L.K.; Lee, J.Y.; Kim, S.; Kim, Y.T.; Kim, S.W.; Nam, E.J. Genetic Profiles Associated with Chemoresistance in Patient-Derived Xenograft Models of Ovarian Cancer. *Cancer Res. Treat.* **2019**, *51*, 1117–1127. [CrossRef] [PubMed]
33. Shin, W.S.; Lee, H.W.; Lee, S.T. Catalytically inactive receptor tyrosine kinase PTK7 activates FGFR1 independent of FGF. *Faseb J.* **2019**, *33*, 12960. [CrossRef]
34. Gammelgaard, K.R.; Vad-Nielsen, J.; Clement, M.S.; Weiss, S.; Daugaard, T.F.; Dagnæs-Hansen, F.; Meldgaard, P.; Sorensen, B.S.; Nielsen, A.L. Up-regulated FGFR1 expression as a mediator of intrinsic TKI resistance in EGFR-mutated NSCLC. *Transl. Oncol.* **2019**, *12*, 432–440. [CrossRef]
35. Chen, T.; Liu, H.; Liu, Z.; Li, K.; Qin, R.; Wang, Y.; Liu, J.; Li, Z.; Gao, Q.; Pan, C.; et al. FGF19 and FGFR4 promotes the progression of gallbladder carcinoma in an autocrine pathway dependent on GPBAR1-cAMP-EGR1 axis. *Oncogene* **2021**, *40*, 4941–4953. [CrossRef]
36. Wang, L.; Ren, Z.; Yu, B.; Tang, J. Development of nomogram based on immune-related gene FGFR4 for advanced non-small cell lung cancer patients with sensitivity to immune checkpoint inhibitors. *J. Transl. Med.* **2021**, *19*, 22. [CrossRef]
37. Shiu, B.; Hsieh, M.H.; Ting, W.; Chou, M.; Chang, L.; Huang, C.; Su, S.; Yang, S. Impact of FGFR4 gene polymorphism on the progression of colorectal cancer. *Diagnostics* **2021**, *11*, 978. [CrossRef] [PubMed]
38. Briukhovetska, D.; Dörr, J.; Endres, S.; Libby, P.; Dinarello, C.A.; Kobold, S. Interleukins in cancer: From biology to therapy. *Nta Rev. Cancer* **2021**, *21*, 481–499. [CrossRef]
39. She, Y.X.; Yu, Q.Y.; Tang, X.X. Role of interleukins in the pathogenesis of pulmonary fibrosis. *Cell Death Discov.* **2021**, *7*, 52. [CrossRef] [PubMed]
40. Tawara, K.; Scott, H.; Emathinger, J.M.; Wolf, C.L.; La Joie, D.; Hedeen, D.S.; Bond, L.; Montgomery, P.G.; Jorcyk, C.L. HIGH expression of OSM and IL-6 are associated with decreased breast cancer survival: Synergistic induction of IL-6 secretion by OSM and IL-1β. *Oncotarget* **2019**, *10*, 2068. [CrossRef]

41. Wu, J.; Gao, F.; Wang, C.; Qin, M.; Han, F.; Xu, T.; Hu, Z.; Long, Y.; He, X.; Deng, X.; et al. IL-6 and IL-8 secreted by tumour cells impair the function of NK cells via the STAT3 pathway in oesophageal squamous cell carcinoma. *JECCR* **2019**, *38*, 1–15. [CrossRef]
42. Chonov, D.C.; Ignatova, M.M.K.; Ananiev, J.R.; Gulubova, M.V. IL-6 activities in the tumour microenvironment. Part 1. *Open Access Maced. J. Med. Sci.* **2019**, *7*, 2391–2398. [CrossRef]
43. Goulet, C.R.; Champagne, A.; Bernard, G.; Vandal, D.; Chabaud, S.; Pouliot, F.; Bolduc, S. Cancer-associated fibroblasts induce epithelial–mesenchymal transition of bladder cancer cells through paracrine IL-6 signalling. *BMC Cancer* **2019**, *19*, 1–13. [CrossRef] [PubMed]
44. Gunassekaran, G.R.; Hong, C.M.; Vadevoo, S.M.; Chi, L.; Guruprasath, P.; Ahn, B.C.; Kim, H.; Kang, T.H.; Lee, B. Non-genetic engineering of cytotoxic T cells to target IL-4 receptor enhances tumor homing and therapeutic efficacy against melanoma. *Biomaterials* **2018**, *159*, 161–173. [CrossRef]
45. Chi, L.; Na, M.H.; Jung, H.K.; Vadevoo, S.M.; Kim, C.; Padmanaban, G.; Park, T.I.; Park, J.; Hwang, I.; Park, K.U.; et al. Enhanced delivery of liposomes to lung tumor through targeting interleukin-4 receptor on both tumor cells and tumor endothelial cells. *J. Control Release* **2015**, *209*, 327–336. [CrossRef] [PubMed]
46. Murugan, P.V.S.; Rangaswamy, G.G.; Lee, B. Interleukin-4 receptor-targeted Abraxane inhibits tumor growth by enhancing drug delivery and reprogramming of M2-type macrophages into M1 phenotype. *Cancer Res.* **2022**, *82*, 245. [CrossRef]
47. Wei, C.Y.; Zhu, M.X.; Zhang, P.F.; Huang, X.Y.; Wan, J.K.; Yao, X.Z.; Hu, Z.T.; Chai, X.Q.; Peng, R.; Yang, X.; et al. PKCα/ZFP64/CSF1 axis resets the tumor microenvironment and fuels anti-PD1 resistance in hepatocellular carcinoma. *J. Hepatol.* **2022**, *77*, 163–176. [CrossRef]
48. Hu, Z.Q.; Zhou, S.L.; Li, J.; Zhou, Z.J.; Wang, P.C.; Xin, H.Y.; Mao, L.; Luo, C.B.; Yu, S.Y.; Huang, X.W.; et al. Circular RNA sequencing identifies CircASAP1 as a key regulator in hepatocellular carcinoma metastasis. *Hepatology* **2020**, *72*, 906–922. [CrossRef]
49. Guo, X.Y.; Zhang, J.Y.; Shi, X.Z.; Wang, Q.; Shen, W.; Zhu, W.W.; Liu, L.K. Upregulation of CSF-1 is correlated with elevated TAM infiltration and poor prognosis in oral squamous cell carcinoma. *Am. J. Transl. Res.* **2020**, *12*, 6235. [PubMed]
50. Sebban, S.; Farago, M.; Rabinovich, S.; Lazer, G.; Idelchuck, Y.; Ilan, L.; Pikarsky, E.; Katzav, S. Vav1 promotes lung cancer growth by instigating tumor-microenvironment cross-talk via growth factor secretion. *Oncotarget* **2014**, *5*, 9214. [CrossRef] [PubMed]
51. Li, H.; Tang, S. Colony stimulating factor-1 and its receptor in gastrointestinal malignant tumors. *J. Cancer* **2021**, *12*, 7111. [CrossRef]
52. Peña-Romero, A.C.; Orenes-Piñero, E. Dual effect of immune cells within tumour microenvironment: Pro-and anti-tumour effects and their triggers. *Cancers* **2022**, *14*, 1681. [CrossRef]
53. Kumar, S.; Sandell, L.L.; Trainor, P.A.; Koentgen, F.; Duester, G. Alcohol and aldehyde dehydrogenases: Retinoid metabolic effects in mouse knockout models. *Biochim. Biophys. Acta* **2012**, *1821*, 198–205. [CrossRef]
54. Chen, M.C.; Hsu, S.L.; Lin, H.; Yang, T. Retinoic acid and cancer treatment. *Biomedicine* **2014**, *4*, 22. [CrossRef]
55. Petkovich, M.; Chambon, P. Retinoic acid receptors at 35 years. *J. Mol. Endocrinol.* **2022**, *69*, T13–T24. [CrossRef] [PubMed]
56. Epplein, M.; Signorello, L.B.; Zheng, W.; Cai, Q.; Hargreaves, M.K.; Michel, A.; Pawlita, M.; Fowke, J.H.; Correa, P.; Blot, W.J. Helicobacter pylori prevalence and circulating micronutrient levels in a low-income United States population. *Cancer Prev. Res.* **2011**, *4*, 871–878. [CrossRef] [PubMed]
57. Bhattacharya, N.; Yuan, R.; Prestwood, T.R.; Penny, H.L.; Dimaio, M.A.; Reticker-Flynn, N.E.; Krois, C.R.; Kenkel, J.A.; Pham, T.D.; Carmi, Y.; et al. Normalizing microbiota-induced retinoic acid deficiency stimulates protective $CD8^+$ T cell-mediated immunity in colorectal cancer. *Immunity* **2016**, *45*, 641–655. [CrossRef]
58. Cai, S.; Chen, M.; Xue, B.; Zhu, Z.; Wang, X.; Li, J.; Wang, H.; Zeng, X.; Qiao, S.; Zeng, X. Retinoic acid enhances ovarian steroidogenesis by regulating granulosa cell proliferation and MESP2/STAR/CYP11A1 pathway. *J. Adv. Res.* **2023**, *13*. [CrossRef]
59. Fonseca, B.M.; Cruz, R.; Pinto, B.; Costa, L.; Felgueira, E.; Oliveira, P.; Casal, S.; Rebelo, I. Retinoic acid (*all-trans*) presents antioxidant properties within human ovary and reduces progesterone production by human granulosa cells. *Syst. Biol. Reprod. Med.* **2022**, *69*, 129–141. [CrossRef]
60. Cui, M.Y.; Yi, X.; Zhu, D.X.; Wu, J. Aberrant lipid metabolism reprogramming and immune microenvironment for gastric cancer: A literature review. *Transl. Cancer Res.* **2021**, *10*, 3829. [CrossRef] [PubMed]
61. Zhang, J.; Song, Y.; Shi, Q.; Fu, L. Research progress on FASN and MGLL in the regulation of abnormal lipid metabolism and the relationship between tumor invasion and metastasis. *Front. Med.* **2021**, *15*, 649–656. [CrossRef]
62. Cheng, H.; Wang, M.; Su, J.; Li, Y.; Long, J.; Chu, J.; Wan, X.; Cao, Y.; Li, Q. Lipid Metabolism and Cancer. *Life* **2022**, *12*, 784. [CrossRef]
63. Guan, S.; Liu, Y.; Guo, Y.; Shen, X.X.; Liu, Y.; Jin, H. Potential biomarkers for clinical outcomes of IVF cycles in women with/without PCOS: Searching with metabolomics. *Front. Endocrinol.* **2022**, *13*, 982200. [CrossRef]
64. Bai, Y.; Zhang, F.; Zhang, H.; Xu, C.; Wu, L.; Xia, C. Follicular Fluid Metabolite Changes in Dairy Cows with Inactive Ovary Identified Using Untargeted Metabolomics. *Biomed. Res. Int.* **2020**, *2020*, 9837543. [CrossRef] [PubMed]
65. Xu, C.; Xia, C.; Sun, Y.; Xiao, X.; Wang, G.; Fan, Z.; Shu, S.; Zhang, H.; Xu, C.; Yang, W. Metabolic profiles using 1H-nuclear magnetic resonance spectroscopy in postpartum dairy cows with ovarian inactivity. *Theriogenology* **2016**, *86*, 1475–1481. [CrossRef] [PubMed]
66. Ridgway, N.D. The role of phosphatidylcholine and choline metabolites to cell proliferation and survival. *Crit. Rev. Biochem. Mol. Biol.* **2013**, *48*, 20–38. [CrossRef] [PubMed]

67. Podo, F.; Paris, L.; Cecchetti, S.; Spadaro, F.; Abalsamo, L.; Ramoni, C.; Ricci, A.; Pisanu, M.E.; Sardanelli, F.; Canese, R.; et al. Activation of phosphatidylcholine-specific phospholipase C in breast and ovarian cancer: Impact on MRS-detected choline metabolic profile and perspectives for targeted therapy. *Front. Oncol.* **2016**, *6*, 171. [CrossRef]
68. Santos, P.H.; Fontes, P.K.; Franchi, F.F.; Nogueira, M.F.; Belaz, K.R.; Tata, A.; Eberlin, M.; Sudano, M.J.; Barros, C.M.; Castilho, A.C. Lipid profiles of follicular fluid from cows submitted to ovarian superstimulation. *Theriogenology* **2017**, *94*, 64–70. [CrossRef]
69. Korbecki, J.; Bosiacki, M.; Gutowska, I.; Chlubek, D.; Baranowska-Bosiacka, I. Biosynthesis and Significance of Fatty Acids, Glycerophospholipids, and Triacylglycerol in the Processes of Glioblastoma Tumorigenesis. *Cancers* **2023**, *15*, 2183. [CrossRef]

genes

MDPI

Article

A Redesigned Method for CNP-Synchronized In Vitro Maturation Inhibits Oxidative Stress and Apoptosis in Cumulus-Oocyte Complexes and Improves the Developmental Potential of Porcine Oocytes

Jinlun Lu [1], Min Guo [1], Xiaodong Wang [1], Rui Wang [1], Guangyin Xi [1], Lei An [1], Jianhui Tian [1] and Meiqiang Chu [1,2,*]

[1] State Key Laboratory of Animal Biotech Breeding, National Engineering Laboratory for Animal Breeding, Key Laboratory of Animal Genetics, Breeding and Reproduction of the Ministry of Agriculture and Rural Affairs, College of Animal Science and Technology, China Agricultural University, No. 2 Yuanmingyuan West Road, Beijing 100193, China

[2] College of Agriculture and Forestry Science, Linyi University, Linyi 276000, China

* Correspondence: chumeiqiang@lyu.edu.cn

Abstract: In vitro embryo production depends on high-quality oocytes. Compared with in vivo matured oocytes, in vitro oocytes undergo precocious meiotic resumption, thus compromising oocyte quality. C-type natriuretic peptide (CNP) is a follicular factor maintaining meiotic arrest. Thus, CNP-pretreatment has been widely used to improve the in vitro maturation (IVM) of oocytes in many species. However, the efficacy of this strategy has remained unsatisfactory in porcine oocytes. Here, by determining the functional concentration and dynamics of CNP in inhibiting spontaneous meiotic resumption, we improved the current IVM system of porcine oocytes. Our results indicate that although the beneficial effect of the CNP pre-IVM strategy is common among species, the detailed method may be largely divergent among them and needs to be redesigned specifically for each one. Focusing on the overlooked role of cumulus cells surrounding the oocytes, we also explore the mechanisms relevant to their beneficial effect. In addition to oocytes per se, the enhanced anti-apoptotic and anti-oxidative gene expression in cumulus cells may contribute considerably to improved oocyte quality. These findings not only emphasize the importance of screening the technical parameters of the CNP pre-IVM strategy for specific species, but also highlight the critical supporting role of cumulus cells in this promising strategy.

Keywords: C-type natriuretic peptide; meiotic arrest; in vitro maturation; developmental potential; porcine oocytes

Citation: Lu, J.; Guo, M.; Wang, X.; Wang, R.; Xi, G.; An, L.; Tian, J.; Chu, M. A Redesigned Method for CNP-Synchronized In Vitro Maturation Inhibits Oxidative Stress and Apoptosis in Cumulus-Oocyte Complexes and Improves the Developmental Potential of Porcine Oocytes. *Genes* **2023**, *14*, 1885. https://doi.org/10.3390/genes14101885

Academic Editor: Bao Yuan

Received: 31 August 2023
Revised: 25 September 2023
Accepted: 26 September 2023
Published: 28 September 2023

1. Introduction

The acquisition of development potential is largely dependent on well-orchestrated oocyte maturation, which involves two relatively independent but related processes, i.e., nuclear and cytoplasmic maturation [1]. The coordination and synchronization of nuclear and cytoplasmic maturation is the prerequisite for acquiring the developmental potential of in vivo oocytes [2,3]. In contrast, the in vitro isolation of oocytes from the natural follicular environment leads to a premature decline in intra-oocyte cyclic adenosine monophosphate (cAMP) concentrations and thus triggers spontaneous and precocious meiotic resumption [4,5], which in turn leads to a nuclear-cytoplasmic asynchrony and ultimately compromises the developmental potential of the oocytes. Thus, spontaneous oocyte meiotic resumption has been thought to be the main obstacles to the success of in vitro embryo production [6]. One widely applied strategy to overcome this asynchrony is to utilize conditions that maintain meiotic arrest and delay spontaneous resumption before in vitro maturation (IVM) [7–10]. A previous study by our research group, using bovine oocytes as the model, has established a natural factor synchronized in vitro oocyte maturation (NF-SOM) system, which recapitulates the in vivo follicular meiotic arrest condition via C-type natriuretic peptide (CNP) pretreatment, thus improving the developmental potential of mature bovine oocytes and resulting in higher efficiency of in vitro embryo production [11].

Over the past decades, CNP has emerged as a critical regulator of the progression of nuclear maturation by maintaining intra-oocyte cAMP levels in many species [12–15]. In growing follicles, CNP secreted from the mural granulosa cells acts on its guanylyl cyclase natriuretic peptide receptor 2 (NPR2), located on the cumulus cells (CCs), and then stimulates the generation of cyclic guanosine monophosphate (cGMP) [12], which is then transferred to the oocyte through gap junctions. Increased intra-oocyte cGMP inhibits the activity of phosphodiesterase 3A (PDE3A), thus resulting in the cAMP hydrolysis inhibition and the maintenance of meiotic arrest until luteinizing hormone-induced oocyte meiotic resumption occurs [12,16].

The CNP-synchronized IVM strategy has been well-accepted and applied to update the conventional IVM method of mouse, bovine, sheep, and goat oocytes [17–20]. Although the CNP-pretreatment strategy has been used to modify the IVM method of porcine oocytes, the beneficial effect was limited to oocytes from small (3–4 mm) and medium (4–6 mm) follicles. Of note, the developmental potential of oocytes from large (6–8 mm) follicles, accounting for a large proportion of the collected oocytes, was unexpectedly decreased due to CNP-pretreatment [21]. This finding seems to challenge the concept of the CNP-synchronized IVM strategy. However, the adverse effect specific to fully grown porcine oocytes led us to question whether or not the detailed method was reasonable and in need of redesign.

Oocyte maturation is a complex process regulated by many internal and external conditions [22]. Among these, the intracellular redox homeostasis and apoptotic regulation in cumulus cells are very important factors for oocyte maturation [23,24]. Dysregulated redox homeostasis and apoptosis were thought to be tightly coupled with lower oocyte quality after IVM [23,25,26]. Thus, the status of cumulus cells is an important indicator of oocyte quality. However, the effect of CNP on the cumulus cell status has been largely overlooked. In addition, a series of heterochromatin-related epigenetic landmarks in oocytes are established upon meiosis, which is critical to safeguard oocyte quality and subsequent embryogenesis [27,28]. Defects in heterochromatin-related epigenetic modifications may contribute to compromised oocyte quality [29,30].

In this study, we tested the efficiency and feasibility of the CNP pre-IVM strategy to improve the porcine oocyte IVM system and explored the possible mechanisms of the beneficial effects. We also focused on the effect of CNP on the status of cumulus cells and heterochromatin-related epigenetic modifications, which are critical to oocyte quality, but which have often been overlooked in previous studies.

2. Materials and Methods

Unless otherwise stated, all chemicals and reagents were purchased from Sigma-Aldrich (St. Louis, MO, USA).

2.1. Isolation and Collection of Cumulus-Oocyte-Complexes (COCs), Denuded Oocytes (DOs), Cumulus Cells (CCs) and Mural Granulosa Cells (MGCs)

Porcine ovaries were acquired from commercially slaughtered prepubertal gilts at a local abattoir. They were promptly transported to the laboratory within 3 h, maintaining a temperature range of 28–30 °C while immersed in a saline solution (0.9% NaCl) containing antibiotics. Healthy follicles of 3–8 mm in diameter were aspirated to recover COCs using an 18-gauge needle attached to a 20 mL syringe. Following a brief incubation at 38.5 °C, the COCs within the follicular fluid were subjected to three washes using HEPES-buffered Tyrodes solution (TL-HEPES) supplemented with 0.3% (*w*/*v*) bovine serum albumin (BSA). Only COCs displaying uniform ooplasm and multilayered cumulus were selected for IVM. DOs were obtained by gentle pipetting into a PBS medium containing 0.5% (*w*/*v*) hyaluronidase. The collected CCs and MGCs were promptly frozen in liquid nitrogen and stored at −80 °C for subsequent analysis.

2.2. In Vitro Maturation of Porcine COCs

To assess the impact of CNP on the nuclear meiosis of porcine oocytes, the dose-response and time-dependent effect of CNP pretreatment on meiotic maturation were analyzed. The pretreatment IVM medium consisted of TCM-199 (Gibco, Grand Island, NY, USA), 10% (*v*/*v*) porcine follicular fluid, 0.6 mM cysteine, 75 mg/L of penicillin, and 50 mg/L of streptomycin. The IVM medium was supplemented with 10 IU/mL of hCG (Chorulon, Intervet Australia Pty Limited, Victoria, Australia), 10 IU/mL of PMSG (Folligon, Intervet Australia Pty Limited), and 10 ng/mL of epidermal growth factor (EGF), based on the pretreatment medium. Selected COCs were subjected to three washes with the specified medium and subsequently transferred to droplets, each containing 60 μL of pre-equilibrated pre-IVM or IVM medium, all situated under 3 mL of mineral oil in a 35 mm Petri culture dish containing 15 to 20 COCs per droplet. The COCs within the droplets were cultured at 38.5 °C in a humidified environment with 5% CO_2. According to the pre-IVM method, COCs in the CNP pre-IVM group underwent culture in IVM droplets for either 17 or 41 h following the initial 24 h pre-IVM phase. In contrast, COCs in the conventional IVM group were cultured in IVM droplets for the full 41 h duration.

2.3. Assessment of Nuclear Maturation

At different timepoints after pre-IVM culture, the oocytes were mechanically denuded. The DOs were then fixed in 4% paraformaldehyde (PFA) for 1 h and permeabilized in 0.5% PBST (0.5% Triton-100 (*v*/*v*) in PBS containing 0.1% (*w*/*v*) polyvinyl alcohol (PVA)). They were then stained with DAPI for 5 min. The stained DOs were placed on a glass slide, which was then covered with a coverslip. Images of the stained DOs were captured and evaluated. The percentages of the oocytes at the germinal vesicle (GV) and meiosis II (MII) stages in each experimental group were calculated.

2.4. Parthenogenetic Activation (PA)

PA served as an alternative to in vitro fertilization to assess the developmental potential of porcine oocytes because of the possible effects of boar semen. After CCs removal, mature MII oocytes underwent three washes in an activation medium (comprising 0.28 mol/L mannitol; 0.01% polyvinyl alcohol; 0.05 mmol/L HEPES; 0.1 mmol/L $CaCl_2 \cdot 2H_2O$; and 0.1 mmol/L $MgCl_2$) before activation. PA was achieved via a 1.5 kV/cm direct pulse for 30 microseconds using a BTX Electro-Cell Manipulator 2001 (BTX, San Diego, CA, USA). Post-activation, the oocytes underwent three washes in PZM-3 medium containing 10 μg/mL cycloheximide (CHX) and 5 μg/mL cytochalasin (CB). Subsequently, the oocytes were cultured in PZM-3 supplemented with CB and CHX for 4 h, followed by three washes and culturing in PZM-3 under mineral oil in a 5% CO_2 environment at 38.5 °C for 7 days. The cleavage and blastocyst rate were measured on days 2 and 7, respectively.

2.5. Immunofluorescence Staining (IF)

The DOs were initially fixed in 4% PFA for 1 h, permeabilized with 0.5% PBST, and subsequently blocked with 1%(*w*/*v*) BSA in 0.5% PBST for 6 h. For NPR2, H3K9me3, or H3K27me3 staining, the oocytes were incubated overnight at 4 °C with specific antibodies (NPR2: 1:50; Santa Cruz Biotechnology, Santa Cruz, CA, USA; H3K9me3 or H3K27me3: 1:1000; Millipore, Bedford, MA, USA). In the case of 5-methyl-cytosine (5mC) or 5-hydroxymethyl-cytosine (5hmC) staining, the permeabilized oocytes underwent denaturation in 4 N HCl for 20 min, neutralization with 100 mM Tris-HCl (pH 8.5) for 15 min, followed by blocking with 1% (*w*/*v*) BSA in 0.5% PBST overnight at 4 °C, and incubated with primary antibody (5mC and 5hmC: 1:250; Active Motif, Carlsbad, CA, USA) for 1 h at room temperature. After three washes, the DOs were incubated with secondary antibodies (488 goat anti-mouse IgG or 594 goat anti-rabbit IgG: 1:1000; Thermo Fisher Scientific

Company, Waltham, MA, USA) for 1 h, with the nuclei counterstained using DAPI. The fluorescence was visualized and quantified using an inverted epifluorescence microscope (IX71; Olympus, Japan), with relative fluorescence intensity analyzed via ImageJ software (Version: 1.53a, National Institutes of Health, Bethesda, MD, USA). The experiments were conducted in triplicate.

2.6. Detection of Intracellular Reactive Oxygen Species (ROS) in the Oocytes

The intracellular ROS content within the oocytes was assessed utilizing a ROS detection assay kit (Beyotime Biotechnology, Shanghai, China), following the manufacturer's instructions. Briefly, the oocytes were incubated in TCM199 medium containing 10 μM 2′,7′-dichlorofluorescin diacetate (DCFH-DA) for 30 min at room temperature. Subsequent to washing with 0.1% PBS-PVA, the oocytes were examined under a fluorescence microscope, with the same scanning settings used for all the groups. The fluorescence intensities of the oocytes were quantified using Image J software. The experiments were repeated at least three times.

2.7. Detection of Early Apoptosis by Annexin-V Staining in the Oocytes

The early apoptosis of the oocytes was detected using an Annexin-V staining kit (Beyotime Biotechnology, Shanghai, China). The viable DOs were washed and stained with 100 μL of a binding buffer containing 5 μL of Annexin-V-FITC for 20 min in the dark, according to the manufacturer's instructions. After washing with PBS-PVA, the fluorescence signals were captured. The experiments were repeated at three times; 10–15 oocytes were included in each detection.

2.8. Terminal Deoxynucleotidyl Transferase-Mediated dUTP Nick-End Labeling (TUNEL) Assay

The apoptotic status of the COCs was evaluated using a TUNEL assay kit (Beyotime Bio-technology, Shanghai, China). The COCs were fixed with 4% PFA, permeabilized in 0.5% PBST for 20 min, and incubated in 0.5% BSA- PBS-PVA at room temperature. The COCs were then incubated in the TUNEL mixture for 1 h at 37 °C in the dark. After staining, the COCs were counterstained with DAPI. The labeled COCs were mounted on glass slides and imaged. The apoptosis rate was calculated as follows: apoptosis rate = (the number of TUNEL-positive cells/total cell number) × 100%. Image J software was employed for analysis. The experiments were conducted in triplicate.

2.9. RNA Extraction and Real-Time Reverse Transcription Quantitative PCR (RT-qPCR) Analysis

Following the manufacturer's recommendations, TRIzol (Invitrogen) was used to extracted the total RNA from the pooled samples from the MGCs, CCs, and DOs, respectively. DNase (Vazyme Biotech, Nanjing, China) was applied to remove the residual genomic DNA before reverse transcription (RT). Approximately 0.5 μg of extracted RNA was used for each RT reaction. The complementary DNA was stored at −20 °C until use. RT-qPCR reactions were performed using a Bio-Rad CFX96 Real-Time PCR System with SosoFast EvaGreen Supermix (Bio-Rad Laboratories). The amplification protocol included an initial denaturation process at 95 °C for 30 s, followed by 40 cycles consisting of 5 s of denaturation at 95 °C and 5 s of annealing/extension at 60 °C. The negative controls were reactions without RT or the substitution of RNA samples with DEPC water (to detect any DNA contamination). The relative fold change of the mRNA transcripts was analyzed using the $2^{-\Delta\Delta Ct}$ method, with *RPL19* or *GAPDH* mRNA serving as reference genes. The primers are listed in Table 1. Each experiment was performed in triplicate.

Table 1. List of primer sequences.

Gene Name	Forward Primers	Reverse Primers
GAPDH	TCGGAGTGAACGGATTTG	CCTGGAAGATGGTGATGG
RPL19	GGAAGGGTACTGCCAATGCT	GTGCTCCATGAGAATCCGCT
CNP	CCGAAGGTCCCTCGAACTC	GGAGTCTTGTCGCCCTTCTT
NPR2	AATGGAGTCTAACGGTCAAG	GGAAGAAGTAGGGTTTATAGGA
DNMT1	GCGTCTTGCAGGCTGGTCAGTA	CTTCTTATCATCGACCACGACGCT
DNMT3A	ATCAGTACGACGATGACGGC	CACCAAGAGATCCACGCACT
DNMT3B	ACCTGTCATCCGACACCTCT	CTCGGCATGAACCCACGTTA
TET3	TCTTCCGTCGTTCAGCTACTACAG	GTGGAGGTCTGGCTTCTTCTCAAA
G9A	GGAGGAGCTGGGGTTTGAC	CAGAGGTGGCTGCTGAGTTG
SUV39H1	GAATCAGCTCCAGGACCTGTGC	CAGGTGCTCTCTGAGTCTGGGTAC
SUV39H2	GCAGGACGAACTCAACAGAA	CAACCAAAGGTGGCTTCATT
SETDB1	CATTGGTTTGGATGCAGCAGC	GATGCATCATCAAAGAGCTGGTC
KDM4A	CTGAAACCTTGAACCCCAGTGC	GATATCGTCATAGGATGCCCGTG
KDM4B	CTGGCCAACAGCGAGAAGTACTG	GATGTTCCACTGGGCCACGTC
KDM4C	TGTGAAAAGCCAGGAGAAGCAAAG	CAGGTTTGGTCAGCCTCGGT
KDM4D	AAGGATGCAGTGTGTGTTGC	CCTGTTCGCGGATCTTTTTA
EZH2	TGCAACACCCAATACTTACAAGC	ACTCTTTTGCTCCCTCCAAGT
KDM6A	GCAGGCTCAGTTGTGTAACC	GGTTTACATGCCTGCTGTGC
BAX	TGCCTCAGGATGCATCTACC	AAGTAGAAAAGCGCGACCAC
BCL2	AATGACCACCTAGAGCCTTG	GGTCATTTCCGACTGAAGAG
CASPASE-3	CCGAGGCACAGAATTGGACT	TCGCCAGGAATAGTAACCAGG
C-MYC	GATAGTGGAAAACCCGGCTGC	CAGATATCCTCGCTGGGTGC
P53	TTTCACCCTCCAGATCCGTG	TTTATGGCGGGAGGGAGACT
SOD1	TCCATGTCCATCAGTTTGGA	AGTCACATTGCCCAGGTCTC
SOD2	AAGCCATCAAACGCGACTTT	CCTTGTTGAAACCGAGCCAA
CAT	ACATGGTCTGGGACTTCTGG	TCATGTGCCTGTGTCCATCT
GPX4	ATTCTCAGCCAAGGACATCG	CCTCATTGAGAGGCCACATT
NRF2	CATAGCAGAGCCCAGTACCA	CACGGTGGTCTTGGTTGAAG

2.10. Western Blot Analysis

The pooled samples from the MGCs, CCs, and DOs were lysed in Laemmli sample buffer (consisting of SDS sample buffer with β-mercaptoethanol) and subsequently heated to 100 °C for 5 min. A total of 10 μg cellular protein from each sample was separated by SDS-PAGE and electrophoretically transferred onto the PVDF membranes. Following transfer, the membranes were blocked using TBST (TBS containing 0.1% Tween 20) containing 5% non-fat milk for 1 h. Subsequently, the membranes were incubated overnight at 4 °C with a rabbit anti-NPR2 antibody (1:300, Santa Cruz Biotechnology, Santa Cruz, CA, USA) and a mouse monoclonal anti-β-GAPDH antibody (1:5000). After three washes with TBST, the membranes were further incubated with HRP-conjugated Goat anti-Rabbit and Goat anti-Mouse IgG (1:8000; ZSGB-Bio, Beijing, China) for 1 h at room temperature. The chemiluminescent signal was captured using a Tanon-5200 chemiluminescence imaging system (Tanon, Shanghai, China). All experiments were conducted independently, no fewer than three times.

2.11. Statistical Analysis

Statistical analyses were performed using SPSS 22.0 (SPSS, Chicago, IL, USA). Differences between groups were assessed using Turkey's test for multiple means comparison. All data are presented as mean ± SEM. A p-value < 0.05 was considered statistically significant.

3. Results

3.1. The Expression Pattern of CNP and NPR2 in Porcine Ovarian Follicles

We first determined the expression patterns of CNP and its exclusive receptor NPR2 in the porcine ovarian follicles. The encoding mRNA of CNP, i.e., natriuretic peptide precursor C (*NPPC*) and *NPR2* were detected in the MGCs, CCs, and oocytes isolated from the preovulatory follicles. The *NPR2* mRNA was preferentially expressed in the CCs, whereas the *NPPC* mRNA level was significantly higher in the MGCs than in the CCs and the oocytes (Figure 1A). Of note, although the previous study by our research group reported that the oocyte membrane-localized NPR2 also contributes to the meiotic arrest of the bovine oocytes [11], the result from the present study indicated that notable NPR2 expression can be detected in the MGCs and CCs (Figure 1B,C), but not in the porcine oocyte membrane (Figure 1B,D), although *NPR2* mRNA was detected at low levels in the oocytes (Figure 1A), which highlighted the importance of NPR2 expression in the CCs for maintaining the meiotic arrest of the porcine oocytes.

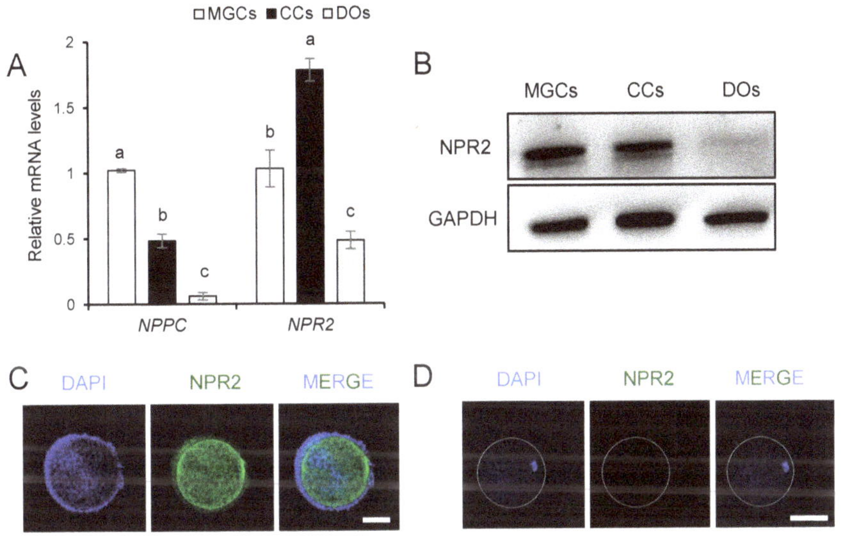

Figure 1. Quantification of CNP and NPR2 expression in porcine MGCs, CCs, and DOs. (**A**) Detection of *NPPC* and *NPR2* by RT-qPCR in isolated MGCs, CCs, and DOs. (**B**) Western blot analysis revealed the presence of NPR2 protein in porcine MGCs and CCs, but not in DOs. (**C**) NPR2 protein in porcine COCs confirmed by fluorescence analysis; scale bar, 100 µm. (**D**) Detection of NPR2 protein in porcine DOs by fluorescence analysis; scale bar, 100 µm. [a–c] Values indicated by different letters are significantly different, $p < 0.05$.

3.2. *CNP Exposure of In Vitro Cultured COCs Maintains the Meiotic Arrest of Porcine Oocytes*

It has been well-established that in vitro cultured oocytes undergo spontaneous and precocious meiotic resumption, thus impairing oocyte maturation. We therefore wanted to determine whether the exogenous supplementation of CNP could recapitulate in an in vivo follicular environment that maintains oocyte meiotic arrest. To test this theory, we exposed in vitro cultured COCs to CNP and measured the percentage of germinal vesicle breakdown (GVBD) that occurred as oocyte meiotic resumption (Figure 2A). Our results showed that CNP inhibited meiotic resumption in a dose-dependent manner, reaching its maximal effect at 200 nM (69.4% vs. 41.9%) (Figure 2B). Furthermore, dynamic analyses showed that 200 nM CNP exhibited a time-dependent inhibitory effect on preserving meiotic arrest up to approximately 24 h (Figure 2C). In addition, we found that vasopressin [31], an effective NPR2 inhibitor, abolished the inhibitory impact of CNP on meiotic resumption in porcine oocytes (Figure 2D), further emphasizing the critical role of NPR2 in mediating CNP-induced meiotic arrest in porcine oocytes.

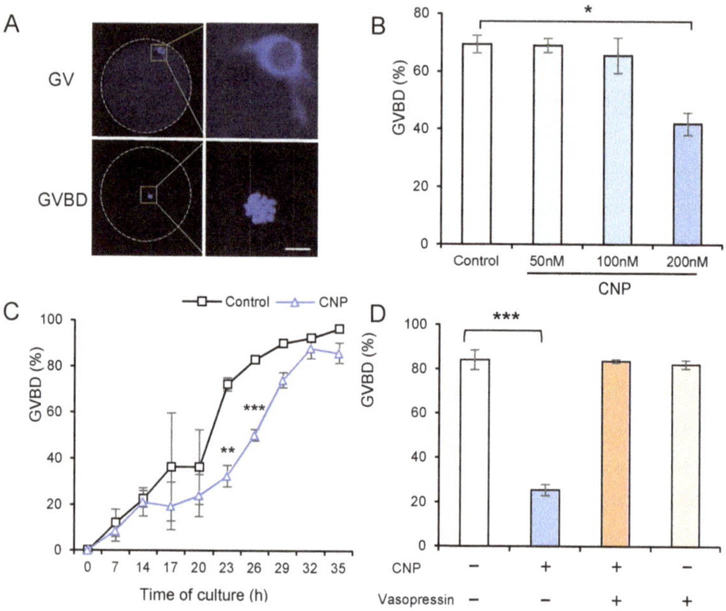

Figure 2. The effect of CNP on the progression of nuclear maturation in porcine oocytes in vitro. (**A**) Representative images of chromatin configurations of porcine oocytes (blue) in different maturation stages (GV: all chromatin is assembled into a compact nuclear circle confined around the nucleolus; GVBD: exhibits a small nucleus and arranged condensed chromosomes). Chromatin was stained with DAPI. Scale bar, 10 μm. (**B**) GVBD rate in porcine CC-enclosed oocytes cultured in medium supplemented with 50, 100, 200 nM CNP for 24 h. (**C**) Kinetics of meiotic resumption of CC-enclosed oocytes cultured in medium supplemented with 200 nM CNP. (**D**) Effect of NPR2 inhibitor (vasopressin) on the nuclear maturation of COCs cultured in medium supplemented with 200 nM CNP for 24 h. * $p < 0.05$; ** $p < 0.01$; *** $p < 0.001$.

3.3. CNP-Based Biphasic IVM System Improves the Developmental Potential of In Vitro Matured Porcine Oocytes

Our group has proposed a CNP-based pre-IVM strategy to improve bovine oocyte developmental potential via enhancing the synchronicity of nuclear and cytoplasmic maturation [11]. Although this method has been well-accepted among different species, the efficacy seems to be unsatisfactory in porcine oocytes. CNP pre-IVM, followed by the conventional IVM, led to a significant decrease in the developmental potential of the oocytes from large follicles (6–8 mm) [21]. This fact, together with the relatively longer duration of conventional porcine IVM per se, led us to postulate that porcine oocytes, especially fully grown oocytes, may be more sensitive to extended IVM duration than those of other species. Therefore, we developed two CNP-based biphasic IVM systems, with or without an extended interval, based on the selected concentration and interval, i.e., the COCs were pretreated with 200 nM CNP for 24 h, followed by 41 h or 17 h (equal to the conventional IVM duration in total) IVM (Figure 3A). Using parthenogenetic embryos, whose developmental performance is largely dependent on oocyte quality, as the model, we tested the preimplantation development following different biphasic IVM systems. Our results showed that oocytes from the CNP-based biphasic IVM system exhibited a higher, although not significant, cleavage rate, as well as a significantly higher blastocyst rate after PA, than those undergoing conventional IVM. In contrast, the extended IVM system, in agreement with previous published CNP pre-IVM system results [21], exhibited no beneficial effects on the oocyte developmental potential. These results indicated that the CNP-based biphasic IVM system was also beneficial to porcine oocytes, similar to the results obtained for previously tested species [11,17–20].

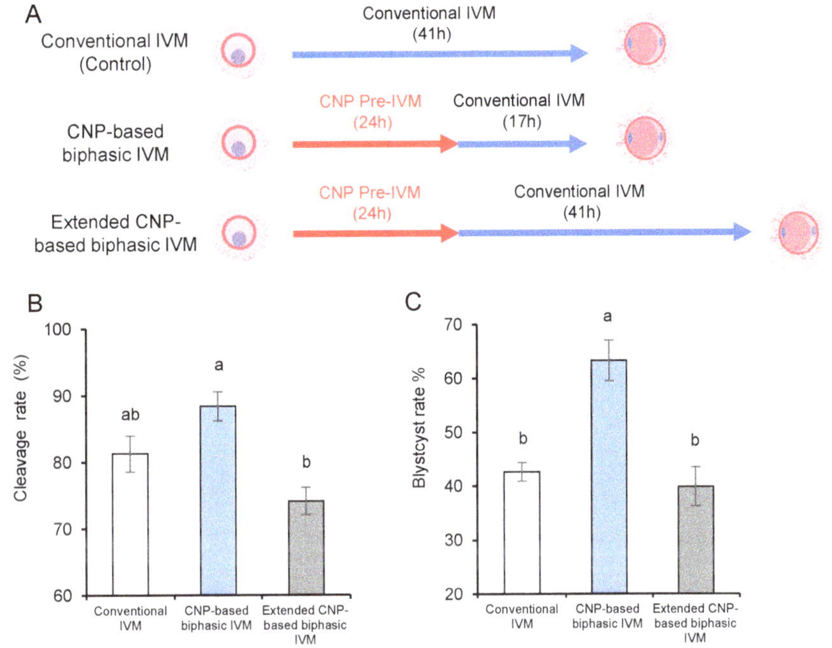

Figure 3. The effect of the CNP-based biphasic IVM system on the development rate of porcine oocytes after parthenogenetic activation. (**A**) Model illustrating the methodology of the CNP-based biphasic IVM system, which consists two steps: a pre-IVM phase based on CNP supplementation, followed by a 17 h or 41 h conventional IVM phase. (**B**,**C**) The effect of CNP-based biphasic IVM on the subsequent cleavage rate (**B**) and blastocyst rate (**C**) after parthenogenetic activation. [a,b] different letters indicate significant differences; $p < 0.05$.

3.4. *CNP Alleviates Oxidative Stress and Early Apoptosis in Porcine Oocytes*

Next, we attempted to determine the mechanism relevant to the beneficial effect of the CNP-based biphasic IVM system. Previous studies have showed that standard in vitro culture conditions lead to oxidative stress and an imbalanced antioxidant defense system [32], and protecting oocytes against oxidative stress during IVM can improve their developmental potential. Therefore, we hypothesized that a CNP-based biphasic IVM system could enhance the resistance of porcine oocytes to oxidative stress. To test this theory, we assessed intracellular ROS levels using DCFH-DA. Quantitative analysis revealed that oocytes subjected to CNP-based biphasic IVM exhibited significantly lower relative intracellular ROS levels compared to those undergoing the 41 h conventional IVM treatment. Of note, the oocytes from extended CNP-based biphasic IVM showed the highest ROS levels (Figure 4A,B). In addition, the mRNA levels of antioxidant enzymes, i.e., *SOD*, *CAT*, *GPX4* and *NRF2*, did not change among the groups (Figure 4C), suggesting that the transcriptional regulation of an enzyme-dependent antioxidative system may not participate in the CNP-induced resistance to oxidative stress in porcine oocytes.

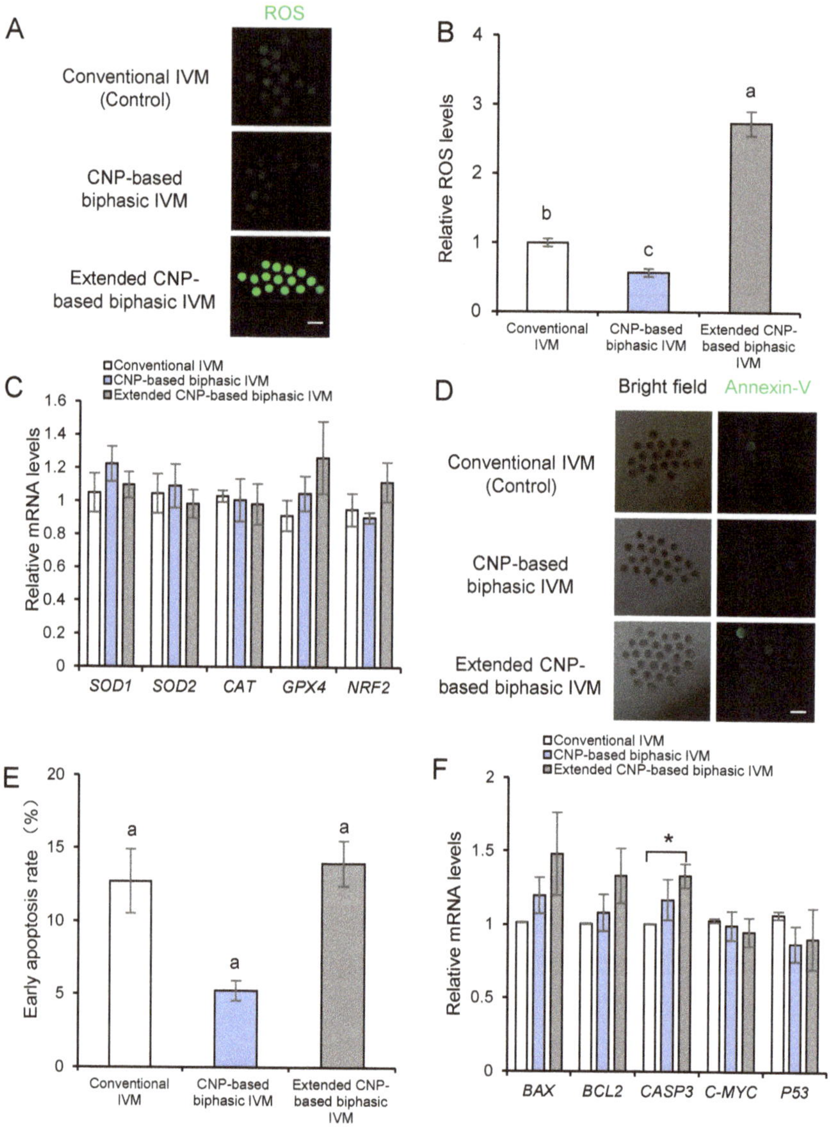

Figure 4. CNP-based biphasic IVM alleviated ROS accumulation and apoptosis initiation in porcine oocytes. (**A**) Illustrative images of DCHF-DA fluorescence (green) in MII oocytes from each group; scale bar, 200 μm. (**B**) Quantitative analysis of ROS fluorescence intensity in MII oocytes from the control IVM (n = 40), CNP-based biphasic IVM (n = 38), and extended CNP-based biphasic IVM (n = 40) groups. (**C**) Relative mRNA expressions of antioxidant enzyme genes in the oocytes among the groups. (**D**) Representative images of annexin-V fluorescence (green) in MII oocytes from each group; scale bar, 200 μm. (**E**) The percentage of annexin-V positive oocytes from each group. (**F**) Relative mRNA expressions of apoptosis-related genes in the oocytes. Experiments were conducted in triplicate, with more than 50 oocytes examined for each group. [a–c] Values indicated by different letters are significantly different; $p < 0.05$. * $p < 0.05$.

Considering that increased intraoocyte ROS accumulation may induce apoptosis, we next examined the initiation of apoptosis in porcine oocytes by using annexin V-FITC staining, a well-accepted early marker of apoptosis. The results showed that the percentage of early apoptosis in the oocytes obtained from CNP-based biphasic IVM, but not extended CNP-based biphasic IVM, was significantly lower than that in oocytes undergoing conventional IVM. (Figure 4D,E). Moreover, expression analyses of apoptosis-related genes showed that the extended biphasic IVM stimulated *CASPASE-3* expression, which always occurs in the late phase of apoptosis (Figure 4F). This finding, together with the increased ROS accumulation, suggested that the extended biphasic IVM system may impair oocyte quality.

3.5. CNP Reduces DNA Damage and Apoptosis in Cumulus Cells

In addition to the apoptotic status in oocytes per se, the apoptotic status of cumulus cells surrounding the oocytes has also been reported to correlate with the developmental potential of the oocytes [25]. Thus, we used a TUNEL assay to evaluate apoptotic DNA fragmentation in the cumulus cells. Porcine COCs were cultured with CNP at different concentrations for 24 h. The results showed that the DNA damage in the cumulus cells was dramatically decreased due to CNP exposure (Figure 5A,B). In addition, CNP exposure significantly downregulated pro-apoptotic genes, i.e., *BAX*, *CASEPASE3*, *C-MYC*, and *P53* (Figure 5C). Next, we analyzed apoptosis in the cumulus cells at the end of different CNP-based biphasic IVM systems. Compared with those undergoing conventional IVM, the cumulus cells from the CNP-based biphasic IVM system, but not the extended CNP-based biphasic IVM system, showed significantly lower apoptotic rates (Figure 5D,E). In line with this, the improved strategy significantly downregulated pro-apoptotic genes, i.e., *BAX*, *C-MYC*, *P53*, and the ratio of *BAX/BCL2*, a key indicator in susceptibility of the cells to apoptosis, in the expanded cumulus cells after oocyte maturation (Figure 5F,G). These results indicated that CNP may be effective in enhancing the anti-apoptotic activity and maintaining the DNA integrity of cumulus cells, thus improving oocyte maturation.

3.6. CNP Does Not Affect Epigenetic Modifications of In Vitro Matured Porcine Oocytes

In addition to the nuclear genome integrity, epigenetic modifications are also essential for oocyte quality and subsequent embryogenesis [27,33]. We next examined some well-established hallmark epigenetic modifications in matured oocytes undergoing different CNP-based biphasic IVM systems. The expression of oocyte-related DNA methyltransferases and demethylases mRNA were first examined by RT-qPCR analyses. Our result showed that there was no significant difference in *DNMT1*, *DNMT3A*, *DNMT3B*, and *TET3* mRNA expression levels among the groups (Figure 6A). Moreover, in line with this, the global 5mC and 5hmC levels were not changed in matured porcine oocytes due to either CNP-based biphasic IVM or extended CNP-based biphasic IVM (Figure 6B,D).

Global transcription repression in matured oocytes was reported to be a prerequisite for normal embryogenesis after fertilization [34]. DNA methylation and repressive histone modifications (H3K9me3, H3K27me3) play essential and coordinated roles in gene silencing. Therefore, we next examined whether CNP is beneficial for the establishment of repressive histone modifications. Expression levels of relevant writers and erasers of H3K9me3 and H3K27me3 were detected by RT-qPCR analysis. The results showed that CNP did not affect either H3K9me3 or H3K27me3 writers and erasers (Figure 6F). The measurement of global H3K9me3 and H3K27me3 showed that H3K9me3 levels were not affected by CNP-based biphasic IVM or extended CNP-based biphasic IVM (Figure 6G,H), but H3K27me3 was significantly decreased in matured porcine oocytes due to extended IVM (Figure 6I,J), also supporting the possible adverse effect of extended IVM.

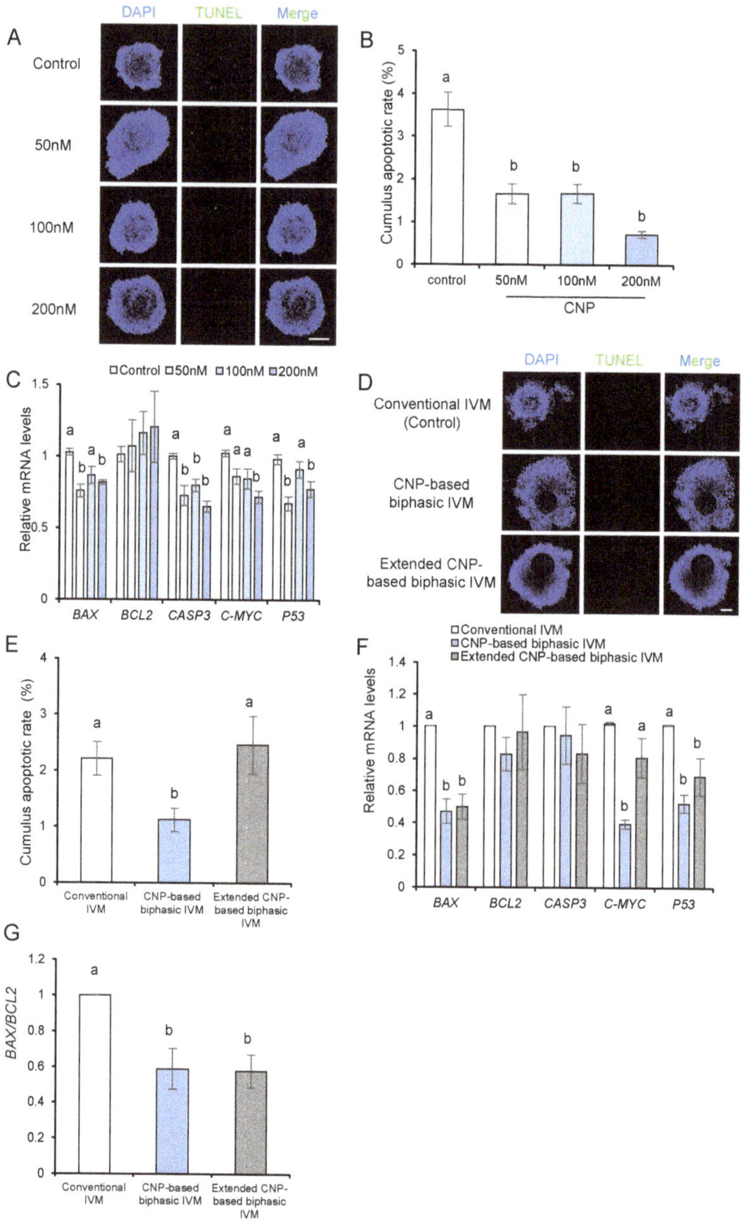

Figure 5. CNP reduced DNA damage and apoptosis in cumulus cells. (**A**) Representative images of TUNEL (green) in COCs from different concentrations of CNP (0, 50, 100, 200 nM); scale bar, 200 μm. (**B**) The percentage of TUNEL positive cumulus cells from the control (n = 61), 50 nM CNP (n = 40), 100 nM CNP (n = 44), and 200 nM CNP (n = 30) groups. (**C**) Relative mRNA expression of apoptosis-related genes, i.e., *BAX*, *BCL2*, *CASPASE-3*, *C-MYC*, *P53*, in cumulus cells. (**D**) Representative images of TUNEL (green) in COCs from each group; scale bar, 100 μm. (**E**) The percentage of TUNEL positive cumulus cells from the control IVM (n = 62), CNP-based biphasic IVM (n = 56), and extended CNP-based biphasic IVM (n = 38) systems. (**F**) Relative mRNA expression of apoptosis-related genes, i.e., *BAX*, *BCL2*, *CASPASE-3*, *C-MYC*, and *P53*, in cumulus cells after IVM. (**G**) The ratio of *BAX/BCL2* in each group; [a,b] Values with different letters are significantly different; $p < 0.05$.

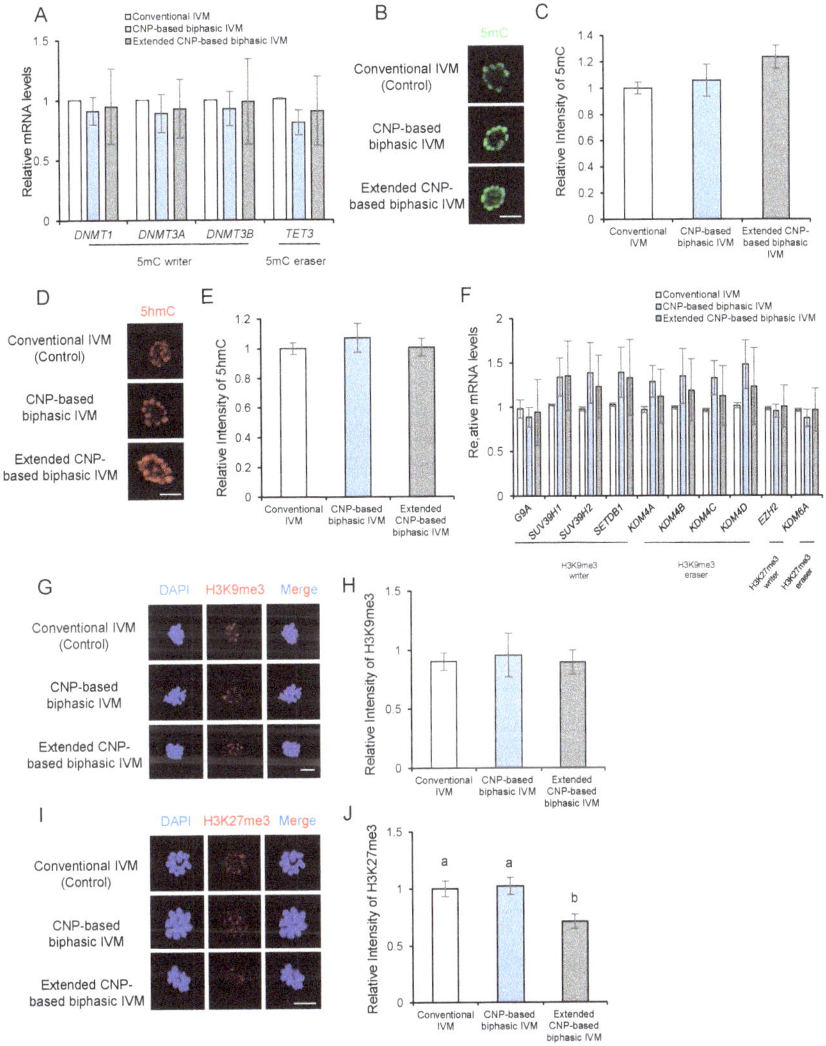

Figure 6. CNP does not affect the epigenetic modifications of in vitro matured porcine oocytes. (**A**) Representative images of 5mC (green) in MII oocytes from each group; scale bar, 10 µm. (**B**) Quantification of relative levels of 5mC in mature oocytes from the control ($n = 67$), CNP-based biphasic IVM ($n = 38$), and extended CNP-based biphasic IVM ($n = 60$) groups; (**C**) Representative images of 5hmC (red) in MII oocytes from each group; scale bar, 10 µm. (**D**) Quantification of the relative levels of 5hmC in mature oocytes from the control ($n = 59$), CNP-based biphasic IVM ($n = 41$), and extended CNP-based biphasic IVM ($n = 68$) groups. (**E**) Expression of DNMTs and TET3 mRNA levels in oocytes. (**F**) Representative images of H3K9me3 marks (red) and nuclear DNA (blue) in MII oocytes from each group; scale bar, 10 µm. (**G**) Quantification of relative levels of H3K9me3 in mature oocytes from the control ($n = 57$), CNP-based biphasic IVM ($n = 43$), and extended CNP-based biphasic IVM ($n = 59$) groups; (**H**) Typical images of H3K27me3 (red) and nuclear DNA (blue) in MII oocytes from each group; scale bar, 10 µm. (**I**) Quantification of relative levels of H3K27me3 in MII oocytes from control IVM ($n = 60$), CNP-based biphasic IVM ($n = 44$), and extended CNP-based biphasic IVM ($n = 55$) groups. (**J**) Expression of writer and eraser mRNA levels of H3K9me3 and H3K27me3 in oocytes. Experiments were repeated at least 3 times, with more than 50 oocytes examined for each experimental condition. [a,b] Values indicated by different letters are significantly different; $p < 0.05$.

4. Discussion

The previous study by our research group has showed that pretreatment with CNP before conventional IVM can improve the developmental potential of in vitro matured bovine oocytes; thus, a novel strategy named the NFSOM system was proposed [11]. Although a pre-IVM phase relying on chemical-extended meiotic arrest has been widely used over past decades, the safety of this method is of primary concern, and evident adverse effects have been frequently reported, e.g., butyrolactone I caused degeneration of cortical granules and peripheral migration of all cytoplasmic organelles, as well as abnormal nuclear morphology [35]; roscovitine disrupted the integrity and subsequent expansion of the cumulus cells and led to mitochondrial swelling in the oocyte cytoplasm [36]; both 6-Dimethylaminopurine (6-DMAP) and cycloheximide induced notable chromosomal abnormality during subsequent embryonic development [37], etc. In contrast, CNP is a natural follicular physiological factor for maintaining meiotic arrest, thus exhibiting advantages over chemical inhibitors in regards to safety and efficacy. Supporting this, the NFSOM system dramatically increased the subsequent developmental rate of IVF bovine embryos [11]. Thus, the strategy of CNP pre-IVM has been well-accepted and widely used to update the current IVM systems in many species [11,17–20]. However, the efficiency and feasibility of CNP-based biphasic IVM system remained unsatisfactory in regards to porcine oocytes.

In the present study, we showed that CNP-based biphasic IVM significantly improved the developmental potential of porcine oocytes, suggesting that the beneficial effect of the CNP pre-IVM strategy is common to various species. However, it should also be noted that the detailed protocol varies among species: in addition to the differences in the duration of the pre-IVM phase, the necessity of extending the IVM phase is totally distinct. In the previous study by our research group [11], using our authorized patent [38], the extended IVM phase, similar to the method used for mouse and sheep oocytes, is indispensable for improving bovine oocyte maturation [17,19]; in contrast, both our own result and previous published data [21] indicate that the extended IVM phase seemed to be detrimental to porcine oocytes; thus, the non-extended system, like that used for goat oocytes, seems to be more suitable [15] (Figure 7). Therefore, our own studies, together with those of other researches, indicate that the CNP pre-IVM strategy is definitely beneficial to oocyte quality, but the suitable technical parameters need to be further determined. Further in-depth studies are needed to explained why porcine oocytes are intolerant towards the extended IVM. A possible explanation may be related to the relatively higher intraoocyte lipid contents in porcine oocytes, which make them highly sensitive to ROS-induced impairments [39]; another possibility may be due to the much longer IVM duration of porcine IVM than those of other livestock species [22].

Having confirmed the beneficial effect of CNP-based biphasic IVM, we next attempted to understand its potential mechanisms. In addition to reduced ROS levels and inhibited early apoptosis in oocytes per se, we also focused on the cumulus cells which surround and support the oocytes. Cumulus cells play vital roles in nurturing oocyte growth, maintaining meiotic arrest, and supporting oocyte cytoplasmic maturation [23]. It has been confirmed that the degree of cumulus apoptosis was negatively correlated with the quality of matured in vitro oocytes [25]. However, whether or not CNP-based biphasic IVM can inhibit apoptosis in cumulus cells has not yet been determined. Our results indicated that CNP-based biphasic IVM enhanced the anti-apoptotic activity and maintained the DNA integrity of the cumulus cells, partly explaining the beneficial effect of CNP in improving the developmental potential of in vitro matured porcine oocytes. Interestingly, it has been reported that abnormalities in cumulus formation were found in the follicles of the *Npr2*-mutant mice [40]. In line with this, the activation of the cGMP pathway, the downstream pathway of CNP-NPR2 signaling, promoted granulosa cell survival by suppressing apoptosis in cultured preantral follicles [41]. Based on our own results and these findings, we speculate that the occurrence of apoptosis in in vitro cumulus cells may be induced by insufficient CNP-NPR2 signaling. To the best of our knowledge, this is the first report that CNP inhibits the apoptosis of ovarian cumulus cells, but further studies are needed to gain in-depth understanding of the underlying mechanisms of this result.

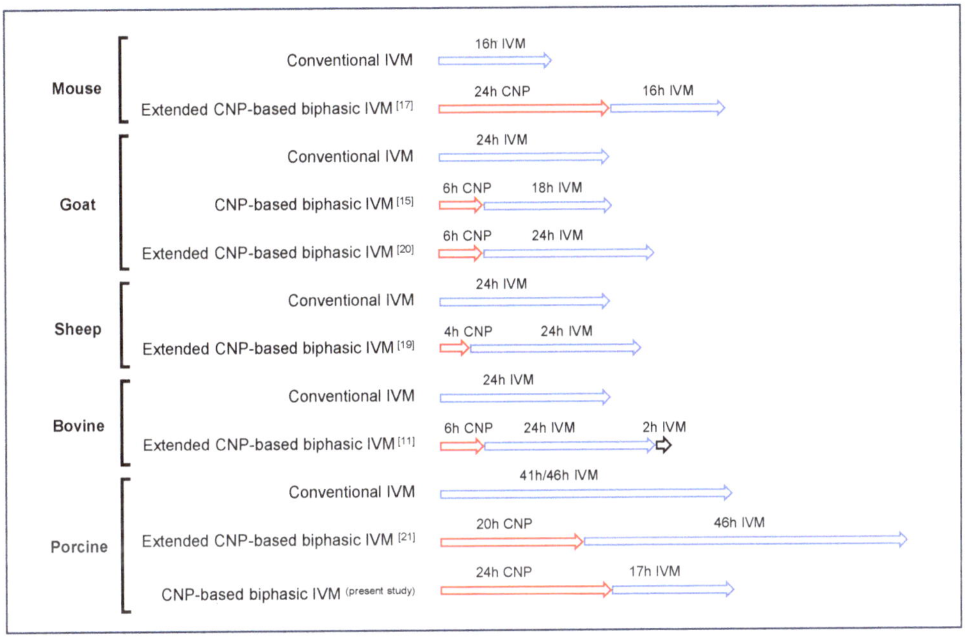

Figure 7. Model illustrating the methodology of the CNP-based biphasic IVM system of different species (mouse [17], goat [15,20], sheep [19], bovine [11], porcine [21]), which consist of two steps: a pre-IVM phase based on CNP treatment, followed by an extended or un-extended IVM phase.

The global transcriptional silencing that occurs in the oocyte genome prior to the resumption of meiosis is crucial to post-fertilization developmental transition, involving DNA methylation and histone modifications [29,42,43]. Despite the essential role of CNP in regulating the oocyte meiotic progression and nuclear maturation, the possible effect of CNP on nuclear epigenetic modifications has not been explored. However, unexpectedly, no obvious changes were detected using our model. This suggests that the beneficial role of CNP in enhancing the developmental potential of porcine oocytes may be independent in regards to the establishment of epigenetic modifications. However, the global level of H3K27me3 was significantly reduced in oocytes undergoing extended IVM, further supporting the notion that suitable technical parameters of biphasic IVM system should be carefully screened to avoid possible adverse effects.

In summary, in the present study, we confirmed the efficiency and feasibility of the CNP pre-IVM strategy to improve the porcine oocyte IVM system (Figure 8). However, distinct from the well-established CNP-based biphasic IVM system, the extended IVM phase is not necessary, but even detrimental to porcine oocytes, emphasizing the importance of exploring the suitable technical parameters of the pre-IVM strategy among species. Concerning the mechanisms relevant to the beneficial effect of CNP pre-IVM, our results suggest that in addition to oocytes per se, the enhanced anti-apoptosis and resistance to oxidative stress in the cumulus cells may contribute considerably to improved oocyte quality, which has been largely overlooked in previous studies.

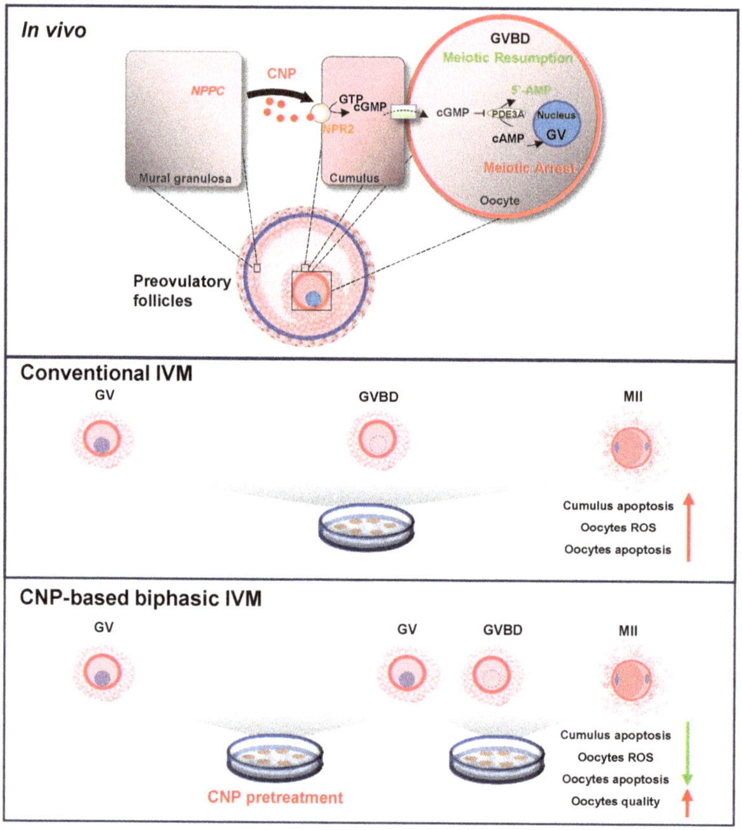

Figure 8. A model illustrating the importance of the CNP-based biphasic IVM system. In the growing follicles, CNP secreted from the mural granulosa cells acts on its specific receptor NPR2 located on the cumulus cells, thus resulting in the maintenance of meiotic arrest until LH-induced oocyte meiotic resumption. In vitro isolation of the oocytes triggers spontaneous and precocious meiotic resumption, ultimately compromising the developmental potential of the oocytes. In the CNP-based biphasic IVM system, exogenous CNP supplementation into the traditional medium not only inhibits the spontaneous meiotic resumption of COCs, but also reduces the apoptosis of cumulus, decreases the ROS and apoptosis of the oocytes, and improves the quality of the MII oocytes.

Author Contributions: Conceptualization, M.C., J.T., L.A., G.X. and J.L.; methodology, M.C., J.T., L.A. and G.X.; software, J.L., M.G. and X.W.; validation, J.L., M.G., X.W. and R.W.; formal analysis, J.L.; investigation, M.C., J.T. and L.A.; data curation, J.L. and R.W.; writing—original draft preparation, J.L., L.A. and M.C.; writing—review and editing, J.L., G.X., L.A., J.T. and M.C.; funding acquisition, J.T., L.A. and G.X. All authors have read and agreed to the published version of the manuscript.

Funding: This work was supported by grants from the National Natural Science Foundation of China (31930103 and 31972573), the Ningbo Major Science and Technology Project (2021Z112), the National Key R&D Program (2022YFD1300301), the Beijing Innovation Consortium of Livestock Research System (grant number BAIC05-2022), and the Linyi University High-level Talents (PhD) Research Start-up Fund (Science and Technology: Z6122060).

Informed Consent Statement: Not applicable.

Data Availability Statement: The data that support the findings of this study are available on request from the corresponding author.

Acknowledgments: All of the contributors of this study have been listed as authors of the manuscript.

Conflicts of Interest: The authors declare no conflict of interest.

References

1. Eppig, J.J. Coordination of nuclear and cytoplasmic oocyte maturation in eutherian mammals. *Reprod. Fertil. Dev.* **1996**, *8*, 485–489. [CrossRef] [PubMed]
2. Josefsberg, L.B.; Galiani, D.; Lazar, S.; Kaufman, O.; Seger, R.; Dekel, N. MPF governs MAPK activation and interphase suppression during meiosis of rat oocytes. *Biol. Reprod.* **2003**, *68*, 1282–1290. [CrossRef] [PubMed]
3. Norris, R.P.; Ratzan, W.J.; Freudzon, M.; Mehlmann, L.M.; Krall, J.; Movsesian, M.A.; Wang, H.; Ke, H.; Nikolaev, V.O.; Jaffe, L.A. Cyclic GMP from the surrounding somatic cells regulates cyclic AMP and meiosis in the mouse oocyte. *Development* **2009**, *136*, 1869–1878. [CrossRef] [PubMed]
4. Edwards, R.G. Maturation in vitro of mouse, sheep, cow, pig, rhesus monkey and human ovarian oocytes. *Nature* **1965**, *208*, 349–351. [CrossRef]
5. Schultz, R.M.; Montgomery, R.R.; Belanoff, J.R. Regulation of mouse oocyte meiotic maturation: Implication of a decrease in oocyte cAMP and protein dephosphorylation in commitment to resume meiosis. *Dev. Biol.* **1983**, *97*, 264–273. [CrossRef]
6. Gilchrist, R.B.; Luciano, A.M.; Richani, D.; Zeng, H.T.; Wang, X.; De Vos, M.; Sugimura, S.; Smitz, J.; Richard, F.J.; Thompson, J.G. Oocyte maturation and quality: Role of cyclic nucleotides. *Reproduction* **2016**, *152*, R143–R157. [CrossRef]
7. Thomas, R.E.; Thompson, J.G.; Armstrong, D.T.; Gilchrist, R.B. Effect of specific phosphodiesterase isoenzyme inhibitors during in vitro maturation of bovine oocytes on meiotic and developmental capacity. *Biol. Reprod.* **2004**, *71*, 1142–1149. [CrossRef]
8. Shu, Y.; Zeng, H.; Ren, Z.; Zhuang, G.; Liang, X.; Shen, H.; Yao, S.; Ke, P.; Wang, N. Effects of cilostamide and forskolin on the meiotic resumption and embryonic development of immature human oocytes. *Hum. Reprod.* **2008**, *23*, 504–513. [CrossRef]
9. Albuz, F.K.; Sasseville, M.; Lane, M.; Armstrong, D.T.; Thompson, J.G.; Gilchrist, R.B. Simulated physiological oocyte maturation (SPOM): A novel in vitro maturation system that substantially improves embryo yield and pregnancy outcomes. *Hum. Reprod.* **2010**, *25*, 2999–3011. [CrossRef]
10. Leal, G.R.; Monteiro, C.A.S.; Souza-Fabjan, J.M.G.; de Paula Vasconcelos, C.O.; Nogueira, L.A.G.; Ferreira, A.M.R.; Serapião, R.V. Role of cAMP modulator supplementations during oocyte in vitro maturation in domestic animals. *Anim. Reprod. Sci.* **2018**, *199*, 1–14. [CrossRef]
11. Xi, G.; An, L.; Jia, Z.; Tan, K.; Zhang, J.; Wang, Z.; Zhang, C.; Miao, K.; Wu, Z.; Tian, J. Natriuretic peptide receptor 2 (NPR2) localized in bovine oocyte underlies a unique mechanism for C-type natriuretic peptide (CNP)-induced meiotic arrest. *Theriogenology* **2018**, *106*, 198–209. [CrossRef] [PubMed]
12. Zhang, M.; Su, Y.; Sugiura, K.; Xia, G.; Eppig, J.J. Granulosa cell ligand NPPC and its receptor NPR2 maintain meiotic arrest in mouse oocytes. *Science* **2010**, *330*, 366–369. [CrossRef] [PubMed]
13. Hiradate, Y.; Hoshino, Y.; Tanemura, K.; Sato, E. C-type natriuretic peptide inhibits porcine oocyte meiotic resumption. *Zygote* **2014**, *22*, 372–377. [CrossRef] [PubMed]
14. Franciosi, F.; Coticchio, G.; Lodde, V.; Tessaro, I.; Modina, S.C.; Fadini, R.; Dal Canto, M.; Renzini, M.M.; Albertini, D.F.; Luciano, A.M. Natriuretic peptide precursor C delays meiotic resumption and sustains gap junction-mediated communication in bovine cumulus-enclosed oocytes. *Biol. Reprod.* **2014**, *91*, 61. [CrossRef]
15. Zhang, J.; Wei, Q.; Cai, J.; Zhao, X.; Ma, B. Effect of C-type natriuretic peptide on maturation and developmental competence of goat oocytes matured in vitro. *PLoS ONE* **2015**, *10*, e132318. [CrossRef]
16. Robinson, J.W.; Zhang, M.; Shuhaibar, L.C.; Norris, R.P.; Geerts, A.; Wunder, F.; Eppig, J.J.; Potter, L.R.; Jaffe, L.A. Luteinizing hormone reduces the activity of the NPR2 guanylyl cyclase in mouse ovarian follicles, contributing to the cyclic GMP decrease that promotes resumption of meiosis in oocytes. *Dev. Biol.* **2012**, *366*, 308–316. [CrossRef]
17. Wei, Q.; Zhou, C.; Yuan, M.; Miao, Y.; Zhao, X.; Ma, B. Effect of C-type natriuretic peptide on maturation and developmental competence of immature mouse oocytes in vitro. *Reprod. Fertil. Dev.* **2017**, *29*, 319–324. [CrossRef]
18. Soto-Heras, S.; Paramio, M.; Thompson, J.G. Effect of pre-maturation with C-type natriuretic peptide and 3-isobutyl-1-methylxanthine on cumulus-oocyte communication and oocyte developmental competence in cattle. *Anim. Reprod. Sci.* **2019**, *202*, 49–57. [CrossRef]
19. Zhang, T.; Fan, X.; Li, R.; Zhang, C.; Zhang, J. Effects of pre-incubation with C-type natriuretic peptide on nuclear maturation, mitochondrial behavior, and developmental competence of sheep oocytes. *Biochem. Biophys. Res. Commun.* **2018**, *497*, 200–206. [CrossRef]
20. Soto-Heras, S.; Menéndez-Blanco, I.; Catalá, M.; Izquierdo, D.; Thompson, J.G.; Paramio, M. Biphasic in vitro maturation with C-type natriuretic peptide enhances the developmental competence of juvenile-goat oocytes. *PLoS ONE* **2019**, *14*, e221663. [CrossRef]
21. Zhang, Y.; Wang, H.; Liu, W.; Yang, Y.; Wang, X.; Zhang, Z.; Guo, Q.; Wang, C.; Xia, G. Natriuretic peptides improve the developmental competence of in vitro cultured porcine oocytes. *Reprod. Biol. Endocrinol.* **2017**, *15*, 41. [CrossRef] [PubMed]
22. Grupen, C.G. The evolution of porcine embryo in vitro production. *Theriogenology* **2014**, *81*, 24–37. [CrossRef] [PubMed]
23. Tatemoto, H.; Sakurai, N.; Muto, N. Protection of porcine oocytes against apoptotic cell death caused by oxidative stress during in vitro maturation: Role of cumulus cells. *Biol. Reprod.* **2000**, *63*, 805–810. [CrossRef] [PubMed]

24. Li, H.J.; Sutton-Mcdowall, M.L.; Wang, X.; Sugimura, S.; Thompson, J.G.; Gilchrist, R.B. Extending prematuration with cAMP modulators enhances the cumulus contribution to oocyte antioxidant defence and oocyte quality via gap junctions. *Hum. Reprod.* **2016**, *31*, 810–821. [CrossRef]
25. Yuan, Y.Q.; Van Soom, A.; Leroy, J.; Dewulf, J.; Van Zeveren, A.; de Kruif, A.; Peelman, L.J. Apoptosis in cumulus cells, but not in oocytes, may influence bovine embryonic developmental competence. *Theriogenology* **2005**, *63*, 2147–2163. [CrossRef]
26. Corn, C.M.; Hauser-Kronberger, C.; Moser, M.; Tews, G.; Ebner, T. Predictive value of cumulus cell apoptosis with regard to blastocyst development of corresponding gametes. *Fertil. Steril.* **2005**, *84*, 627–633. [CrossRef]
27. Gu, L.; Wang, Q.; Sun, Q. Histone modifications during mammalian oocyte maturation: Dynamics, regulation and functions. *Cell Cycle* **2010**, *9*, 1942–1950. [CrossRef]
28. Racedo, S.E.; Wrenzycki, C.; Lepikhov, K.; Salamone, D.; Walter, J.; Niemann, H. Epigenetic modifications and related mRNA expression during bovine oocyte in vitro maturation. *Reprod. Fertil. Dev.* **2009**, *21*, 738–748. [CrossRef]
29. Andreu-Vieyra, C.V.; Chen, R.; Agno, J.E.; Glaser, S.; Anastassiadis, K.; Stewart, A.F.; Matzuk, M.M. MLL2 is required in oocytes for bulk histone 3 lysine 4 trimethylation and transcriptional silencing. *PLoS Biol.* **2010**, *8*, e1000453. [CrossRef]
30. Xiong, X.; Zhang, X.; Yang, M.; Zhu, Y.; Yu, H.; Fei, X.; Mastuda, F.; Lan, D.; Xiong, Y.; Fu, W. Oocyte-Specific Knockout of Histone Lysine Demethylase KDM2a Compromises Fertility by Blocking the Development of Follicles and Oocytes. *Int. J. Mol. Sci.* **2022**, *23*, 12008. [CrossRef]
31. Abbey, S.E.; Potter, L.R. Vasopressin-dependent inhibition of the C-type natriuretic peptide receptor, NPR-B/GC-B, requires elevated intracellular calcium concentrations. *J. Biol. Chem.* **2002**, *277*, 42423–42430. [CrossRef] [PubMed]
32. Combelles, C.M.; Gupta, S.; Agarwal, A. Could oxidative stress influence the in-vitro maturation of oocytes? *Reprod. Biomed. Online* **2009**, *18*, 864–880. [CrossRef] [PubMed]
33. Beaujean, N. Epigenetics, embryo quality and developmental potential. *Reprod. Fertil. Dev.* **2015**, *27*, 53–62. [CrossRef]
34. Liu, H.; Aoki, F. Transcriptional activity associated with meiotic competence in fully grown mouse GV oocytes. *Zygote* **2002**, *10*, 327–332. [CrossRef] [PubMed]
35. Fair, T.; Hyttel, P.; Motlik, J.; Boland, M.; Lonergan, P. Maintenance of meiotic arrest in bovine oocytes in vitro using butyrolactone I: Effects on oocyte ultrastructure and nucleolus function. *Mol. Reprod. Dev. Inc. Gamete Res.* **2002**, *62*, 375–386. [CrossRef]
36. Lonergan, P.; Faerge, I.; Hyttel, P.M.; Boland, M.; Fair, T. Ultrastructural modifications in bovine oocytes maintained in meiotic arrest in vitro using roscovitine or butyrolactone. *Mol. Reprod. Dev.* **2003**, *64*, 369–378. [CrossRef]
37. Alexander, B.; Coppola, G.; Di Berardino, D.; Rho, G.J.; St John, E.; Betts, D.H.; King, W.A. The effect of 6-dimethylaminopurine (6-DMAP) and cycloheximide (CHX) on the development and chromosomal complement of sheep parthenogenetic and nuclear transfer embryos. *Mol. Reprod. Dev. Inc. Gamete Res.* **2006**, *73*, 20–30. [CrossRef] [PubMed]
38. Tian, J.; Jia, Z.; Zhang, J.; An, L.; Wu, Z. Method for In Vitro Oocyte Maturation. Google Patents US10011818B2, 3 July 2018.
39. Scheuerer, B. Factors and methods of pig oocyte and embryo quality improvement and their application in reproductive biotechnology. *Reprod. Biol.* **2009**, *9*, 97–112. [CrossRef]
40. Kiyosu, C.; Tsuji, T.; Yamada, K.; Kajita, S.; Kunieda, T. NPPC/NPR2 signaling is essential for oocyte meiotic arrest and cumulus oophorus formation during follicular development in the mouse ovary. *Reproduction* **2012**, *144*, 187. [CrossRef]
41. Mcgee, E.; Spears, N.; Minami, S.; Hsu, S.; Chun, S.; Billig, H.; Hsueh, A.J. Preantral ovarian follicles in serum-free culture: Suppression of apoptosis after activation of the cyclic guanosine 3′, 5′-monophosphate pathway and stimulation of growth and differentiation by follicle-stimulating hormone. *Endocrinology* **1997**, *138*, 2417–2424. [CrossRef]
42. De La Fuente, R. Chromatin modifications in the germinal vesicle (GV) of mammalian oocytes. *Dev. Biol.* **2006**, *292*, 1–12. [CrossRef] [PubMed]
43. Kageyama, S.; Liu, H.; Kaneko, N.; Ooga, M.; Nagata, M.; Aoki, F. Alterations in epigenetic modifications during oocyte growth in mice. *Reproduction* **2007**, *133*, 85–94. [CrossRef] [PubMed]

genes

MDPI

Article

miR-302d Targeting of CDKN1A Regulates DNA Damage and Steroid Hormone Secretion in Bovine Cumulus Cells

Jianbo Liu [1,2,†], Jiabao Zhang [1,†], Yi Zheng [1], Guokun Zhao [1], Hao Jiang [1] and Bao Yuan [1,*]

1 Department of Laboratory Animals, College of Animal Sciences, Jilin University, Changchun 130062, China; liu8541@126.com (J.L.); zjb@jlu.edu.cn (J.Z.); zhaogk20@mails.jlu.edu.cn (G.Z.); jhhaojiang@jlu.edu.cn (H.J.)
2 Experimental Testing Center, Jilin Agricultural Science and Technology University, Jilin 132101, China
* Correspondence: yuan_bao@jlu.edu.cn
† These authors contributed equally to this work.

Abstract: (1) Background: DNA damage in cumulus cells hinders oocyte maturation and affects steroid hormone secretion. It is crucial to identify the key factors that regulate cellular DNA damage and steroid hormone secretion. (2) Methods: Treatment of bovine cumulus cells with bleomycin to induce DNA damage. The effects of DNA damage on cell biology were determined by detecting changes in DNA damage degree, cell cycle, viability, apoptosis, and steroid hormones. It was verified that mir-302d targeted regulation of CDKN1A expression, and then affected DNA damage and steroid hormone secretion in cumulus cells. (3) Results: Bleomycin induced increased DNA damage, decreased G1-phase cells, increased S-phase cells, inhibited proliferation, promoted apoptosis, affected E_2 and P_4 secretion, increased *CDKN1A* expression, and decreased miR-302d expression. Knockdown of *CDKN1A* reduced DNA damage, increased G1-phase cells, decreased G2-phase cells, promoted proliferation, inhibited apoptosis, increased E_2 and P_4 secretion, and increased the expression of *BRCA1*, *MRE11*, *ATM*, *CDK1*, *CDK2*, *CCNE2*, *STAR*, *CYP11A1*, and *HSD3B1*. The expression of *RAD51*, *CCND1*, *p53*, and *FAS* was decreased. Overexpression of CDKN1A resulted in the opposite results. miR-302d targets *CDKN1A* expression to regulate DNA damage and then affects the cell cycle, proliferation, apoptosis, steroid hormone secretion, and the expression of related genes. (4) Conclusions: miR-302d and CDKN1A were candidate molecular markers for the diagnosis of DNA damage in bovine cumulus cells.

Keywords: bovine; cumulus cells; CDKN1A; miR-302d; DNA damage; steroid hormone

Citation: Liu, J.; Zhang, J.; Zheng, Y.; Zhao, G.; Jiang, H.; Yuan, B. miR-302d Targeting of CDKN1A Regulates DNA Damage and Steroid Hormone Secretion in Bovine Cumulus Cells. *Genes* **2023**, *14*, 2195. https://doi.org/10.3390/genes14122195

Academic Editor: Shaochen Sun

Received: 8 November 2023
Revised: 5 December 2023
Accepted: 7 December 2023
Published: 10 December 2023

1. Introduction

Cumulus cell–oocyte complexes (COCs) are recognized as the functionally responsible unit of the germ cell in female mammals [1]. Gap junctions between cumulus cells (CCs) and oocytes modulate the expression of hormonal factors and related regulators in oocytes, thereby affecting oocyte maturation [2]. DNA damage can be caused by diseases or environmental changes [3]. Before ovulation, oocytes need to maintain corresponding physiological functions for a long time in the body and will accumulate some damage, including DNA damage. CCs have a protective effect on oocytes and can reduce damage to oocytes from harmful substances during oocyte maturation. The expression of genes in CCs is highly correlated with the developmental competence of oocytes, mainly in DNA damage, cell cycle, apoptosis, metabolism, meiosis, and cell signaling pathways [4–6]. Animals need an efficient DNA damage recognition and repair (DDR) mechanism to ensure the reproductive capacity of oocytes in the ovary. Failure to properly activate DDR mechanisms can result in an inability of oocytes to mature and fertilize properly, leading to diseases or mutations in offspring [7,8]. After DNA damage occurs, chromatin reorganization, cell cycle arrest, apoptosis, or other forms of cell death may occur if the DDR is not activated correctly [9–11]. DNA damage in CCs could inhibit the resumption of oocyte meiosis through gap junctions [12]. DNA damage has been linked to abnormal levels of estrogen(E_2) and progesterone(P_4) in women [13].

MicroRNAs (miRNAs) play a critical role in the post-transcriptional modulation of gene expression and have become a focus of research on reproductive regulation mechanisms [14]. Studies have shown that DNA damage is regulated by miRNAs [15]. miRNA affects estradiol and progesterone secretion in CCs [16,17]. The abnormal expression of miR-302d is related to the productivity, migration, and apoptosis of endometrial cancer cells [18]. miR-302d can affect the cell viability and apoptosis of chondrocytes by regulating target genes [19]. Earlier studies reported that CDKN1A is a protein of 165 amino acids and belongs to the CIP/Kip group of CDK inhibitors [20]. The carboxyl group of CDKN1A reacts with PCNA to inhibit DNA damage [21]. Mammalian cell cycle progression is regulated by CDKs and the regulatory subunit cyclins, cell cycle progression is triggered by partial phosphorylation of Rb by CDK-Cyclin, and CDKN1A disrupts this interaction and affects cell cycle progression and cell proliferation [22,23]. Earlier studies showed that miR-302d and CDKN1A were abnormally expressed during DNA damage in bovine cumulus cells [24]. This indicated that CDKN1A and miR-302d may be potential markers of DNA damage. However, it is unclear whether miR-302d or CDKN1A is involved in the regulation of DNA damage and steroid hormone production in bovine cumulus cells. Bleomycin (BLM) is a chemotherapeutic agent that is often used to induce DNA damage in cells [25]. After DNA damage, γH2AX is immediately recruited to the lesion site and is the most effective biomarker to detect DNA damage [26]. Therefore, this study investigated the regulatory effects of miR-302d and CDKN1A on DNA damage in bovine cumulus cells and their effects on steroid hormone secretion.

2. Materials and Methods

2.1. Collection and Culture

Collection and culture of bovine CCs were performed in the same way as previously described [24]. In short, the ovaries were taken from healthy cows from Changchun Haoyue Halal Meat Products Co., Ltd.* (Changchun, Jilin, China) and transferred to the laboratory within 1 h. Follicular fluid containing COCs in follicles 3–8 mm in diameter was obtained using a 10 mL syringe. After three washes with HEPES (Gibco, Paisley, Scotland, UK), more than five layers of tightly packed CCs were collected under a microscope. CCs were subsequently isolated from oocytes using 0.1% hyaluronidase. CCs were then collected after centrifugation for 5 min and washed twice with phosphate-buffered saline (PBS). CCs were then cultivated in DMEM/F12 (Gibco, Grand Island, NY, USA). The culture medium was supplemented with 1% penicillin, streptomycin (HyClone, Logan, UT, USA), and 10% fetal bovine serum (Biological Industries, Kibbutz Beit Haemek, Israel), and placed in an incubator at 38.5 °C under 5% CO_2.

2.2. Cumulus Cell Treatment

After CCs were cultured in 6-well plates (5×10^5 cells/well) for 6 h, the original culture medium was removed. A culture medium containing 0 µM BLM or 200 µM BLM (Thermo, Carlsbad, CA, USA, diluted with cell culture medium) was added to the culture medium and treated for 3 h.

2.3. Cell Transfection

After culturing the cells in a 6-well plate with 70% cell fusion, 100 nM si-*CDKN1A* was transfected for the *CDKN1A* knockdown assay in accordance with the instructions of the RiboFECT CP Transfection Reagent (Ribo, Guangzhou, China). Cells were transfected with 500, 1000, and 1500 ng/mL *CDKN1A* overexpression plasmids for the *CDKN1A* overexpression assay. miRNA regulation experiments were performed by transfecting 100 nM miR-302d mimics or miR-302d inhibitor. The specific operations were as follows: (a) Diluted si-*CDKN1A*, *CDKN1A* overexpression plasmid, or miR-302d mimics/inhibitor were mixed with RiboFECT CP Transfection Reagent (Ribo, China), and the mixture was incubated indoors for 15 min. (b) The transfection mixture was added dropwise to the cells, and the plates were incubated in an oven at 38.5 °C and 5% CO_2 for 48 h (miRNA-24 h). siRNA, overexpression plasmids, or miR302d mimics/inhibitor were synthesized by GenePharma (GenePharma, Suzhou, China). For the siRNA sequences, see Supplementary Table S1.

2.4. RNA Extraction and RT–qPCR

After transfection or treatment of CCs, total RNA was isolated with TriPure (Roche, Basel, Switzerland), and different reagents were used according to the manufacturer's instructions. All RNA samples were tested for integrity and purity. An Agilent 2100 device (Agilent, Karlsruhe, Baden-Württemberg, Germany) was used to detect integrity, and a NanoDrop ND-2000 Spectrophotometer (Thermo, Waltham, MA, USA) was used to detect concentration and purity. RNAs with integrity meeting RIN value $\geq$ 8.0, 28S/18S $\geq$ 1.5, normal 5S peak, and purity meeting OD 260/280 $\geq$ 1.8 and OD 260/230 $\geq$ 2.2 were used for further experiments. Reverse transcription of RNA was performed with a reverse transcription kit (Tiangen, Beijing, China), which was used for RT–qPCR. Power SYBR Green (Tiangen, China) was used for real-time PCR assays. The overall 20 µL RT–qPCR mixture consisted of 8 µL deionized water, 10 µL SuperReal PreMix Plus (Tiangen, China), 1 µL of cDNA, and 0.5 µL of forward and reverse primers (10 mM). All real-time PCR programs consisted of 120 s of denaturation at 95 °C and 45 cycles (15 s of denaturation at 95 °C, 15 s of annealing at 60 °C, and 30 s of extension at 72 °C). Treatments were repeated at least three times in each group. Comparative expression was calculated and normalized by the $2^{-\Delta\Delta Ct}$ method with respect to *GAPDH (miRNA-U6)*. The sequences of the primers used are shown in Supplementary Table S1.

2.5. γH2AX Detection

After transfection or treatment, 500 µL of trypsin was used to dissociate the cells for 3 min. Cells were collected in a 1.5 mL centrifuge tube, then centrifuged at 500 g for 5 min and washed three times with PBS. The cells were then fixed with 100 µL (4%) paraformaldehyde solution for 10 min. Next, 900 µL of pre-cooled methanol was directly added, mixed gently, and incubatedon ice for 30 min. After centrifugation, the solution was removed, and 1 mL of incubation solution (500 g BSA in 100 mL PBS) was added to the centrifuge tube and mixed well. The mixture was centrifuged at 500× *g* for 5 min and the incubation solution was removed. Cells were treated with 100 µL γH2AX antibody (CST, Danvers, MA, USA) diluent for 1 h at room temperature. The ratio of antibody to incubation solution was 1:50. Afterwards, the cells were washed three times with incubation solution. Finally, cytosolic cells were resolved and analyzed in 200 µL of PBS with the use of a flow cytometer (CytoFLEX, Beckman Coulter, Brea, CA, USA).

2.6. Cell Proliferation Assays

A CCK-8 kit (Beyotime, Shanghai, China) was used to measure the proliferation of bovine CCs. Briefly, 96-well plates contained 1×10^4 CCs per well in 100 µL of medium and were incubated at 38.5 °C and 5% CO_2 for 12 h. After culturing the bovine CCs for 24 h, the cells were treated for 0 h, 24 h, 48 h, or 72 h. A solution of 10 µL CCK-8 was then injected into every well, and the incubation was carried out at 38.5 °C for 3 h. An enzyme marker (BioTek Instruments, Winooski, VT, USA) was used to measure absorbance at 450 nm.

2.7. Cell Cycle Analysis

Determination of the cell cycle was performed under strict adherence to the instructions of the Cycle Kit (Beyotime, China). In brief, 6-well plates were inoculated with 5×10^5 CCs per well. Following transfection or treatment, CCs were collected into 1.5 mL centrifuge tubes and centrifuged at 500× *g* over 5 min. The cells were frozen in 70% ethanol for 24 h. After cleaning with PBS, the cells were re-suspended in 500 µL of dye buffer containing PI (50 µg/mL) and RNaseA (1 mg/mL) and incubated at 37 °C for 30 min away from light. The suspension was resuspended in 100 µL of PBS. Cell cycle distributions were detected in a flow cytometer (CytoFLEX, Beckman, USA), and data were processed using MODFIT software (v. 2.0, Verity Software House, Topsham, ME, USA).

2.8. Apoptosis Analysis

An assay for apoptosis was performed in strict accordance with the directions of the Apoptosis Detection Kit (Beyotime, China). Bovine CCs (5×10^5 cells per well) were inoculated into 6-well plates. After treatment, in each centrifuge tube, 100 μL of PBS was added, followed by 5 μL of FITC solution and 5 μL of propidium iodide (20 μg/mL). It was then grown in the shade at room temperature for 15 min. Measurement of samples was performed within 2 h by a flow cytometer (CytoFLEX, Beckman Coulter, USA). Propidium iodide and FITC staining were used to distinguish apoptotic cells from dead cells.

2.9. Analysis of Steroid Hormone Secretion

After the cells were transfected, the levels of estradiol or progesterone secreted by bovine cumulus cells were measured by ELISA kits (Elabscience, Wuhan, Hubei, China). Briefly, (a) 100 μL of cell culture solution was centrifuged at 500× *g* over 5 min, and the supernatant was placed in a centrifuge tube. (b) An amount of 50 μL standard solution of different concentrations per well was added to the standard well. (c) A total of 100 μL 1 × HRP (10 μL 100 × HRP concentrate + 990 μL diluent)-labeled estradiol or progesterone antibody was added to the sample wells and standard wells and incubated at 37 °C for 1 h. (d) The liquid was emptied from each reaction hole, and 50 μL of substrate A and B were added to each reaction well and incubated at 37 °C for 15 min under dark conditions. (e) Fifty microliters of termination solution were added to every reaction well, and the OD value of each reaction well was measured at 450 nm within 15 min using an enzyme-labeling instrument (BioTek Instruments, USA).

2.10. DualLuciferase Reporter Gene Analysis

Within the dual luciferase reporter gene test, the *CDKN1A*3′UTR was cloned and inserted into the pmiR-RB REPORT™ plasmid to construct the primr-CDKN1A-3′UTR-WT plasmid. Next, miR-302d mimics and NC, primr-CDKN1A-3′UTR-WT plasmid, or pmiR-RB REPORT™ plasmid were transfected into cumulus cells. CCs were cultured in 96-well plates (1×10^4 cells/well) and transfected with Ribo transfection reagent (Ribo, China) over 48 h. Then, the fluorescence intensity was measured by a fluorescence microplate reader. Reporter plasmid construction and luciferase reporter gene assays were performed by Guangzhou Ribo Biotechnology Co., Ltd. (Guangzhou, China, www.ribobio.com, accessed on 1 December 2023).

2.11. Statistical Analysis

Statistical analysis was performed with SPSS 23.0 (IBM, Armonk, NY, USA) [27]. Data analysis between the two groups used an unpaired *t*-test. Data were shown as mean ± standard deviation (SD). Significant differences are denoted by * ($p \leq 0.05$) and ** ($p < 0.01$).

3. Results

3.1. Effect of DNA Damage on Cumulus Cells in Bovines

In this study, a flow cytometry assay after treating bovine cumulus cells with 200 μM BLM revealed that the rate of γH2AX-positive cells was observed to be higher in the BLM group than in the NC group ($p < 0.01$, Figure 1a). Since γH2AX is a marker of DNA damage, BLM successfully induced DNA damage in cumulus cells. In this study, the changes in CDKN1A expression after DNA damage occurred in the cells were analyzed by RT–qPCR, and we found that the expression of *CDKN1A* was significantly higher in the BLM group than in the control group ($p < 0.01$), which was in accordance with the results of a previous study (Figure 1b). We clarified the variations in the proliferation and cell cycle of CCs after DNA damage occurred. In this study, after treating the cells with BLM, a flow cytometry assay revealed that the CCs in the BLM group passed through the G1 phase rapidly and were significantly blocked in the S phase ($p \leq 0.05$), but no significant difference was observed in the G2 phase ($p > 0.05$, Figure 1c). CCK-8 detection demonstrated that the cell proliferation level in the BLM group was obviously lower than that in the NC group at 24 h ($p < 0.01$), 48 h ($p < 0.01$), and 72 h ($p \leq 0.05$) after DNA damage occurred (Figure 1d). Flow cytometry detection of apoptosis revealed that the relative apoptosis level in the NC group was notably lower than that in the BLM group ($p \leq 0.05$, Figure 1e). To verify the effect of DNA damage on steroid hormone secretion in cumulus cells, the levels of E_2 and P_4 in cell cultures were measured by ELISA. The results showed that the secretion of E_2 and P_4 in cumulus cells was significantly reduced after the occurrence of DNA damage ($p \leq 0.05$, Figure 1f, please check raw data in Table S2).

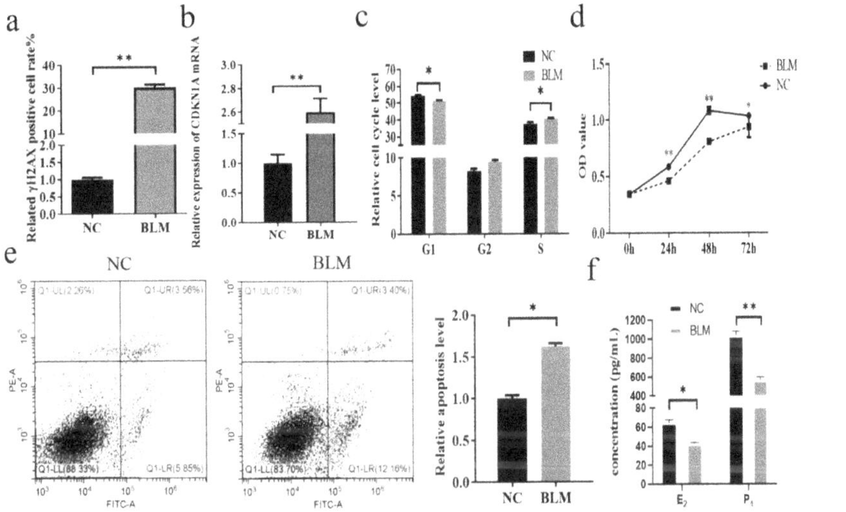

Figure 1. Effect of DNA damage on bovine cumulus cells. Significant differences are denoted by * ($p \leq 0.05$) and ** ($p < 0.01$). After 200 μM BLM treatment of bovine cumulus cells for 3 h, the γH2AX-positive cell rate, cell cycle, and apoptosis were detected by flow cytometry; cell proliferation by CCK-8; *CDKN1A* expression by RT–qPCR; and E_2 and P_4 by ELISA. (**a**) The rate of γH2AX-positive cells was significantly higher than that in the NC after BLM treatment ($p \leq 0.05$). (**b**) DNA damage caused a marked rise in *CDKN1A* mRNA expression ($p \leq 0.05$). (**c**) After DNA damage, compared to the NC group, the BLM group had fewer cells in the G1 phase ($p \leq 0.05$), clearly more cells in the S phase ($p \leq 0.05$), and more cells in the G2 phase, but the difference was not marked ($p > 0.05$). (**d**) After the onset of DNA damage, the cell proliferation levels at 24 h ($p < 0.01$), 48 h ($p < 0.01$), and 72 h ($p \leq 0.05$) in the BLM group were significantly lower than those in the NC group, while the difference was not significant at 0 h ($p > 0.05$). (**e**) The relative apoptotic level of cells was significantly increased after DNA damage occurred ($p \leq 0.05$); left upper/Q1-UL: dead cells, right upper/Q1-UR: advanced apoptotic cells, right lower/Q1-LR: earlier apoptotic cells, left lower/Q1-LL: nonexpired cells. (**f**) After DNA damage occurred, the E_2 and P_4 contents in the cell cultures of the BLM group were significantly reduced relative to those of the NC group ($p \leq 0.05$).

3.2. *Effect of CDKN1A on DNA Damage and Steroid Hormone Secretion in Bovine Cumulus Cells*

To elucidate the regulatory role of *CDKN1A* in DNA damage and steroid hormone secretion in bovine cumulus cells, *CDKN1A* was knocked down and overexpressed in the cells. After transfection of siCDKN1A-1, siCDKN1A-2, and siCDKN1A-3 into cumulus cells, RT–qPCR showed that the expression of CDKN1A was significantly decreased in the siCDKN1A-1 ($p \leq 0.05$) and siCDKN1A-2 ($p < 0.01$) groups, while the expression was increased in the siCDKN1A-3 group (Figure 2a). Therefore, siCDKN1A-2 was used as a siRNA that specifically interferes with CDKN1Ain this paper. Next, *CDKN1A* overexpression plasmids were constructed. The results of RT–qPCR showed that the optimal treatment concentration was 500 ng/mL ($p < 0.01$), and the subsequent *CDKN1A* overexpression plasmids were treated with this concentration (Figure 2b). Figure 2c shows that inhibition of *CDKN1A* significantly attenuated the γH2AX-positive cell rate, and overexpression of *CDKN1A* significantly enhanced the γH2AX-positive cell rate ($p \leq 0.05$). Moreover, remarkable changes in the expression of DNA damage-related genes and the expression of *BRCA1*, *MRE11*, and *ATM* were significantly increased after *CDKN1A* inhibition, while the expression of *RAD51* was markedly diminished ($p \leq 0.05$). The promotion of *CDKN1A* expression resulted in suppression of *BRCA1*, *MRE11*, and *ATM* expression ($p \leq 0.05$) and elevation of *RAD51* expression ($p < 0.01$). Cell cycle analysis by flow cytometry showed that the inhibition of *CDKN1A* blocked cell passage through the G1 ($p \leq 0.05$) and S phases ($p \leq 0.05$) and promoted rapid cell passage through the G2 phase ($p \leq 0.05$, Figure 2e). The CCK-8 assay of proliferation levels after cell transfection showed that inhibition of *CDKN1A* significantly promoted proliferation (24 h, 48 h) and that overexpression of *CDKN1A* significantly inhibited cell proliferation levels ($p \leq 0.05$, Figure 2f). The results of the RT–qPCR assay of cell cycle- and proliferation-related genes showed that knockdown of *CDKN1A* caused a significant increase in the expression of *CDK1* ($p < 0.01$), *CDK2* ($p \leq 0.05$), and *CCNE2* ($p \leq 0.05$) and a significant decrease in the expression of *CCND1* ($p \leq 0.05$, Figure 2h). Overexpression of *CDKN1A* caused a marked diminish in the expression of *CDK1*, *CDK2*, and *CCNE2* ($p \leq 0.05$) and a dramatic increase in the expression of *CCND1* ($p < 0.01$). Apoptosis assays showed that inhibition of *CDKN1A* substantially reduced the level of apoptosis ($p \leq 0.05$) and that overexpression of *CDKN1A* induced apoptosis ($p < 0.01$, Figure 2g,i). The results of the apoptosis-related gene expression profile assay showed that knockdown of *CDKN1A* caused a drastic reduction in the *BAX*/*BCL2* ratio ($p \leq 0.05$), *p53* ($p \leq 0.05$) and *FAS* ($p < 0.01$), and overexpression of *CDKN1A* promoted the *BAX*/*BCL2* ratio ($p < 0.01$), *p53* ($p \leq 0.055$), and *FAS* ($p < 0.01$, Figure 2j,k). ELISA results for E_2 and P_4 levels in cell cultures showed that E_2 ($p \leq 0.05$) and P_4 ($p < 0.01$) levels increased significantly after *CDKN1A* inhibition, and overexpression of *CDKN1A* significantly promoted the secretion of E_2 and P_4 ($p < 0.01$, Figure 2l). Analysis of steroid hormone secretion-related gene expression showed that the knockdown of *CDKN1A* distinctly facilitated the expression of *STAR* ($p < 0.01$), *CYP11A1* ($p < 0.01$), and *HSD3B1* ($p \leq 0.05$). In contrast, *CDKN1A* overexpression markedly inhibited the expression of *STAR* ($p \leq 0.05$), *CYP11A1* ($p < 0.01$), and *HSD3B1* ($p < 0.01$, Figure 2m).

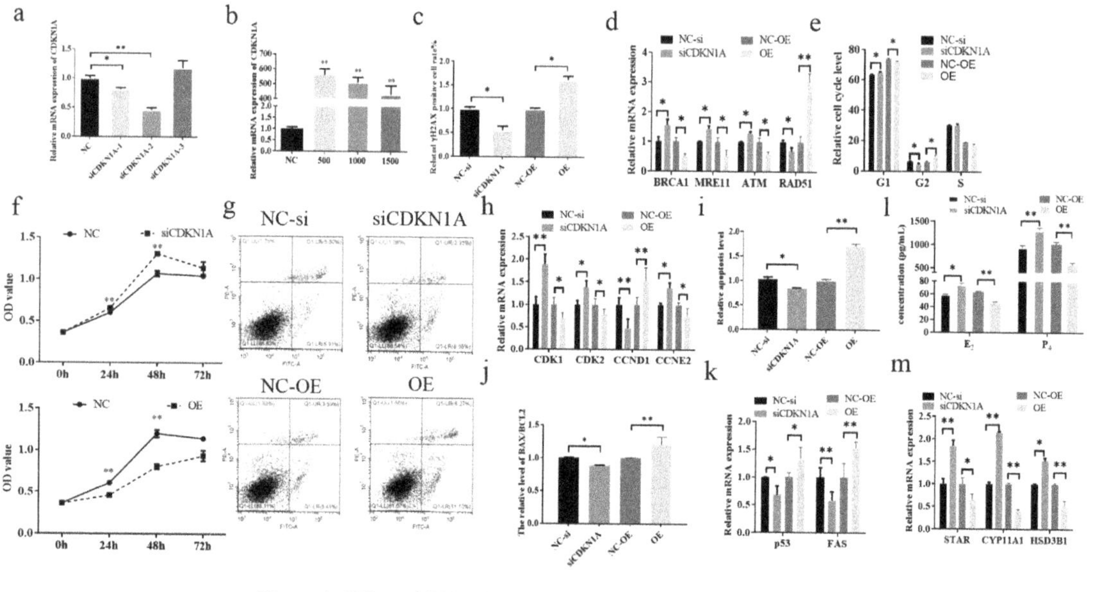

Figure 2. Effect of DNA damage on bovine cumulus cells. Significant differences are denoted by * ($p \leq 0.05$) and ** ($p < 0.01$). After knocking down or overexpressing *CDKN1A*, the cell cycle, γH2AX-

positive cell rate, and apoptosis were detected by flow cytometry; cell proliferation was detected by CCK-8; E_2 and P_4 content in cell culture medium was detected by ELISA; and related gene expression was detected by RT–qPCR. (**a**) After siRNA interference of the cells, the siCDKN1A-1 group ($p \leq 0.05$) and siCDKN1A-2 ($p < 0.01$) group showed a marked decrease, and the siCDKN1A-3 group showed an increase ($p > 0.05$). (**b**) The expression level of *CDKNA1* substantially rose after transfection with the overexpression plasmid ($p < 0.01$), and CDKNA1 expression was the highest in group 500. (**c**) The γH2AX-positive cell rate was significantly reduced after knocking down *CDKN1A* in cumulus cells. After overexpression of *CDKN1A*, the γH2AX-positive cell rate ($p \leq 0.05$) was significantly increased. (**d**) Inhibition of *CDKN1A* promoted the expression of *BRCA1*, *MRE11*, and *ATM* and suppressed the expression of *RAD51* ($p \leq 0.05$). After overexpression of *CDKN1A*, the expression of *BRCA1*, *MRE11*, and *ATM* was suppressed ($p \leq 0.05$), while *RAD51* expression was markedly increased ($p < 0.01$). (**e**) Transfection with siCDKN1A markedly increased the number of cells in G1 phase ($p \leq 0.05$), significantly declined the number of CCs in G2 phase ($p \leq 0.05$), and minimized the number of cells in S phase, but the change was not significant ($p > 0.05$). Overexpression of *CDKN1A* declined the number of cells in G1 phase and aggrandized the number of cells in G2 phase ($p \leq 0.05$), and the change in S phase was not marked ($p > 0.05$). (**f**) After *CDKN1A* was knocked down, the cell proliferation level at 24 h and 48 h was significantly higher than that of the NC group ($p < 0.01$), and the difference between 0 h and 72 h was not significant ($p > 0.05$). *CDKN1A* overexpression significantly inhibited cell proliferation at 24 h and 48 h ($p < 0.01$), and the variations between 0 h and 72 h were not significant ($p > 0.05$). (**g**) The results of the apoptosis level assay. After transfection of cells, left upper/Q1-UL: dead cells, right upper/Q1-UR: advanced apoptotic cells, right lower/Q1-LR: earlier apoptotic cells, left lower/Q1-LL: nonexpired cells. (**h**) Inhibition of *CDKN1A* significantly promoted the expression of *CDK1* ($p < 0.01$), *CDK2* ($p \leq 0.05$), and *CCNE2* ($p \leq 0.05$) and significantly inhibited the expression of *CCND1* ($p \leq 0.05$). After overexpression of *CDKN1A*, the expression of *CDK1*, *CDK2*, and *CCNE2* was substantially reduced ($p \leq 0.05$), and *CCND1* expression was considerably increased ($p < 0.01$). (**i**) Inhibition of *CDKN1A* markedly declined the level of apoptosis ($p \leq 0.05$). The relative apoptotic level of cells was markedly enhanced after overexpression of *CDKN1A* ($p < 0.01$). (**j**,**k**) Inhibition of *CDKN1A* brought about a substantial diminished in the *BAX*/*BCL2* ratio ($p \leq 0.05$) and *p53* ($p \leq 0.05$) and *FAS* ($p < 0.01$) expression. Overexpression of *CDKN1A* induced a strong increase in the *BAX*/*BCL2* ratio ($p < 0.01$) and *p53* ($p \leq 0.05$) and *FAS* ($p < 0.01$) expression. (**l**) Inhibition of *CDKN1A* promoted the production of E_2 ($p \leq 0.05$) and P_4 ($p < 0.01$) by cumulus cells, and overexpression of *CDKN1A* inhibited the secretion of E_2 and P_4 ($p < 0.01$). (**m**) Inhibition of *CDKN1A* induced the expression of *STAR* ($p < 0.01$), *CYP11A1* ($p < 0.01$) and *HSD3B1* ($p \leq 0.05$), and upregulation of *CDKN1A* expression inhibited *STAR* ($p \leq 0.05$), *CYP11A1* ($p < 0.01$), and *HSD3B1* ($p < 0.01$) expression.

3.3. *miR-302d Targeted Binding to the* CDKN1A *3′UTR and Inhibited Its Expression*

To identify miRNAs that target and regulate *CDKN1A* during DNA damage in cumulus cells, this study utilized the TargetScan database (www.targetscan.org, accessed on 2 April 2023) in conjunction with data from earlier studies to screen miRNAs. TargetScan database predictions illustrate that miR-302d targets binding to the *CDKN1A* 3′UTR at positions 614 to 620 (Figure 3a). BLM treatment of bovine cumulus cells resulted in a significant down regulation of miR-302d expression ($p < 0.01$, Figure 3b), which was opposite to the trend of *CDKN1A* expression (Figure 1b) and was also in agreement with the results of earlier studies [24]. Transfection of bovine cumulus cells with miR-302d mimics enhanced miR-302d expression and significantly diminished *CDKN1A* expression ($p < 0.01$, Figure 3c,d). The miR-302d inhibitor dramatically inhibited miR-302dexpression while promoting *CDKN1A* expression ($p \leq 0.05$). The dual luciferase reporter assay showed that miR-302d mimics remarkably inhibited the expression of $CDKN1A^{WT}$ (Figure 3e).

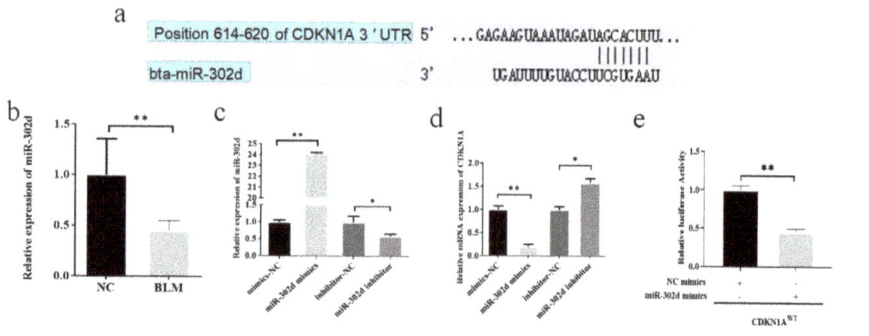

Figure 3. miR-302d targeted binding to the *CDKN1A* 3′UTR and inhibited its expression. Significant differences are denoted by * ($p \leq 0.05$) and ** ($p < 0.01$). (**a**) Predictive discovery of miR-302d targeting binding to the *CDKN1A* 3′UTR. (**b**) BLM treatment of cumulus cells significantly inhibited miR-302d expression ($p < 0.01$). (**c**) Transfection of miR-302d mimics significantly increased miR-302d expression levels ($p < 0.01$), and the inhibitor significantly inhibited miR-302d expression ($p \leq 0.05$). (**d**) Transfection of miR-302d mimics remarkably inhibited *CDKN1A* expression levels ($p < 0.01$), and the inhibitor significantly promoted *CDKN1A* expression ($p \leq 0.05$). (**e**) Transfection with miR-302d mimics substantially lowered the relative activity of the wild-type reporter vector luciferase ($p < 0.01$).

3.4. miR-302d Regulated DNA Damage and Steroid Hormone Secretion in Bovine CCs

In the present study, we clarified that miR-302d in bovine cumulus cells regulates the expression of *CDKN1A*. The expression of γH2AX in bovine CCs transfected with miR-302d mimics or inhibitors was examined by flow cytometry, and the results showed that the rate of γH2AX-positive cells was reduced by miR-302d overexpression, and inhibition of miR-302d promoted the γH2AX-positive cell rate ($p \leq 0.05$, Figure 4a). The results of RT–qPCR detection of DNA damage-related genes showed that transfection of miR-302d mimics led to a pronounced increase in the expression of *BRCA1*, *RAD51*, *MRE11* and *ATM* in cells ($p \leq 0.05$), while the inhibitor resulted in a distinct decrease in the expression of *BRCA1* and *RAD51* ($p \leq 0.05$), a reduction in *MRE11* expression but with a nonsignificant difference ($p > 0.05$), and a slight but nonsignificant difference in the expression of *ATM* ($p > 0.05$, Figure 4b). miR-302d overexpression caused a noticeable decline in the number of cells in the G1 phase ($p \leq 0.05$), a visible increase in the S phase ($p \leq 0.05$), and an expansion in the number of cells in the G2 phase, but the difference was not significant ($p > 0.05$). Suppression of miR-302d drastically blocked the process of the G1 phase and caused cells to pass through the S phase rapidly ($p \leq 0.05$), and the quantities of cells in phase G2 decreased, but the difference was not significant ($p \leq 0.05$, Figure 4c). miR-302d overexpression led to an appreciable rise in the expression of *CDK1* ($p \leq 0.05$), *CCND1* ($p < 0.01$) and *CCNE2* ($p < 0.01$) and a significant decrease in the expression of *CDK2* ($p \leq 0.05$) in cumulus cells (Figure 4d). Inhibition of miR-302d resulted in a significant decrease in *CDK1* and *CCND1* ($p < 0.01$), an increase in *CCNE2* expression, but the variation was not marked ($p > 0.05$), and an increase in *CDK2* expression, but the difference was not obvious ($p > 0.05$). The results showed that overexpression of miR-302d significantly promoted cell proliferation (24 h and 48 h, $p \leq 0.05$), and interference with miR-302d expression resulted in significant inhibition of proliferation (24 h ($p \leq 0.05$), 48 h ($p < 0.01$), Figure 4e). Cell transfection with miR-302d mimics inhibited the relative level of apoptosis and inhibitor-induced apoptosis ($p \leq 0.05$, Figure 4f,g). After cell transfection, RT–qPCR analysis of apoptosis-related genes showed that miR-302d overexpression dramatically reduced the *BAX*/*BCL2* ratio ($p \leq 0.05$) and *FAS* expression ($p < 0.01$) and promoted *p53* expression ($p < 0.01$, Figure 4g–i). A decrease in miR-302d expression caused a significant elevation in the *BAX*/*BCL2* ratio and *FAS* expression and suppressed *p53* expression ($p \leq 0.05$). ELISA analysis showed that an increase in miR-302d expression significantly promoted the generation of E_2 ($p \leq 0.05$) and P_4 ($p < 0.01$), while a reduction in miR-302d expression was associated with a notable decline in the synthesis of E_2 and P_4 ($p \leq 0.05$, Figure 4j). RT–qPCR detection of steroid hormone-related genes showed that miR-302d overexpression dramatically promoted the expression of *STAR*, *CYP11A1*, and *HSD3B1* ($p < 0.01$), and inhibition of miR-302d significantly interfered with the expression of *STAR* ($p \leq 0.05$), *CYP11A1* ($p < 0.01$), and *HSD3B1* ($p \leq 0.05$, Figure 4k).

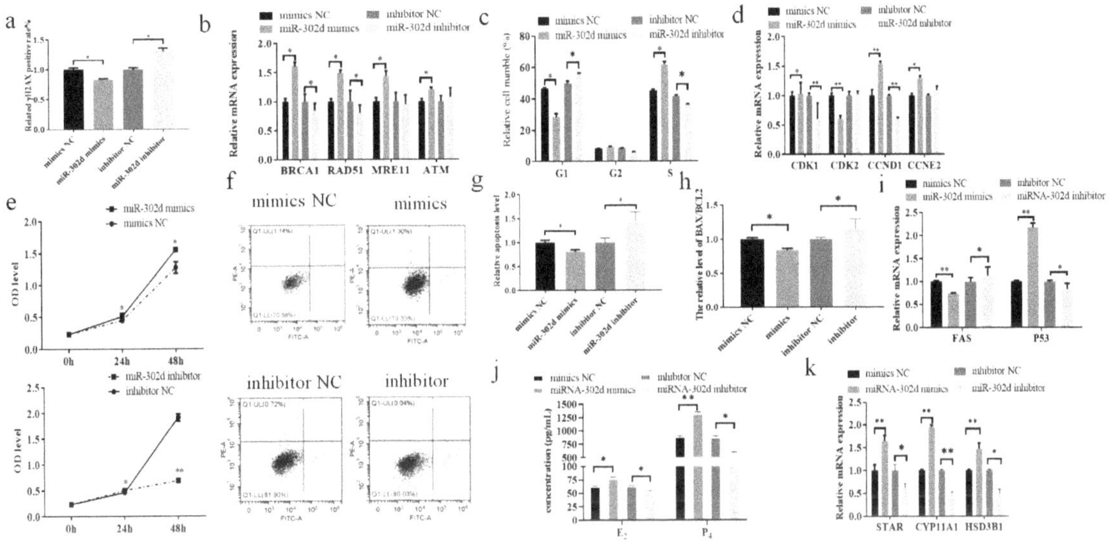

Figure 4. miR-302d regulated DNA damage and steroid hormone secretion in CCs. Significant differences are denoted by * ($p \leq 0.05$) and ** ($p < 0.01$). Upon transfection of cells with miR-302d mimics or inhibitors, flow cytometry was employed to examine the rate of γH2AX-positive cells, cell cycle, and apoptosis, CCK-8 was used to detect cell proliferation, ELISA was used to analyze the content of E_2 and P_4 in the culture fluid, and expression levels of relevant genes were analyzed by RT–qPCR. (**a**) Transfection of mimics led to a marked decline in the rate of γH2AX-positive cells, and the inhibitor resulted in a significant increase ($p \leq 0.05$). (**b**) The expression of *BRCA1, RAD51, MRE11*, and *ATM* was dramatically enhanced after transfection with miR-302d mimics ($p \leq 0.05$); after transfection with miR-302d inhibitors, the expression of *BRCA1* and *RAD51* was greatly diminished ($p \leq 0.05$), the expression of *MRE11* was decreased, and the expression of *ATM* was enlarged ($p > 0.05$). (**c**) A statistically significant decline in the number of cells in the G1 phase ($p \leq 0.05$), a remarkable enlargement in the count of cells in the S phase ($p \leq 0.05$), and an insignificant variation in the cell number in the G2 phase were caused by transfection with miR-302d mimics ($p > 0.05$). Upon transfection with the miR-302d inhibitor, the number of cells in the G1 phase was remarkably higher ($p \leq 0.05$), the number of S phase cells was reduced considerably ($p \leq 0.05$), and the number of G2 phase cells was diminished, but the effect of the difference was inconsequential ($p > 0.05$). (**d**) *CDK1* ($p \leq 0.05$), *CCND1* ($p < 0.01$), and *CCNE2* ($p < 0.01$) expression was obviously elevated after transfection of miR-302d mimics and *CDK2* expression was notably diminished ($p \leq 0.05$) in bovine CCs. Transfection of miR-302d inhibitor brought about a notable decline in *CDK1* and *CCND1* ($p < 0.01$), a rise in *CCNE2* expression, but the variety was insignificant ($p > 0.05$), and a rise in *CDK2* expression, but the variety was not obvious ($p > 0.05$). (**e**) miR-302d overexpression promoted cell proliferation at 24 h and 48 h ($p \leq 0.05$), and inhibition of miR-302d interfered with cell proliferation at 24 h ($p \leq 0.05$) and 48 h ($p < 0.01$). (**f**) The results of the apoptosis level assay after transfection of cells, left upper/Q1-UL: dead cells, right upper/Q1-UR: advanced apoptotic cells, right lower/Q1-LR: earlier apoptotic cells, left lower/Q1-LL: nonexpired cells. (**g**) Transfection of miR-302d mimics produced dramatic declines in relative apoptosis ($p \leq 0.05$), and miR-302d inhibitor contributed to a noticeable improvement in the relative apoptosis levels ($p \leq 0.05$). (**h**,**i**) *BAX/BCL2* expression was visibly displayed by transfection of the miR-302d mimic; the ratio of *BAX/BCL2* expression to that of *FAS* was visibly elevated by transfection of the miR-302d inhibitor, whereas the expression of *p53* was

considerably minimized ($p \leq 0.05$). (**j**) The levels of E_2 and P_4 were dramatically enlarged ($p \leq 0.05$) after the miR-302d mimic was transfected, whereas the levels of E_2 and P_4 were diminished ($p \leq 0.05$) after the miR-302d inhibitor was transfected. (**k**) The expression of *STAR*, *CYP11A1*, and *HSD3B1* was dramatically augmented after miR-302d mimic transfection ($p < 0.01$); the expression of *STAR* ($p \leq 0.055$), *CYP11A1* ($p < 0.01$) and *HSD3B1* ($p \leq 0.05$) prominently declined after transfection with miR-302d inhibitor.

4. Discussion

BLM-induced cellular DNA damage had an effect on the consequent production of γH2AX in a dose-dependent/time-dependent manner [28]. γH2AX not only indicates the extent of DNA damage but also serves as a signal transduction factor for DDR [29]. DNA damage in CCs causes COC communication disorder, which is dedicated to the inability of oocytes to initiate DDR, and increased DNA damage, bringing about the withdrawal of oocytes from meiosis [30,31]. DNA damage causes cell cycle arrest and induces apoptosis [10,11]. DNA damage has been linked to abnormal levels of estrogen and progesterone in women [13]. These results suggest that BLM causes DNA damage, induces apoptosis, arrests the cell cycle, and affects steroid hormone secretion. We found that the rate of γ-H2AX-positive cells increased significantly after BLM treatment, indicating that BLM successfully induced DNA damage in bovine cumulus cells. DNA damage in cumulus cells resulted in a significant decrease in cell abundance in the G1 phase and an obvious increase in cell abundance in the S phase, inhibited cell proliferation, increased relative apoptosis, and significantly reduced E_2 and P_4 secretion. During DNA damage, the expression of *CDKN1A* was apparently observed to expand, which was consistent with the results of previous studies [24], demonstrating that *CDKN1A* may be a key gene in the cellular DNA damage process.

DNA damage affects cellular physiological functions, cell lifespan depends on the effectiveness of DDR [32], and revitalization of the DDR pathway contributes to in vitro fertilization and embryo transfer [33]. *CDKN1A* inhibits DNA damage repair [21]. γH2AX is formed at DNA damage sites and is a biomarker of DNA damage that enhances cascading signaling during the DDR [26]. In this study, inhibition of *CDKN1A* resulted in a substantial decline in the positive rate of γH2AX in cells and changes in the expression of related DDR factors. DNA damage is able to activate ataxia-telangiectasia mutated (ATM), and once activated, ATM is phosphorylated and recruited to damaged DNA to initiate the DDR process [34]. Breast cancer susceptibility gene 1 (BRCA1) is a key member of the ATM-mediated DDR [35] and directs the homologous recombination repair (HR) pathway [36], and mutations in this gene are associated with ovarian cancers [37]. MRE11 promotes ATM activation, homologous recombination repair of DNA damage, and genome stability [38]. The expression of DNA damage repair-related genes is reduced [39], such as RAD51 [40], and the expression of RAD51 varies widely between species [41]. These findings support our hypothesis that inhibition of CDKN1A leads to a substantial expansion in *ATM*, *BRCA1*, and *MRE11* mRNA expression and a substantial reduction in *RAD51* expression. These changes activate the DDR pathway (HR pathway), reducing the γH2AX-positive rate and repairing DNA damage.

The stability and order of the cell cycle are foundations for the proliferation of cells and biological function performance. The phosphorylation of CDK2 and CCNE2 reaction products is related to the number of G1 phase cells [42]. Inhibition of CDKN1A leads to the upregulation of CDK2 and CCNE2, which enables CCs to pass through the G1 phase rapidly, promoting substance metabolism and RNA and protein synthesis. The interaction of CDKN1A and CDK1 leads to G2 phase abnormalities in the cell cycle [43]. Overexpression of CCND1 can lead to uncontrolled cell proliferation and corresponding cell cycle disorder in the G1 phase [44]. In this study, inhibition of CDKN1A altered the expression of cell cycle-related genes. Our study further confirmed that inhibition of CDKN1A led to up-regulation of CDK1 expression and down-regulation of CCND1 expression, which enabled a large number of cumulus cells to rapidly pass the G2 phase. The HR mechanism favors sister chromatids rather than homologous chromosomes as templates for DDR and is most active during the G2 phase when sister chromatids are available [45]. We hypothesize that inhibition of CDKN1A activates DNA damage repair through the HR pathway, leading to abnormal expression of cell cycle-related factors, leading to changes in cell cycle and proliferation. As a highly modulated program of cell death, apoptosis is a sensible and aggressive determination to compromise selected cells for the betterment of the biological organism and is a conventional and necessary physiological process in multicellular organisms [46]. Apoptosis plays an important role in cell renewal and embryonic development. Early studies confirmed that apoptosis is beneficial to multicellular organisms in a coordinated manner, allowing organisms to balance and fine-tune their life cycle [47]. Relevant studies have shown that CDKN1A allows DNA damage repair while inhibiting apoptosis, has the ability to activate apoptosis protein BAX to promote apoptosis, and can also promote apoptosis by relying on the p53 pathway [48]. Signals such as DNA damage and oncogene activation induce elevated levels of cellular *p53* and apoptosis [49]. In addition, FAS and FASL binding is also closely related to apoptosis [48–50], indicating that inhibition of *CDKN1A* activates the DDR mechanism and causes abnormal apoptotic gene expression, leading to declining apoptosis in cumulus cells.

DNA damage can lead to infertility and related endocrine disorders [13,51]. The abnormal secretion of estrogen and progesterone in cumulus cells results in low fecundity and weakened follicle development [52–54]. Downregulating 3β-hydroxysteroid dehydrogenase (HSD3B1) expression inhibited the synthesis of progesterone [55]. Cytochrome P450 family 11 subfamily A member 1 (CYP11A1) and steroidogenic acute regulatory protein (STAR) participate in cholesterol transport across mitochondrial membranes and cholesterol decomposition into pregnenolone, respectively [56]. In combination with BLM for cancer, abnormal expression of STAR, CYP11A1, and HSD3B1 led to hormone secretion disorder [55]. The secreted progesterone and estradiol of CCs are concerned with the expression of STAR, CYP11A1, and HSD3B1 [57,58]. In this study, the inhibition of CDKN1A significantly increased the mRNA levels of STAR, CYP11A1, and HSD3B1 in bovine CCs and significantly promoted the secretion of E_2 and P_4. These results indicated that CDKN1A could negatively regulate the secretion of steroid hormones in CCs. Therefore, the damage of CCs to estradiol and progesterone secretion caused by DNA damage may be caused by abnormal expression of steroid hormone marker genes. CDKN1A has become a focus of research on the regulatory mechanisms of reproduction as well as a critical factor in the post-transcriptional modulation of gene development [14]. miRNAs can affect DNA damage, proliferation, the cell cycle, and apoptosis by regulating target genes [15,59]. miRNA affected E_2 and P_4 secretion in CCs [16,17]. Disturbance of the pathway affected E_2 and P_4 synthesis and DNA damage in cumulus cells [52,60,61]. BLM treatment of CCs (Figure 3) caused a significant reduction in miR-302d expression, which was opposite to *CDKN1A* expression. miR-302d regulates the expression of *CDKN1A*. Combined with the results of the dual luciferase reporter gene, it was proven that miR-302d targeted *CDKN1A* and regulated its expression. In addition (Figure 4), the expression of *CDKN1A* in bovine cumulus cells transfected with miR-302d mimics decreased, the rate of γH2AX-positive cells decreased significantly, the cell cycle changed, proliferation was promoted, apoptosis was inhibited, and the secretion levels of E_2 and P_4 were increased. RAD51 can regulate the DDR through multiple pathways [62,63]. The expression of several targeted genes is controlled by a single miRNA [64]. The inconsistent expression of *RAD51* in cells transfected with miR-302d or siCDKN1A may be due to the change in the *RAD51* pathway to repair DNA damage or miR-302d regulation of other target genes to affect the expression of *RNA51*. miR-302d mimics upregulated *CCND1* expression and downregulated *CDK2* expression, causing a pronounced G2 phase cell count depression and an S phase cell count enlargement, which was inconsistent with the results of *CDKN1A* knockdown. It is possible that miR-302d regulates the expression of other genes involved in the cell cycle.

Oocyte quality is a key limiting factor for female fertility, but little is known about what constitutes oocyte quality or the mechanisms that control it. The developmental capacity of oocytes is mainly evaluated by the cumulus cell-mediated ovarian follicular microenvironment and the growth and development of oocytes [65]. DNA damage in cumulus cells is regarded as an important indicator for evaluating oocyte quality [66]. Different levels of cumulus cell DNA damage can reflect the developmental capacity of oocytes [67]. COCs coped with DNA damage better than denuded oocytes [8]. Our study found that inhibition of CDKN1A could activate the DDR pathway to reduce the degree of DNA damage in cumulus cells, suggesting that inhibition of CDKN1A could improve oocyte quality. The *p53* gene in cumulus cells can affect ovarian function by regulating the quality of mouse oocytes [68]. Specific inhibition of ATM may affect meiosis and cytoplasmic maturation in porcine oocytes by reducing their sensitivity to DNA damage [69]. In this study, after inhibiting *CDKN1A*, *p53* expression was downregulated and *ATM* gene expression was upregulated, which indicates that *CDKN1A* can improve oocyte quality and further regulate ovarian function by regulating the expression of cumulus cell genes.

Cumulus cells regulate the expression of hormonal factors and related regulatory factors in oocytes through the gap junction with oocytes, thereby affecting oocyte maturation [2]. DNA damage disrupts bovine COC communication, leading to oocyte withdrawal from meiosis [30], which in turn affects blastocyst production, hatchability, and embryo quality [70,71]. These results suggest that DNA damage in cumulus cells may affect the quality of oocytes and early embryos. Our study found that miR-302d targeted regulation of CDKN1A affected DNA damage and steroid hormone secretion in bovine cumulus cells. These results indicated that miR-302d and CDKN1A were candidate molecular markers for the diagnosis of DNA damage in bovine cumulus cells.

5. Conclusions

miR-302d targets *CDKN1A* and regulates its expression, resulting in reduced DNA damage levels, disordered cell cycle progression, increased proliferation levels, inhibited relative apoptosis levels, and increased steroid hormone secretion in bovine cumulus cells. Our current findings provided molecular targets for a more effective study of DNA damage in bovine cumulus cells.

Supplementary Materials: The following supporting information can be downloaded at: https://www.mdpi.com/article/10.3390/genes14122195/s1, Table S1: Primer and siRNA sequences used in this study. Table S2: The raw data of Figures 1–4.

Author Contributions: Conceptualization, B.Y. and H.J.; methodology, H.J.; software, Y.Z.; validation, J.L., J.Z. and G.Z.; formal analysis, Y.Z.; investigation, J.L., J.Z. and Y.Z.; resources, Y.Z.; data curation, J.L. and J.Z.; writing—original draft preparation, J.L. and J.Z.; writing—review and editing, B.Y. and H.J.; visualization, Y.Z.; supervision, B.Y. and H.J.; project administration, B.Y. and H.J.; funding acquisition, B.Y. and H.J. All authors have read and agreed to the published version of the manuscript.

Funding: This research was funded by the National Natural Science Foundation of China (31972570) and the China Agriculture Research System of MOF and MARA (CARS-37).

Institutional Review Board Statement: The experiments were strictly performed according to the guidelines of the Guide for the Care and Use of Laboratory Animals of Jilin University. In addition, all experimental protocols were approved by the Institutional Animal Care and Use Committee of Jilin University (Permit Number: 201801026).

Informed Consent Statement: Not applicable.

Data Availability Statement: The data presented in this study are available in Supplementary Material.

Conflicts of Interest: The authors declare that they have no competing interests.

References

1. Su, Y.Q.; Sugiura, K.; Eppig, J.J. Mouse oocyte control of granulosa cell development and function: Paracrine regulation of cumulus cell metabolism. *Semin. Reprod. Med.* **2009**, *27*, 32–42. [CrossRef] [PubMed]

2. Jiao, X.; Ding, Z.; Meng, F.; Zhang, X.; Wang, Y.; Chen, F.; Duan, Z.; Wu, D.; Zhang, S.; Miao, Y.; et al. The toxic effects of Fluorene-9-bisphenol on porcine oocyte in vitro maturation. *Environ. Toxicol.* **2020**, *35*, 152–158. [CrossRef] [PubMed]
3. De Bont, R.; van Larebeke, N. Endogenous DNA damage in humans: A review of quantitative data. *Mutagenesis* **2004**, *19*, 169–185. [CrossRef] [PubMed]
4. Shen, Q.; Chen, M.; Zhao, X.; Liu, Y.; Ren, X.; Zhang, L. Versican expression level in cumulus cells is associated with human oocyte developmental competence. *Syst. Biol. Reprod. Med.* **2020**, *66*, 176–184. [CrossRef] [PubMed]
5. Xiong, X.R.; Lan, D.L.; Li, J.; Yin, S.; Xiong, Y.; Zi, X.D. Identification of differential abundances of mRNA transcript in cumulus cells and CCND1 associated with yak oocyte developmental competence. *Anim. Reprod. Sci.* **2019**, *208*, 106135. [CrossRef] [PubMed]
6. Raman, R.S.; Chan, P.J.; Corselli, J.U.; Patton, W.C.; Jacobson, J.D.; Chan, S.R.; King, A. Comet assay of cumulus cell DNA status and the relationship to oocyte fertilization via intracytoplasmic sperm injection. *Hum. Reprod.* **2001**, *16*, 831–835. [CrossRef] [PubMed]
7. Tuppi, M.; Kehrloesser, S.; Coutandin, D.W.; Rossi, V.; Luh, L.M.; Strubel, A.; Hotte, K.; Hoffmeister, M.; Schafer, B.; De Oliveira, T.; et al. Oocyte DNA damage quality control requires consecutive interplay of CHK2 and CK1 to activate p63. *Nat. Struct. Mol. Biol.* **2018**, *25*, 261–269. [CrossRef] [PubMed]
8. Marangos, P.; Carroll, J. Oocytes progress beyond prophase in the presence of DNA damage. *Curr. Biol.* **2012**, *22*, 989–994. [CrossRef]
9. Pandita, T.K.; Richardson, C. Chromatin remodeling finds its place in the DNA double-strand break response. *Nucleic Acids Res.* **2009**, *37*, 1363–1377. [CrossRef]
10. van Gent, D.C.; Hoeijmakers, J.H.; Kanaar, R. Chromosomal stability and the DNA double-stranded break connection. *Nat. Rev. Genet.* **2001**, *2*, 196–206. [CrossRef]
11. Bohgaki, T.; Bohgaki, M.; Hakem, R. DNA double-strand break signaling and human disorders. *Genome Integr.* **2010**, *1*, 15. [CrossRef] [PubMed]
12. Wang, L.; Xu, X.; Teng, M.; Zhao, G.; Lei, A. Coping with DNA Double-Strand Breaks via ATM Signaling Pathway in Bovine Oocytes. *Int. J. Mol. Sci.* **2020**, *21*, 8892. [CrossRef] [PubMed]
13. Maneschi, F.; Benedetti-Panici, P.; Scambia, G.; Salerno, M.G.; D'Agostino, G.; Mancuso, S. Menstrual and hormone patterns in women treated with high-dose cisplatin and bleomycin. *Gynecol. Oncol.* **1994**, *54*, 345–348. [CrossRef] [PubMed]
14. Harries, L. MicroRNAs as Mediators of the Ageing Process. *Genes* **2014**, *5*, 656–670. [CrossRef]
15. Wei, S.; Xue, J.; Sun, B.; Zou, Z.; Chen, C.; Liu, Q.; Zhang, A. miR-145 via targeting ERCC2 is involved in arsenite-induced DNA damage in human hepatic cells. *Toxicol. Lett.* **2018**, *295*, 220–228. [CrossRef]
16. Pan, B.; Toms, D.; Shen, W.; Li, J. MicroRNA-378 regulates oocyte maturation via the suppression of aromatase in porcine cumulus cells. *Am. J. Physiol. Endocrinol. Metab.* **2015**, *308*, E525–E534. [CrossRef]
17. Andreas, E.; Pandey, H.O.; Hoelker, M.; Salilew-Wondim, D.; Gebremedhn, S.; Schellander, K.; Tesfaye, D. The regulatory role of miR-20a in bovine cumulus cells and its contribution to oocyte maturation. *Zygote* **2021**, *29*, 435–444. [CrossRef]
18. Li, Y.; Huo, J.; Pan, X.; Wang, C.; Ma, X. MicroRNA 302b-3p/302c-3p/302d-3p inhibits epithelial-mesenchymal transition and promotes apoptosis in human endometrial carcinoma cells. *Onco Targets Ther.* **2018**, *11*, 1275–1284. [CrossRef]
19. Yang, Y.; Sun, Z.; Liu, F.; Bai, Y.; Wu, F. FGD5-AS1 Inhibits Osteoarthritis Development by Modulating miR-302d-3p/TGFBR2 Axis. *Cartilage* **2021**, *13*, 1412S–1420S. [CrossRef]
20. Gartel, A.L. Is p21 an oncogene? *Mol. Cancer Ther.* **2006**, *5*, 1385–1386. [CrossRef]
21. Tom, S.; Ranalli, T.A.; Podust, V.N.; Bambara, R.A. Regulatory roles of p21 and apurinic/apyrimidinic endonuclease 1 in base excision repair. *J. Biol. Chem.* **2001**, *276*, 48781–48789. [CrossRef]
22. Wang, Y.; Fisher, J.C.; Mathew, R.; Ou, L.; Otieno, S.; Sublet, J.; Xiao, L.; Chen, J.; Roussel, M.F.; Kriwacki, R.W. Intrinsic disorder mediates the diverse regulatory functions of the Cdk inhibitor p21. *Nat. Chem. Biol.* **2011**, *7*, 214–221. [CrossRef]
23. Chan, T.A.; Hwang, P.M.; Hermeking, H.; Kinzler, K.W.; Vogelstein, B. Cooperative effects of genes controlling the G(2)/M checkpoint. *Genes Dev.* **2000**, *14*, 1584–1588. [CrossRef]
24. Liu, J.B.; Zhang, J.B.; Yan, X.M.; Xie, P.G.; Fu, Y.; Fu, X.H.; Sun, X.L.; Han, D.X.; Li, S.P.; Zheng, Y.; et al. DNA Double-Strand Break-Related Competitive Endogenous RNA Network of Noncoding RNA in Bovine Cumulus Cells. *Genes* **2023**, *14*, 290. [CrossRef] [PubMed]
25. Todorova, T.; Miteva, D.; Chankova, S. DNA susceptibility of Saccharomyces cerevisiae to Zeocin depends on the growth phase. *Int. Microbiol.* **2019**, *22*, 419–428. [CrossRef] [PubMed]
26. Tu, W.Z.; Li, B.; Huang, B.; Wang, Y.; Liu, X.D.; Guan, H.; Zhang, S.M.; Tang, Y.; Rang, W.Q.; Zhou, P.K. gammaH2AX foci formation in the absence of DNA damage: Mitotic H2AX phosphorylation is mediated by the DNA-PKcs/CHK2 pathway. *FEBS Lett.* **2013**, *587*, 3437–3443. [CrossRef] [PubMed]
27. Cappelen, K.; Aase, K.; Storm, M.; Hetland, J.; Harris, A. Psychometric properties of the Nursing Home Survey on Patient Safety Culture in Norwegian nursing homes. *BMC Health Serv. Res.* **2016**, *16*, 446. [CrossRef] [PubMed]
28. Zhang, T.; Zhang, G.L.; Ma, J.Y.; Qi, S.T.; Wang, Z.B.; Wang, Z.W.; Luo, Y.B.; Jiang, Z.Z.; Schatten, H.; Sun, Q.Y. Effects of DNA damage and short-term spindle disruption on oocyte meiotic maturation. *Histochem. Cell Biol.* **2014**, *142*, 185–194. [CrossRef]
29. Stucki, M.; Jackson, S.P. gammaH2AX and MDC1: Anchoring the DNA-damage-response machinery to broken chromosomes. *DNA Repair* **2006**, *5*, 534–543. [CrossRef] [PubMed]

30. Yamamoto, T.; Iwata, H.; Goto, H.; Shiratuki, S.; Tanaka, H.; Monji, Y.; Kuwayama, T. Effect of maternal age on the developmental competence and progression of nuclear maturation in bovine oocytes. *Mol. Reprod. Dev.* **2010**, *77*, 595–604. [CrossRef]
31. Sun, M.H.; Yang, M.; Xie, F.Y.; Wang, W.; Zhang, L.; Shen, W.; Yin, S.; Ma, J.Y. DNA Double-Strand Breaks Induce the Nuclear Actin Filaments Formation in Cumulus-Enclosed Oocytes but Not in Denuded Oocytes. *PLoS ONE* **2017**, *12*, e0170308. [CrossRef] [PubMed]
32. Best, B.P. Nuclear DNA damage as a direct cause of aging. *Rejuvenation Res.* **2009**, *12*, 199–208. [CrossRef] [PubMed]
33. Sun, X.L.; Jiang, H.; Han, D.X.; Fu, Y.; Liu, J.B.; Gao, Y.; Hu, S.M.; Yuan, B.; Zhang, J.B. The activated DNA double-strand break repair pathway in cumulus cells from aging patients may be used as a convincing predictor of poor outcomes after in vitro fertilization-embryo transfer treatment. *PLoS ONE* **2018**, *13*, e0204524. [CrossRef] [PubMed]
34. Ismail, I.H.; Nystrom, S.; Nygren, J.; Hammarsten, O. Activation of ataxia telangiectasia mutated by DNA strand break-inducing agents correlates closely with the number of DNA double strand breaks. *J. Biol. Chem.* **2005**, *280*, 4649–4655. [CrossRef] [PubMed]
35. Rzepka-Gorska, I.; Tarnowski, B.; Chudecka-Glaz, A.; Gorski, B.; Zielinska, D.; Toloczko-Grabarek, A. Premature menopause in patients with BRCA1 gene mutation. *Breast Cancer Res. Treat.* **2006**, *100*, 59–63. [CrossRef] [PubMed]
36. Tarsounas, M.; Sung, P. The antitumorigenic roles of BRCA1-BARD1 in DNA repair and replication. *Nat. Rev. Mol. Cell Biol.* **2020**, *21*, 284–299. [CrossRef] [PubMed]
37. Oktay, K.; Turan, V.; Titus, S.; Stobezki, R.; Liu, L. BRCA Mutations, DNA Repair Deficiency, and Ovarian Aging. *Biol. Reprod.* **2015**, *93*, 67. [CrossRef] [PubMed]
38. Wang, Z.; Gong, Y.; Peng, B.; Shi, R.; Fan, D.; Zhao, H.; Zhu, M.; Zhang, H.; Lou, Z.; Zhou, J.; et al. MRE11 UFMylation promotes ATM activation. *Nucleic Acids Res.* **2019**, *47*, 4124–4135. [CrossRef]
39. Govindaraj, V.; Keralapura Basavaraju, R.; Rao, A.J. Changes in the expression of DNA double strand break repair genes in primordial follicles from immature and aged rats. *Reprod. Biomed. Online* **2015**, *30*, 303–310. [CrossRef]
40. Suwaki, N.; Klare, K.; Tarsounas, M. RAD51 paralogs: Roles in DNA damage signalling, recombinational repair and tumorigenesis. *Semin. Cell Dev. Biol.* **2011**, *22*, 898–905. [CrossRef]
41. Bilotto, S.; Boni, R.; Russo, G.L.; Lioi, M.B. Meiosis progression and donor age affect expression profile of DNA repair genes in bovine oocytes. *Zygote* **2015**, *23*, 11–18. [CrossRef] [PubMed]
42. Caruso, J.A.; Duong, M.T.; Carey, J.P.W.; Hunt, K.K.; Keyomarsi, K. Low-Molecular-Weight Cyclin E in Human Cancer: Cellular Consequences and Opportunities for Targeted Therapies. *Cancer Res.* **2018**, *78*, 5481–5491. [CrossRef]
43. Niculescu, A.B., 3rd; Chen, X.; Smeets, M.; Hengst, L.; Prives, C.; Reed, S.I. Effects of p21(Cip1/Waf1) at both the G1/S and the G2/M cell cycle transitions: pRb is a critical determinant in blocking DNA replication and in preventing endoreduplication. *Mol. Cell Biol.* **1998**, *18*, 629–643. [CrossRef] [PubMed]
44. Tchakarska, G.; Sola, B. The double dealing of cyclin D1. *Cell Cycle* **2020**, *19*, 163–178. [CrossRef]
45. Huertas, P.; Cortes-Ledesma, F.; Sartori, A.A.; Aguilera, A.; Jackson, S.P. CDK targets Sae2 to control DNA-end resection and homologous recombination. *Nature* **2008**, *455*, 689–692. [CrossRef] [PubMed]
46. Kerr, J.F. History of the events leading to the formulation of the apoptosis concept. *Toxicology* **2002**, *181–182*, 471–474. [CrossRef]
47. King, K.L.; Cidlowski, J.A. Cell cycle regulation and apoptosis. *Annu. Rev. Physiol.* **1998**, *60*, 601–617. [CrossRef]
48. Abbas, T.; Dutta, A. p21 in cancer: Intricate networks and multiple activities. *Nat. Rev. Cancer* **2009**, *9*, 400–414. [CrossRef]
49. Vogelstein, B.; Lane, D.; Levine, A.J. Surfing the p53 network. *Nature* **2000**, *408*, 307–310. [CrossRef]
50. Chauhan, P.; Sodhi, A.; Tarang, S. Cisplatin-treated murine peritoneal macrophages induce apoptosis in L929 cells: Role of Fas-Fas ligand and tumor necrosis factor-tumor necrosis factor receptor 1. *Anti-Cancer Drugs* **2007**, *18*, 187–196. [CrossRef]
51. Jahnukainen, K.; Ehmcke, J.; Hou, M.; Schlatt, S. Testicular function and fertility preservation in male cancer patients. *Best Pract. Res. Clin. Endocrinol. Metab.* **2011**, *25*, 287–302. [CrossRef] [PubMed]
52. Yamashita, Y.; Kawashima, I.; Gunji, Y.; Hishinuma, M.; Shimada, M. Progesterone is essential for maintenance of Tace/Adam17 mRNA expression, but not EGF-like factor, in cumulus cells, which enhances the EGF receptor signaling pathway during in vitro maturation of porcine COCs. *J. Reprod. Dev.* **2010**, *56*, 315–323. [CrossRef] [PubMed]
53. Krege, J.H.; Hodgin, J.B.; Couse, J.F.; Enmark, E.; Warner, M.; Mahler, J.F.; Sar, M.; Korach, K.S.; Gustafsson, J.A.; Smithies, O. Generation and reproductive phenotypes of mice lacking estrogen receptor β. *Proc. Natl. Acad. Sci. USA* **1998**, *95*, 15677–15682. [CrossRef] [PubMed]
54. Cheng, G.; Weihua, Z.; Makinen, S.; Makela, S.; Saji, S.; Warner, M.; Gustafsson, J.A.; Hovatta, O. A role for the androgen receptor in follicular atresia of estrogen receptor β knockout mouse ovary. *Biol. Reprod.* **2002**, *66*, 77–84. [CrossRef]
55. Al-Bader, M.; Kilarkaje, N. Effects of bleomycin, etoposide and cisplatin treatment on Leydig cell structure and transcription of steroidogenic enzymes in rat testis. *Eur. J. Pharmacol.* **2015**, *747*, 150–159. [CrossRef] [PubMed]
56. Lehmann, K.P.; Phillips, S.; Sar, M.; Foster, P.M.; Gaido, K.W. Dose-dependent alterations in gene expression and testosterone synthesis in the fetal testes of male rats exposed to di (n-butyl) phthalate. *Toxicol. Sci.* **2004**, *81*, 60–68. [CrossRef] [PubMed]
57. Mateo-Otero, Y.; Yeste, M.; Roca, J.; Llavanera, M.; Bucci, D.; Galeati, G.; Spinaci, M.; Barranco, I. Seminal extracellular vesicles subsets modulate gene expression in cumulus cells of porcine in vitro matured oocytes. *Sci. Rep.* **2022**, *12*, 19096. [CrossRef]
58. Caillaud, M.; Gerard, N. In vivo and in vitro effects of interleukin-1beta on equine oocyte maturation and on steroidogenesis and prostaglandin synthesis in granulosa and cumulus cells. *Reprod. Fertil. Dev.* **2009**, *21*, 265–273. [CrossRef]
59. Fu, X.; He, Y.; Song, J.; Wang, L.; Guo, P.; Cao, J. MiRNA-181b-5p Modulates Cell Proliferation, Cell Cycle, and Apoptosis by Targeting SSX2IP in Acute Lymphoblastic Leukemia. *Turk. J. Haematol.* **2022**, *39*, 160–169. [CrossRef]

60. Qiao, G.Y.; Dong, B.W.; Zhu, C.J.; Yan, C.Y.; Chen, B.L. Deregulation of WNT2/FZD3/β-catenin pathway compromises the estrogen synthesis in cumulus cells from patients with polycystic ovary syndrome. *Biochem. Biophys. Res. Commun.* **2017**, *493*, 847–854. [CrossRef]
61. Maidarti, M.; Anderson, R.A.; Telfer, E.E. Crosstalk between PTEN/PI3K/Akt Signalling and DNA Damage in the Oocyte: Implications for Primordial Follicle Activation, Oocyte Quality and Ageing. *Cells* **2020**, *9*, 200. [CrossRef]
62. Kwon, Y.; Rosner, H.; Zhao, W.; Selemenakis, P.; He, Z.; Kawale, A.S.; Katz, J.N.; Rogers, C.M.; Neal, F.E.; Badamchi Shabestari, A.; et al. DNA binding and RAD51 engagement by the BRCA2 C-terminus orchestrate DNA repair and replication fork preservation. *Nat. Commun.* **2023**, *14*, 432. [CrossRef]
63. Bobyk, L.; Vianna, F.; Martinez, J.S.; Gruel, G.; Benderitter, M.; Baldeyron, C. Differential Recruitment of DNA Repair Proteins KU70/80 and RAD51 upon Microbeam Irradiation with α-Particles. *Biology* **2022**, *11*, 1652. [CrossRef] [PubMed]
64. Vishnoi, A.; Rani, S. miRNA Biogenesis and Regulation of Diseases: An Updated Overview. *Methods Mol. Biol.* **2023**, *2595*, 1–12. [PubMed]
65. Gilchrist, R.B.; Lane, M.; Thompson, J.G. Oocyte-secreted factors: Regulators of cumulus cell function and oocyte quality. *Hum. Reprod. Update* **2008**, *14*, 159–177. [CrossRef]
66. Lopez, A.; Betancourt, M.; Ducolomb, Y.; Rodriguez, J.J.; Casas, E.; Bonilla, E.; Bahena, I.; Retana-Marquez, S.; Juarez-Rojas, L.; Casillas, F. DNA damage in cumulus cells generated after the vitrification of in vitro matured porcine oocytes and its impact on fertilization and embryo development. *Porc. Health Manag.* **2021**, *7*, 56. [CrossRef] [PubMed]
67. Barcena, P.; Lopez-Fernandez, C.; Garcia-Ochoa, C.; Obradors, A.; Vernaeve, V.; Gosalvez, J.; Vassena, R. Detection of DNA damage in cumulus cells using a chromatin dispersion assay. *Syst. Biol. Reprod. Med.* **2015**, *61*, 277–285. [PubMed]
68. Haraguchi, H.; Hirota, Y.; Saito-Fujita, T.; Tanaka, T.; Shimizu-Hirota, R.; Harada, M.; Akaeda, S.; Hiraoka, T.; Matsuo, M.; Matsumoto, L.; et al. Mdm2-p53-SF1 pathway in ovarian granulosa cells directs ovulation and fertilization by conditioning oocyte quality. *FASEB J.* **2019**, *33*, 2610–2620. [CrossRef]
69. Lin, Z.L.; Kim, N.H. Role of ataxia-telangiectasia mutated (ATM) in porcine oocyte in vitro maturation. *Cell Biol. Int.* **2015**, *39*, 710–720. [CrossRef]
70. Sun, M.H.; Zheng, J.; Xie, F.Y.; Shen, W.; Yin, S.; Ma, J.Y. Cumulus Cells Block Oocyte Meiotic Resumption via Gap Junctions in Cumulus Oocyte Complexes Subjected to DNA Double-Strand Breaks. *PLoS ONE* **2015**, *10*, e0143223. [CrossRef]
71. Sirini, M.A.; Anchordoquy, J.M.; Anchordoquy, J.P.; Pascua, A.M.; Nikoloff, N.; Carranza, A.; Relling, A.E.; Furnus, C.C. The presence of acylated ghrelin during in vitro maturation of bovine oocytes induces cumulus cell DNA damage and apoptosis, and impairs early embryo development. *Zygote* **2017**, *25*, 601–611. [CrossRef] [PubMed]

genes

Communication

Efficient and Specific Generation of *MSTN*-Edited Hu Sheep Using C-CRISPR

Rihong Guo [1,2,†], Huili Wang [1,2,†], Chunhua Meng [1,2], Hongbing Gui [1], Yinxia Li [1,2], Fang Chen [1,2], Chenjian Zhang [1], Han Zhang [1], Qiang Ding [1,2], Jianli Zhang [1,2], Jun Zhang [1,2], Yong Qian [1,2], Jifeng Zhong [1,2] and Shaoxian Cao [1,2,3,*]

1 Jiangsu Provincial Engineering Research Center for Precision Animal Breeding, Nanjing 210014, China; rhguo@jaas.ac.cn (R.G.)
2 Institute of Animal Science, Jiangsu Academy of Agricultural Sciences, Nanjing 210014, China
3 Key Laboratory of Crop and Animal Integrated Farming, Ministry of Agriculture and Rural Affairs, Nanjing 210014, China
* Correspondence: sxcao@jaas.ac.cn
† These authors contributed equally to this work.

Abstract: Hu sheep, an indigenous breed in China known for its high fecundity, are being studied to improve their growth and carcass traits. MSTN is a negative regulator of muscle development, and its inactivation results in muscularity. The C-CRISPR system, utilizing multiple neighboring sgRNAs targeting a key exon, has been successfully used to generate genes for complete knockout (KO) monkeys and mice in one step. In this study, the C-CRISPR system was used to generate *MSTN*-edited Hu sheep; 70 embryos injected with Cas9 mRNA and four sgRNAs targeting exon 3 of sheep MSTN were transferred to 13 recipients. Out of 10 lambs born from five recipients after full-term pregnancies, nine had complete *MSTN* KO with various mutations. No off-target effects were found. These *MSTN*-KO Hu sheep showed a double-muscled (DM) phenotype, characterized by a higher body weight at 3 and 4 months old, prominent muscular protrusion, clearly visible intermuscular groves, and muscle hypertrophy. The molecular analysis indicated enhanced AKT and suppressed ERK1/2 signaling in the gluteus muscle of the edited Hu sheep. In conclusion, *MSTN* complete KO Hu sheep with a DM phenotype were efficiently and specifically generated using C-CRISPR, and the C-CRISPR method is a promising tool for farm animal breeding.

Keywords: CRISPR/Cas9; Hu sheep; MSTN; double muscling phenotype; off-target effects

Citation: Guo, R.; Wang, H.; Meng, C.; Gui, H.; Li, Y.; Chen, F.; Zhang, C.; Zhang, H.; Ding, Q.; Zhang, J.; et al. Efficient and Specific Generation of *MSTN*-Edited Hu Sheep Using C-CRISPR. *Genes* **2023**, *14*, 1216. https://doi.org/10.3390/genes14061216

Academic Editors: Bao Yuan and Martien A. M. Groenen

Received: 24 March 2023
Revised: 11 May 2023
Accepted: 30 May 2023
Published: 2 June 2023

1. Introduction

Hu sheep are a Chinese indigenous breed well known for its high fecundity, early maturity, year-round estrus, high tolerance to heat stress, and suitability for stabling. Hu sheep are introduced to northern areas of China and used as dam lines for commercial breeds. Attempts have been made to cross Hu sheep with meat sheep to breed so-called high-fecundity meat sheep [1], but this is costly and time consuming. Moreover, the cross may introduce undesirable genes and dilute the desired characteristics [2] of Hu sheep. Fortunately, the development of genome editing technology, especially CRISPR (Clustered Regularly Interspaced Short Palindromic Repeats)/Cas9 (CRISPR-associated protein) [3,4], and the identification of myostatin (*MSTN*) as a major gene controlling animal muscularity offer a convenient alternative approach to achieve the breeding goal.

MSTN belongs to the TGFβ superfamily; it negatively regulates skeletal muscle development [5]. Natural mutations that either inactivate the encoded protein or suppress its quantity cause enhanced muscling in humans [6], sheep [7–9], cattle [10–12], pigs [13], and dogs [14]. In Texel sheep, an SNP c.*1232G>A in *MSTN* 3′-UTR creating an illegitimate microRNA site reduces 2/3 of MSTN protein level in serum and results in muscularity [8].

In addition to the SNP c.*1232G>A, a frameshift *MSTN* mutation, c.960delG, has been identified in Norwegian white sheep; both SNPs reduced fatness and increased muscle mass in Norwegian white sheep [15]. *MSTN* knockout also results in muscularity in goats [16,17], sheep [18–20], cattle [19], and pigs [21–23]. Wang et al. [20], Zhou et al. [24], and Li et al. [18] reported that *MSTN* edition in Tan sheep and small tail Han sheep significantly enhanced body weight and muscularity. Consequently, *MSTN* is an ideal target gene to improve the growth and carcass traits of Hu sheep.

CRISPR/Cas9 is an efficient and precision genome editor [3,4] and has been extensively used to generate healthier and more productive farm animals [25,26]. Microinjection of Cas9 mRNA and sgRNA into one-cell-stage embryos generates frameshift mutations at target sites. However, animals generated by this method show gene-functional mosaicism, with gene disruption occurring in some cells but not others [27]. One or two generations are required to produce heterozygous or homozygous edited animals for trait evaluation. However, generation intervals are always long in farm animals, and crossbreeding to generate homozygous individuals is laborious. Using somatic cell nuclear cloning (SCNT) to construct cloning embryos with cells harboring verified mutations can overcome gene-functional mosaicism, and this method has been extensively utilized in farm animal genome editing [18,21,28–32]. Unfortunately, SCNT is hindered by low efficiency and health issues in founder animals. Zuo et al. [33] proposed an alternative method called C-CRISPR, which was successfully used to generate complete gene knockout mice and monkeys in one step. Unlike the standard CRISPR method, C-CRISPR simultaneously uses three to four neighboring gRNAs spaced 10–200 bp apart to target a single key exon of the target gene. This makes C-CRISPR a promising technique for farm animal breeding.

In this study, we used the C-CRISPR method to generate *MSTN*-edited Hu sheep and evaluated the off-target effects. We also analyzed the phenotypic and molecular changes resulting from genome editing.

2. Materials and Methods

2.1. Animals

The Hu sheep were maintained at the Luhe Animal Scientific Base of the Jiangsu Academy of Agricultural Sciences in Jiangsu province. The experimental procedures were approved by the Research Committee of the Jiangsu Academy of Agricultural Sciences and conducted with adherence to the Regulations for the Administration of Affairs Concerning Experimental Animals (Decree No. 63 of the Jiangsu Academy of Agricultural Science on 8 July 2014).

2.2. Preparation of Cas9 mRNA and sgRNA

The sgRNA design and in vitro transcription (IVT) of Cas9 mRNA and sgRNA were described in our previous study [17]. Briefly, four sgRNAs (Supplementary Table S1) targeting sheep *MSTN* exon 3 were designed with CRISPR tools (https://benchling.com/, accessed on 15 September 2020). The oligos for each sgRNA (Supplementary Table S1) were annealed and cloned into the pX330 plasmid. The IVT templates for Cas9 and sgRNAs were amplified using the T7 promotor-appended primers (Supplementary Table S2) and were gel-purified using a QiaQuick Spin Column (28104Qiagen, Shanghai, China). The Cas9 IVT template (400 ng for a 20 μL reaction) was subjected to a T7 Ultra Kit (AM1345, Ambion, Shanghai, China), and sgRNA IVT templates (200 ng for a 20 μL reaction) were transcribed using the MEGAshortscript Kit (AM1354, Ambion) in vitro. The Cas9 mRNA and sgRNAs were purified using the MEGAclear Kit (AM1908, Ambion). The purified RNAs were quantified using NanoDrop 2000 and then subjected to electrophoresis in 1.5% agarose gel (Supplementary Figure S1). The high-quality RNAs were stored at −80 °C before use.

2.3. Manipulation of Sheep Embryos

Sheep superovulation and estrous synchronization, embryo collection, injection, and transfer were performed as previously described [17] with modifications (Supplementary Figure S2). In brief, healthy donors (n = 5) and recipients (n = 13) with normal estrous cycles were intravaginally implanted with progesterone sponges (EAZI-BREED CIDR Sheep and Goat Devices, ZOETIS, Rhodes, NSW, Australia) for 13 days, followed by administration of 0.1 mg PGF2α analogues (Cloprostenol Sodium Injection) at the time of sponge removal. The donors received 675 IU FSH twice daily in a decreasing dose over 3.5 days (200/150, 100/75, 75/50, and 25 IU) starting 72 h before sponge removal. Then donors were naturally mated twice at 36 and 48 h after sponge removal. The recipients received 400 IU PMSG 72 h before sponge removal. All hormones or analogues used in this study were provided by Sansheng Pharmaceutics (Ningbo, China).

Twenty hours after the last insemination, the one-cell-stage embryos were flushed from the oviducts, and the collected zygotes were injected with a mixture of Cas9 mRNA (100 ng/μL) and sgRNAs (50 ng/μL for each sgRNA). The injected zygotes were cultured in M2 media containing 10% FBS (26140, Gibco, Shanghai, China) at 37 °C for 2 to 4 h, and then live embryos were transferred to estrous-synchronized recipient sheep.

2.4. Detection of Genome Editing at the Target Sites and POTSs

Blood or ear tissue genomic DNA of the newborn lambs was extracted using a DNA extraction kit (DP348, Tiangen, Beijing, China). The genomic regions surrounding exon 3 were amplified by PCR using the primers listed in Supplementary Table S3. The purified PCR amplicons were subjected to Sanger sequencing and subcloned into the pMD-19T vector (Takara, D103A), with 10–20 colonies randomly selected from each sample for sequencing.

The potential off-target sites (POTSs) of sgC1-C4 were computationally predicted using Cas-OFFinder according to the Hu sheep genome assembly GCA_011170295.1_ASM1117029v1. The POTS sites were PCR amplified and subjected to Sanger sequencing. The POTSs and primers used for the amplification of the target sites and POTSs are listed in Supplementary Table S4.

2.5. Western Blotting and Histological Analysis

Gluteus muscle biopsies were surgically collected from a control and an *MSTN*-edited Hu sheep (#521) 12 months of age. The aliquots of the muscle biopsies were either fixed by 4% paraformaldehyde for hematoxylin and eosin (HE) staining or immediately immersed into liquid nitrogen for storage.

For the Western blotting analysis, the total protein of the frozen muscle sample was extracted using RIPA lysis buffer (P0013B, Beyotime Biotechnology, Nantong, China) with phosphatase inhibitor cocktail C (P1091, Beyotime Biotechnology). Then, the lysates were boiled in the gel-loading buffer, and 30 μg of protein was separated by SDS-PAGE in each lane of a 12% gel. The proteins were subsequently transferred to a polyvinylidene fluoride membrane (Millipore, Darmstadt, Germany) and probed with primary antibodies against GAPDH (60004-1-Ig, Proteintech, Wuhan, China), MSTN (19142-1-AP, Proteintech), ERK1/2 (4695T, Cell Signaling Technology, Danvers, MA, USA), p-ERK1/2 (4370T, Cell Signaling Technology), AKT (9272S, Cell Signaling Technology), p-AKT (4060S, Cell Signaling Technology), P38 (9272S, Cell Signaling Technology), and p-P38 (8690S, Cell Signaling Technology). The chemiluminescence was detected by an ECL kit from Pierce Chemical (Dallas, TX, USA) and visualized through Image Quant LAS 4000 (Fujifilm, Tokyo, Japan). The band intensity was quantified with ImageJ software (NIH, Bethesda, MA, USA).

For the histological analysis, the fixed muscle samples were dehydrated in a graded alcohol series and then embedded in paraffin. The paraffin-embedded tissues were sectioned at 3–5 μm. The slices were then stained with an HE kits (C0105S, Beyotime Biotechnology). The stained slides were viewed via microscopy (Leica, Wetzlar, Germany). The myofiber sizes from five randomly selected fields of view in each slice were analyzed using the Image Pro Plus software (Media Cybernetics, Silver Spring, MD, USA).

2.6. Data Analysis

Data were presented as mean ± SEM. The differences in body weight and myofiber size were analyzed using t-tests. Statistical analyses were performed using GraphPad Prism (Version 9.0). A *p*-value < 0.05 was considered significant.

3. Results

3.1. Efficient Generation of MSTN-Edited Hu Sheep Using C-CRISPR

Four sgRNAs (sgC1–C4) targeting the sheep *MSTN* exon 3 were designed (Figure 1A, Supplementary Table S1). The efficiencies of the sgRNAs were not verified in cells or in vitro cultured embryos. A total of 83 embryos were collected from five donors and injected with a mixture of Cas9 mRNA, sgC1–C4. After culturing for 1–2 h, 70 live embryos were transferred to 13 recipients. Five recipients delivered 10 lambs after full-term pregnancies. (Table 1).

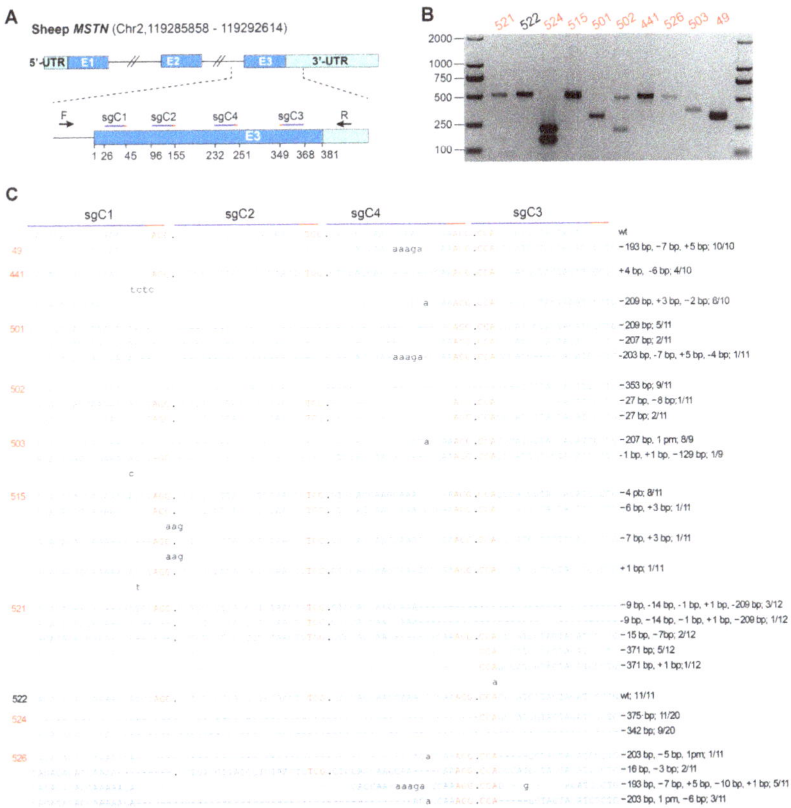

Figure 1. Generation of *MSTN*-edited Hu sheep using C-CRISPR. (**A**) Four sgRNAs (sgC1–C4) targeting sheep *MSTN* exon 3 were designed, and the numbers below the panel indicate the position in the 381 bp exon 3. F/R were primers used for amplification of exon3. (**B**) The target site amplification result of the ten newborn lambs. The nine edited lambs are highlighted in red. (**C**) TA clone sequencing results. The mutated and inserted bases are indicated in lowercase, and the deletions are indicated by "−". WT: wild type; −: deletion; +: insertion; N/N indicates the positive colonies of the total colonies sequenced.

Table 1. Summary of the generation of *MSTN*-edited sheep using C-CRISPR.

sgRNAs	Donors Superovulated	Embryo Injected/Embryos Transferred	Delivered Recipients /Total Recipients	Edited Lambs /Total Lambs
sgC1–C4	5	83/70	5/13 (38.46%)	9/10 (90.00%)

Sanger sequencing of the target sites showed that 9 out of the 10 lambs were edited (Figure 1B,C). The TA clone sequencing results indicated that the mutation ratios in all the edited sheep were 100%, and each lamb had one to five types of indels, which were −371 bp ~ +1 bp. Indels at sgC1 target sites were present in all nine edited lambs, and all mutations harbored indels at sgC1 sites except lamb #515, whose indel ratio at sgC1 sites was 27.27%. For sgC4, eight out of the nine lambs had indels and were 100% edited except lamb #515 (the mutation rate at sgC4 sites was 72.72%). Five lambs had indels at sgC3 and none at sgC2. The indel ratio at the sgC3 target site varied from 9.09% to 100%. By combining the indels of sgC1 and sgC4, *MSTN* in eight out of the nine lambs was 100% edited. By combining the indels of sgC1, sgC3, and sgC4, *MSTN* in the nine lambs was 100% edited. Notably, there were long-range deletions across sgC1 and sgC4 (~190 bp) and sgC4 and sgC3 (~209 bp), as well as sgC1 and sgC3 (~353 bp), with the most frequent type of long-range deletion across sgC1 and sgC4 in eight out of the nine edited lambs.

3.2. Specificity of Genome Editing Using C-CRISPR

To evaluate the off-target effects, potential off-target sites (POTSs) with PAM NGG of these four sgRNAs were predicted using Cas-OFFinder [34] according to the reference genome. The details of these POTSs are listed in Supplementary Table S4, with six to seven POTSs detected for each sgRNA. No de novo mutations were found among these POTSs after PCR and Sanger sequencing. Multiple alignments of the Sanger sequencing results are presented in Supplementary Figure S3.

3.3. Phenotype Analysis of MSTN-Edited Hu Sheep

MSTN-KO Hu sheep displayed double-muscled (DM) phenotype (Figure 2A). This was particularly evident in the proximal regions of both the forequarters and hindquarters, where there was a prominent muscular protrusion. The intermuscular boundaries and grooves were also clearly visible beneath the skin.

The body weights of the edited lambs (n = 9) and the control lambs (n = 8) were monitored from birth to 8 months old (Figure 2B). At birth, the edited lambs showed a trend toward higher body weights compared with the control group (p = 0.09). This difference became significant at 3 and 4 months old. However, as the lambs grew older, the difference in body weight between the two groups gradually diminished and was no longer significant from 5 months old (p = 0.053).

The histological analysis of the gluteus muscle indicated muscle hypertrophy (Figure 2C). The cross-sectional area of the muscle fibers in the edited Hu sheep increased 51.4% compared with the control Hu sheep (630.00 ± 78.56 vs. 416 ± 35.19 μm^2, $p < 0.001$) (Figure 2D).

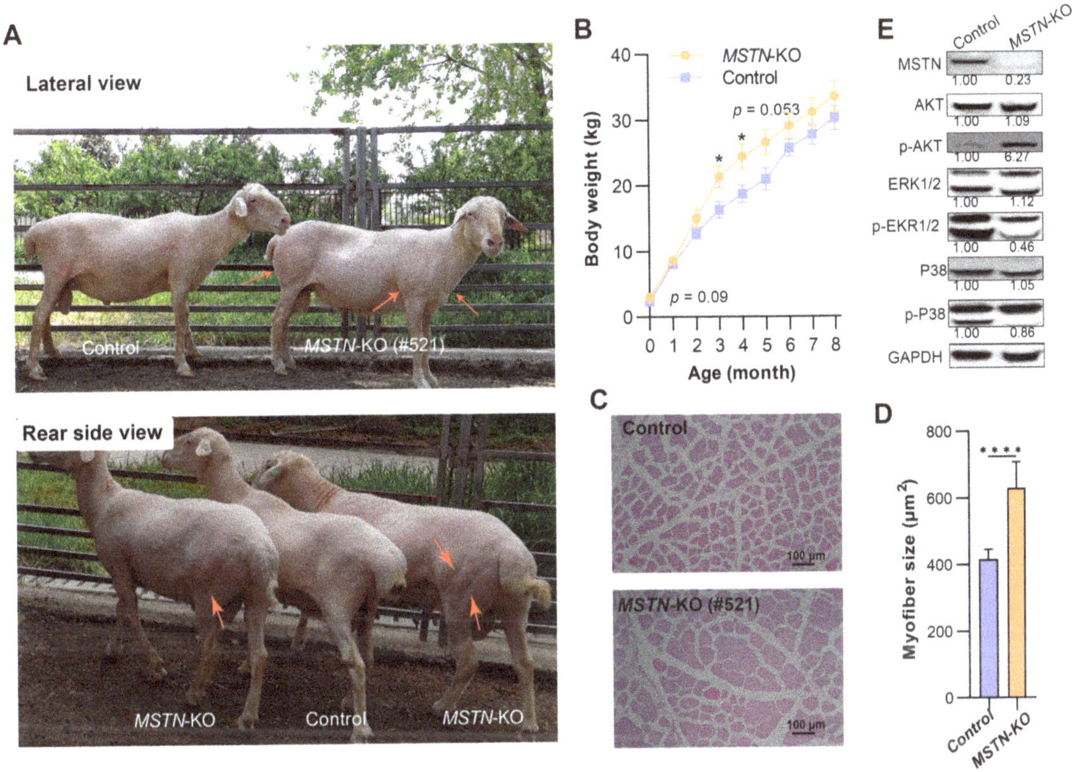

Figure 2. Phenotype and molecular analysis of *MSTN*-edited Hu sheep. (**A**) Pictures of *MSTN*-KO and control male sheep. The top and bottom pictures were taken from the lateral view and the rear side view, respectively. Red arrows indicate muscular protrusion, intermuscular boundaries, and grooves in the forequarters and hindquarters of *MSTN*-KO sheep. (**B**) Growth curve of *MSTN*-KO and control Hu sheep from birth to 8 months. Shown are mean values ± SEM. * $p < 0.05$. (**C**,**D**) HE-stained cross section and myofiber size of gluteus muscles from *MSTN*-KO (#521) and control sheep. **** $p < 0.001$. (**E**) Western blot analysis of MSTN, AKT, p-AKT, ERK1/2, p-EKR1/2, p38, p-P38 and GAPDH. The number under the blot picture is the relative protein expression relative to the control group.

3.4. Molecular Analysis of MSTN-Edited Hu Sheep

Constant with the genome editing at the *MSTN* locus, mature MSTN protein of the gluteus muscle decreased 77% in the edited Hu sheep compared with the control Hu sheep (Figure 2E). The remaining MSTN protein in edited sheep had a slightly lower molecular weight, indicating truncation of the MSTN protein. Among the downstream signaling mediators of MSTN (AKT, ERK1/2, and P38), p-AKT increased by 6.27-fold, while p-ERK1/2 decreased by 0.46-fold. p-P38 remained unchanged (0.86-fold) (Figure 2E). These results indicated that the activation of AKT signaling and repression of ERK1/2 signaling contributed to muscle hypertrophy in the *MSTN*-KO Hu sheep.

4. Discussion

In China, intense cross-breeding efforts have been afforded to improve the carcass performance of Hu sheep. However, the grading-up process is labor intensive and time consuming, and other desired genes may be diluted. In this study, we successfully achieved complete *MSTN* gene KO in 90% of newborn lambs using the C-CRISPR method. This high level of gene KO efficiency has never been reported before in genome editing of farm animals. Furthermore, the *MSTN*-KO Hu sheep founders exhibited the typical DM phenotype, with obvious intermuscular boundaries and grooves in both forequarters and hindquarters. These results indicate that the C-CRISPR method is a promising tool for breeding farm animals.

The C-CRISPR method was first proposed by [33]; they generated gene total KO founder animals with adjacent sgRNAs in one step. The high efficiencies of C-CRISPR can be easily understood. As the efficiency of each sgRNA is less than 100%, the gene editing must result in gene function mosaicisms when one sgRNA is used. For example, in our *MSTN*-edited Haimen goats and rabbits and *CLPG1*-edited rabbits, the efficiencies of the three sgRNAs targeting *MSTN* were from 11.1 to 100% [17], and the efficiencies of the two sgRNAs targeting *CLPG1* were from 20.0 to 100% [35]. In Wang et al. [20], five sgRNAs were used to edit the sheep genome; the efficiencies of these sgRNAs were as follows: *MSTN* sg1, 4.3 to 100%; *MSTN* sg2, 16.6 to 100%; *ASIP* sg1 + sg2, 11.1 to 60%; and *BCO2* sg1 + sg2, 9.5% to 100%. When two or more sgRNAs are used, the total genome editing efficiency can be calculated as $p = 1 - (1 - p_1)(1 - p_2) \ldots (1 - p_N)$, while $P_1, P_2 \ldots P_N$ are the genome editing efficiencies of the each sgRNA, and N is the number of sgRNAs used in C-CRISPR. The p value will gradually increase to 100% as the number of sgRNAs increases and as the values of $p_1, p_2 \ldots p_N$ increase. The present study achieved total gene knockout mediated by two (sgC1 and sgC4) and three sgRNAs (sgC1, sgC3, and sgC4). These results support that C-CRISPR is efficient, even with sgRNAs whose efficiencies have not been validated in advance, and it could generate genetic function homozygous KO farm animals in one step.

CRISPR/Cas9 has been proven to be high fidelity in several in vivo studies [36–38] and stem cell research [39–41] using whole-genome sequencing; rare but bona fide off-target mutations were reported in founder mice [42,43]. In our current study, no de novo mutations were found in the founders. The high fidelity of our genome editing in Hu sheep is in line with Wang et al. [20] and Wang et al. [44], who proved the genome editing specificity in sheep by genotyping POTSs predicted in silico [20] or by whole-genome sequencing [44].

Our *MSTN*-KO Hu sheep founders exhibited the DM phenotype characterized by prominent higher body weight, muscular protrusion, clearly visible intermuscular groves, and muscle hypertrophy. This was consistent with the muscularity in other sheep with *MSTN* natural mutations [8,9,15,45] or *MSTN* KO [18,20,46]. The average fiber size increased 51.4% in our *MSTN*-KO Hu sheep, and the muscle hypertrophy was consistent with [18,24], which may contribute to their muscularity. The body weight of the edited Hu sheep significantly increased at 3 and 4 months and tended to be higher at 0 d or 5 months, which was in line with the effects of *MSTN* knockout on sheep growth performance. The enhancement effects of *MSTN* knockout on body weight vary in the literature. Crispo et al. [46] generated homozygous and heterozygous *MSTN*-edited sheep guided by sgRNA targeting exon 1 and found that the body weight of homozygous *MSTN* knockout sheep was significantly higher from 15 d to 30 d old and tended to be higher at 60 d old. Wang et al. [20] generated *MSTN*-knockout Tan sheep by microinjection of Cas9 mRNA and two sgRNAs targeting exon 2 and exon 3, the efficiencies ranging from 4.34% to 100%, and the body weight of the edited sheep was higher at birth, 1 month old, and 5 months old. Li et al. [18] generated *MSTN*-edited small tail Han sheep using TALEN by targeting exon 1 of ovine *MSTN*, and the body weights were significantly higher from 1 to 7 months old.

The effects of *MSTN* edition on phenotype are target location, breed, and editing efficiency dependent. First, *MSTN* editing has a target location effect [47]. For example, all $MSTN^{-/-}$ large white pigs with exon 3 mutations showed lameness and only survived for 1–2 days, while $MSTN^{-/-}$ piglets with exon 1 mutations were healthy [23]. Additionally, the well-known sheep *MSTN* mutations c.*1232G>A and c.960delG lead to the typical double muscle phenotype and exist in Norwegian white sheep. However, after 30 years of breeding for improved carcass quality, only the frequencies of c.*1232G>A increased [7]. This is because most homozygous c.960delG lambs die after birth for unknown reasons [7], while homozygous c.*1232G>A sheep are healthy. Second, the effects of *MSTN* mutation are breed dependent. As mentioned above, $MSTN^{-/-}$ large white pigs are born with health problems, while $MSTN^{-/-}$ Meishan pigs with mutations at exon 3 are healthy. Last, the effects of MSTN are dosage dependent; the genome editing efficiencies are key factors influencing the growth performance. Thus, the differences in sgRNA used (different target sites), the sheep breed, and also editing efficiencies may all contribute to the variation in body weight and other phenotypes between studies.

MSTN functions as a negative regulator of muscle development [5] and satellite cell proliferation and differentiation [48,49]. The mechanisms are not fully understood, but it is known that MSTN acts through several downstream mediators including the TGFβ signaling pathway, TGFβ signaling pathway [5,50], AKT [51–53], P38 [54], and ERK1/2 [55]. In in vitro cultured myoblasts, the repression of AKT signaling was involved in MSTN-inhibited, both in differentiation in myoblasts and hypertrophy in myotubes [52,53]; in vivo overexpression of *MSTN* resulted in muscle loss through inhibiting AKT signaling [49]. In contrast, AKT signaling was enhanced in myostatin-deficient mice [56] and pigs [29]. Our results also showed a 6.27-fold increase in AKT activation. The ERK1/2 cascade has been reported to be activated by myostatin in C2C12 myoblasts and blocking ERK1/2 activation attenuates MSTN-suppressed myotube fusion and differentiation [55]. ERK1/2 has also been reported to be involved in MSTN-suppressed progesterone secretion in ovary granulosa cells [57] and preadipocyte differentiation [58,59]. In our study, we found that ERK1/2 activation decreased by 0.57-fold. P38 has also been shown to be activated by MSTN in C2C12 myoblasts, and treatment with P38 inhibitors reduces the MSTN-induced inhibition of proliferation [54]. However, P38 was only slightly changed in our study. Our results emphasize the importance of AKT and ERK1/2 signaling in MSTN-regulated muscle function in Hu sheep.

Recently, PRLR knockout SLICK cattle were approved by the FDA as a human food source after strict molecular characterization and assessment of the phenotypic data and animal safety, human food safety, and environmental risks [60]. This has paved the way to utilize genome editors as domestic animal precision breeding tools. Accordingly, *MSTN*-edited Hu sheep have the potential to be approved as a human food source after molecular characterization and evaluation of phenotypic data, animal safety, human food safety, and environmental risks. In the future, other phenotypic data such as carcass traits and meat traits will be evaluated after preparing enough offspring, along with an assessment of the environmental risks.

In conclusion, our study indicated that C-CRISPR is a promising method in farm animal breeding. In addition, *MSTN*-edited Hu sheep with the DM phenotype were efficiently and precisely generated.

Supplementary Materials: The following supporting information can be downloaded at: https://www.mdpi.com/article/10.3390/genes14061216/s1, Supplemental Figure S1. Electrophoresis results of IVT templates and RNAs for Cas9 and sgRNAs. Supplemental Figure S2. Protocols for superovulation and estrous synchronization of Hu sheep. Supplemental Figure S3. Detection of the off-target effects by Sanger sequencing of the POTS for sgC1, sgC2, sgC3, and sgC4. Supplemental Table S1. Information of sgRNA and oligos for sgRNA construction. Supplemental Table S2. Primers for IVT template amplification. Supplemental Table S3. Primers for exon3 amplification. Supplemental Table S4. Information on the potential off-target sites (POTs) for sgC1-C4 and the primers for PCR amplification of the sites.

Author Contributions: Conceptualization, R.G., S.C. and H.W.; methodology, R.G., H.W. and S.C.; investigation, R.G., H.W., S.C., C.M., H.G., Y.L., F.C., C.Z., H.Z., Q.D., J.Z. (Jianli Zhang) and J.Z. (Jun Zhang); data curation, S.C.; writing—original draft preparation, R.G., F.C., H.W. and S.C.; writing—review and editing, R.G., S.C., H.W. and Q.D.; visualization, R.G.; supervision, Y.Q. and J.Z. (Jifeng Zhong); project administration, S.C., J.Z. (Jianli Zhang) and J.Z. (Jun Zhang); funding acquisition, S.C. and R.G. All authors have read and agreed to the published version of the manuscript.

Funding: This study was supported by the Jiangsu Provincial Key R&D Program–Modern Agriculture (grant no. BE2019373) and the Jiangsu Province Agricultural Science and Technology Independent Innovation Fund Project (grant no. CX(21)3128).

Institutional Review Board Statement: The experimental procedures were approved by the Research Committee of the Jiangsu Academy of Agricultural Sciences and conducted with adherence to the Regulations for the Administration of Affairs Concerning Experimental Animals (Decree No. 63 of the Jiangsu Academy of Agricultural Science on 8 July 2014).

Informed Consent Statement: Not applicable.

Data Availability Statement: The data presented in this study are available on request from the corresponding author.

Acknowledgments: We would like to express our gratitude to Ma Baohua from Northwest A&F University for providing suggestions on sheep ovulation.

Conflicts of Interest: The authors declare no competing financial interests.

References

1. CRISPR beef cattle get FDA green light. *Nat. Biotechnol.* **2022**, *40*, 448. [CrossRef] [PubMed]
2. Zhang, Y.; Guo, J.; Yao, X.; Wang, F. Research progress on regulation mechanism of reproductive traits of Hu sheep. *J. Nanjing Agric. Univ.* **2022**, *45*, 1032–1040.
3. Chebo, C.; Betsha, S.; Melesse, A. Chicken genetic diversity, improvement strategies and impacts on egg productivity in Ethiopia: A review. *Worlds Poult. Sci. J.* **2022**, *78*, 803–821. [CrossRef]
4. Cong, L.; Ran, F.A.; Cox, D.; Lin, S.; Barretto, R.; Habib, N.; Hsu, P.D.; Wu, X.; Jiang, W.; Marraffini, L.A.; et al. Multiplex Genome Engineering Using CRISPR/Cas Systems. *Science* **2013**, *339*, 819–823. [CrossRef] [PubMed]
5. Hsu, P.D.; Lander, E.S.; Zhang, F. Development and Applications of CRISPR-Cas9 for Genome Engineering. *Cell* **2014**, *157*, 1262–1278. [CrossRef]
6. Bogdanovich, S.; Krag, T.O.B.; Barton, E.R.; Morris, L.D.; Whittemore, L.-A.; Ahima, R.S.; Khurana, T.S. Functional improvement of dystrophic muscle by myostatin blockade. *Nature* **2002**, *420*, 418–421. [CrossRef]
7. Schuelke, M.; Wagner, K.R.; Stolz, L.E.; Hübner, C.; Riebel, T.; Kömen, W.; Braun, T.; Tobin, J.F.; Lee, S.J. Myostatin mutation associated with gross muscle hypertrophy in a child. *N. Engl. J. Med.* **2004**, *350*, 2682–2688. [CrossRef]
8. Boman, I.A.; Klemetsdal, G.; Nafstad, O.; Blichfeldt, T.; Våge, D.I. Selection based on progeny testing induces rapid changes in myostatin allele frequencies—A case study in sheep. *J. Anim. Breed. Genet. Z. Tierz. Zucht.* **2011**, *128*, 52–55. [CrossRef]
9. Clop, A.; Marcq, F.; Takeda, H.; Pirottin, D.; Tordoir, X.; Bibé, B.; Bouix, J.; Caiment, F.; Elsen, J.M.; Eychenne, F.; et al. A mutation creating a potential illegitimate microRNA target site in the myostatin gene affects muscularity in sheep. *Nat. Genet.* **2006**, *38*, 813–818. [CrossRef]
10. Boman, I.A.; Klemetsdal, G.; Blichfeldt, T.; Nafstad, O.; Våge, D.I. A frameshift mutation in the coding region of the myostatin gene (MSTN) affects carcass conformation and fatness in Norwegian White Sheep (Ovis aries). *Anim. Genet.* **2009**, *40*, 418–422. [CrossRef]
11. Grobet, L.; Martin, L.J.; Poncelet, D.; Pirottin, D.; Brouwers, B.; Riquet, J.; Schoeberlein, A.; Dunner, S.; Ménissier, F.; Massabanda, J.; et al. A deletion in the bovine myostatin gene causes the double-muscled phenotype in cattle. *Nat. Genet.* **1997**, *17*, 71–74. [CrossRef] [PubMed]
12. Kambadur, R.; Sharma, M.; Smith, T.P.; Bass, J.J. Mutations in myostatin (GDF8) in double-muscled Belgian Blue and Piedmontese cattle. *Genome Res.* **1997**, *7*, 910–916. [CrossRef] [PubMed]
13. McPherron, A.C.; Lee, S.J. Double muscling in cattle due to mutations in the myostatin gene. *Proc. Natl. Acad. Sci. USA* **1997**, *94*, 12457–12461. [CrossRef]
14. Matika, O.; Robledo, D.; Pong-Wong, R.; Bishop, S.C.; Riggio, V.; Finlayson, H.; Lowe, N.R.; Hoste, A.E.; Walling, G.A.; Del-Pozo, J.; et al. Balancing selection at a premature stop mutation in the myostatin gene underlies a recessive leg weakness syndrome in pigs. *PLoS Genet.* **2019**, *15*, e1007759. [CrossRef] [PubMed]
15. Mosher, D.S.; Quignon, P.; Bustamante, C.D.; Sutter, N.B.; Mellersh, C.S.; Parker, H.G.; Ostrander, E.A. A mutation in the myostatin gene increases muscle mass and enhances racing performance in heterozygote dogs. *PLoS Genet.* **2007**, *3*, e79. [CrossRef]
16. Boman, I.A.; Klemetsdal, G.; Nafstad, O.; Blichfeldt, T.; Våge, D.I. Impact of two myostatin (MSTN) mutations on weight gain and lamb carcass classification in Norwegian White Sheep (Ovis aries). *Genet. Sel. Evol. GSE* **2010**, *42*, 4. [CrossRef]
17. Wang, X.; Yu, H.; Lei, A.; Zhou, J.; Zeng, W.; Zhu, H.; Dong, Z.; Niu, Y.; Shi, B.; Cai, B.; et al. Generation of gene-modified goats targeting MSTN and FGF5 via zygote injection of CRISPR/Cas9 system. *Sci. Rep.* **2015**, *5*, 13878. [CrossRef]
18. Guo, R.; Wan, Y.; Xu, D.; Cui, L.; Deng, M.; Zhang, G.; Jia, R.; Zhou, W.; Wang, Z.; Deng, K.; et al. Generation and evaluation of Myostatin knock-out rabbits and goats using CRISPR/Cas9 system. *Sci. Rep.* **2016**, *6*, 29855. [CrossRef]
19. Li, H.; Wang, G.; Hao, Z.; Zhang, G.; Qing, Y.; Liu, S.; Qing, L.; Pan, W.; Chen, L.; Liu, G.; et al. Generation of biallelic knock-out sheep via gene-editing and somatic cell nuclear transfer. *Sci. Rep.* **2016**, *6*, 33675. [CrossRef]

20. Proudfoot, C.; Carlson, D.F.; Huddart, R.; Long, C.R.; Pryor, J.H.; King, T.J.; Lillico, S.G.; Mileham, A.J.; McLaren, D.G.; Whitelaw, C.B.; et al. Genome edited sheep and cattle. *Transgenic Res.* **2015**, *24*, 147–153. [CrossRef]
21. Wang, X.; Niu, Y.; Zhou, J.; Yu, H.; Kou, Q.; Lei, A.; Zhao, X.; Yan, H.; Cai, B.; Shen, Q.; et al. Multiplex gene editing via CRISPR/Cas9 exhibits desirable muscle hypertrophy without detectable off-target effects in sheep. *Sci. Rep.* **2016**, *6*, 32271. [CrossRef] [PubMed]
22. Bi, Y.; Hua, Z.; Liu, X.; Hua, W.; Ren, H.; Xiao, H.; Zhang, L.; Li, L.; Wang, Z.; Laible, G.; et al. Isozygous and selectable marker-free MSTN knockout cloned pigs generated by the combined use of CRISPR/Cas9 and Cre/LoxP. *Sci. Rep.* **2016**, *6*, 31729. [CrossRef] [PubMed]
23. Ren, H.; Xiao, W.; Qin, X.; Cai, G.; Chen, H.; Hua, Z.; Cheng, C.; Li, X.; Hua, W.; Xiao, H.; et al. Myostatin regulates fatty acid desaturation and fat deposition through MEF2C/miR222/SCD5 cascade in pigs. *Commun. Biol.* **2020**, *3*, 612. [CrossRef] [PubMed]
24. Fan, Z.; Liu, Z.; Xu, K.; Wu, T.; Ruan, J.; Zheng, X.; Bao, S.; Mu, Y.; Sonstegard, T.; Li, K. Long-term, multidomain analyses to identify the breed and allelic effects in MSTN-edited pigs to overcome lameness and sustainably improve nutritional meat production. *Sci. China Life Sci.* **2022**, *65*, 362–375. [CrossRef] [PubMed]
25. Zhou, S.; Kalds, P.; Luo, Q.; Sun, K.; Zhao, X.; Gao, Y.; Cai, B.; Huang, S.; Kou, Q.; Petersen, B.; et al. Optimized Cas9:sgRNA delivery efficiently generates biallelic MSTN knockout sheep without affecting meat quality. *BMC Genom.* **2022**, *23*, 348. [CrossRef]
26. Tait-Burkard, C.; Doeschl-Wilson, A.; McGrew, M.J.; Archibald, A.L.; Sang, H.M.; Houston, R.D.; Whitelaw, C.B.; Watson, M. Livestock 2.0—Genome editing for fitter, healthier, and more productive farmed animals. *Genome Biol.* **2018**, *19*, 204. [CrossRef]
27. Kalds, P.; Gao, Y.; Zhou, S.; Cai, B.; Huang, X.; Wang, X.; Chen, Y. Redesigning small ruminant genomes with CRISPR toolkit: Overview and perspectives. *Theriogenology* **2020**, *147*, 25–33. [CrossRef]
28. Niu, Y.; Shen, B.; Cui, Y.; Chen, Y.; Wang, J.; Wang, L.; Kang, Y.; Zhao, X.; Si, W.; Li, W.; et al. Generation of gene-modified cynomolgus monkey via Cas9/RNA-mediated gene targeting in one-cell embryos. *Cell* **2014**, *156*, 836–843. [CrossRef]
29. Dingwei, P.; Ruiqiang, L.; Wu, Z.; Min, W.; Xuan, S.; Zeng, J.; Liu, X.; Chen, Y.; He, Z. Editing the cystine knot motif of MSTN enhances muscle development of Liang Guang Small Spotted pigs. *Hereditas* **2021**, *43*, 261–270. [CrossRef]
30. Li, R.; Zeng, W.; Ma, M.; Wei, Z.; Liu, H.; Liu, X.; Wang, M.; Shi, X.; Zeng, J.; Yang, L.; et al. Precise editing of myostatin signal peptide by CRISPR/Cas9 increases the muscle mass of Liang Guang Small Spotted pigs. *Transgenic Res.* **2020**, *29*, 149–163. [CrossRef]
31. Zhu, X.X.; Zhan, Q.M.; Wei, Y.Y.; Yan, A.F.; Feng, J.; Liu, L.; Lu, S.S.; Tang, D.S. CRISPR/Cas9-mediated MSTN disruption accelerates the growth of Chinese Bama pigs. *Reprod. Domest. Anim. Zuchthyg.* **2020**, *55*, 1314–1327. [CrossRef] [PubMed]
32. Moro, L.N.; Viale, D.L.; Bastón, J.I.; Arnold, V.; Suvá, M.; Wiedenmann, E.; Olguín, M.; Miriuka, S.; Vichera, G. Generation of myostatin edited horse embryos using CRISPR/Cas9 technology and somatic cell nuclear transfer. *Sci. Rep.* **2020**, *10*, 15587. [CrossRef] [PubMed]
33. Wang, K.; Ouyang, H.; Xie, Z.; Yao, C.; Guo, N.; Li, M.; Jiao, H.; Pang, D. Efficient Generation of Myostatin Mutations in Pigs Using the CRISPR/Cas9 System. *Sci. Rep.* **2015**, *5*, 16623. [CrossRef] [PubMed]
34. Zuo, E.; Cai, Y.J.; Li, K.; Wei, Y.; Wang, B.A.; Sun, Y.; Liu, Z.; Liu, J.; Hu, X.; Wei, W.; et al. One-step generation of complete gene knockout mice and monkeys by CRISPR/Cas9-mediated gene editing with multiple sgRNAs. *Cell Res.* **2017**, *27*, 933–945. [CrossRef]
35. Bae, S.; Park, J.; Kim, J.S. Cas-OFFinder: A fast and versatile algorithm that searches for potential off-target sites of Cas9 RNA-guided endonucleases. *Bioinformatics* **2014**, *30*, 1473–1475. [CrossRef]
36. Wan, Y.; Guo, R.; Deng, M.; Liu, Z.; Pang, J.; Zhang, G.; Wang, Z.; Wang, F. Efficient generation of CLPG1-edited rabbits using the CRISPR/Cas9 system. *Reprod. Domest. Anim. Zuchthyg.* **2019**, *54*, 538–544. [CrossRef] [PubMed]
37. Kilian, N.; Goryaynov, A.; Lessard, M.D.; Hooker, G.; Toomre, D.; Rothman, J.E.; Bewersdorf, J. Assessing photodamage in live-cell STED microscopy. *Nat. Methods* **2018**, *15*, 755–756. [CrossRef]
38. Iyer, V.; Boroviak, K.; Thomas, M.; Doe, B.; Riva, L.; Ryder, E.; Adams, D.J. No unexpected CRISPR-Cas9 off-target activity revealed by trio sequencing of gene-edited mice. *PLoS Genet.* **2018**, *14*, e1007503. [CrossRef]
39. Wilson, C.J.; Fennell, T.; Bothmer, A.; Maeder, M.L.; Reyon, D.; Cotta-Ramusino, C.; Fernandez, C.A.; Marco, E.; Barrera, L.A.; Jayaram, H.; et al. Response to "Unexpected mutations after CRISPR-Cas9 editing in vivo". *Nat. Methods* **2018**, *15*, 236–237. [CrossRef]
40. Smith, C.; Gore, A.; Yan, W.; Abalde-Atristain, L.; Li, Z.; He, C.; Wang, Y.; Brodsky, R.A.; Zhang, K.; Cheng, L.; et al. Whole-genome sequencing analysis reveals high specificity of CRISPR/Cas9 and TALEN-based genome editing in human iPSCs. *Cell Stem Cell* **2014**, *15*, 12–13. [CrossRef]
41. Suzuki, K.; Yu, C.; Qu, J.; Li, M.; Yao, X.; Yuan, T.; Goebl, A.; Tang, S.; Ren, R.; Aizawa, E.; et al. Targeted gene correction minimally impacts whole-genome mutational load in human-disease-specific induced pluripotent stem cell clones. *Cell Stem Cell* **2014**, *15*, 31–36. [CrossRef]
42. Veres, A.; Gosis, B.S.; Ding, Q.; Collins, R.; Ragavendran, A.; Brand, H.; Erdin, S.; Cowan, C.A.; Talkowski, M.E.; Musunuru, K. Low incidence of off-target mutations in individual CRISPR-Cas9 and TALEN targeted human stem cell clones detected by whole-genome sequencing. *Cell Stem Cell* **2014**, *15*, 27–30. [CrossRef] [PubMed]
43. Aryal, N.K.; Wasylishen, A.R.; Lozano, G. CRISPR/Cas9 can mediate high-efficiency off-target mutations in mice in vivo. *Cell Death Dis.* **2018**, *9*, 1099. [CrossRef] [PubMed]

44. Iyer, V.; Shen, B.; Zhang, W.; Hodgkins, A.; Keane, T.; Huang, X.; Skarnes, W.C. Off-target mutations are rare in Cas9-modified mice. *Nat. Methods* **2015**, *12*, 479. [CrossRef] [PubMed]
45. Wang, X.; Liu, J.; Niu, Y.; Li, Y.; Zhou, S.; Li, C.; Ma, B.; Kou, Q.; Petersen, B.; Sonstegard, T.; et al. Low incidence of SNVs and indels in trio genomes of Cas9-mediated multiplex edited sheep. *BMC Genom.* **2018**, *19*, 397. [CrossRef] [PubMed]
46. Wang, J.; Zhou, H.; Hu, J.; Li, S.; Luo, Y.; Hickford, J.G. Two single nucleotide polymorphisms in the promoter of the ovine myostatin gene (MSTN) and their effect on growth and carcass muscle traits in New Zealand Romney sheep. *J. Anim. Breed. Genet. Z. Tierz. Zucht.* **2016**, *133*, 219–226. [CrossRef]
47. Crispo, M.; Mulet, A.P.; Tesson, L.; Barrera, N.; Cuadro, F.; dos Santos-Neto, P.C.; Nguyen, T.H.; Creneguy, A.; Brusselle, L.; Anegon, I.; et al. Efficient Generation of Myostatin Knock-Out Sheep Using CRISPR/Cas9 Technology and Microinjection into Zygotes. *PLoS ONE* **2015**, *10*, e0136690. [CrossRef]
48. Wang, X.; Petersen, B. More abundant and healthier meat: Will the MSTN editing epitome empower the commercialization of gene editing in livestock? *Sci. China Life Sci.* **2022**, *65*, 448–450. [CrossRef]
49. McCroskery, S.; Thomas, M.; Maxwell, L.; Sharma, M.; Kambadur, R. Myostatin negatively regulates satellite cell activation and self-renewal. *J. Cell Biol.* **2003**, *162*, 1135–1147. [CrossRef]
50. Thomas, M.; Langley, B.; Berry, C.; Sharma, M.; Kirk, S.; Bass, J.; Kambadur, R. Myostatin, a negative regulator of muscle growth, functions by inhibiting myoblast proliferation. *J. Biol. Chem.* **2000**, *275*, 40235–40243. [CrossRef]
51. Sartori, R.; Milan, G.; Patron, M.; Mammucari, C.; Blaauw, B.; Abraham, R.; Sandri, M. Smad2 and 3 transcription factors control muscle mass in adulthood. *Am. J. Physiol. Cell Physiol.* **2009**, *296*, C1248–C1257. [CrossRef] [PubMed]
52. Amirouche, A.; Durieux, A.C.; Banzet, S.; Koulmann, N.; Bonnefoy, R.; Mouret, C.; Bigard, X.; Peinnequin, A.; Freyssenet, D. Down-regulation of Akt/mammalian target of rapamycin signaling pathway in response to myostatin overexpression in skeletal muscle. *Endocrinology* **2009**, *150*, 286–294. [CrossRef] [PubMed]
53. Trendelenburg, A.U.; Meyer, A.; Rohner, D.; Boyle, J.; Hatakeyama, S.; Glass, D.J. Myostatin reduces Akt/TORC1/p70S6K signaling, inhibiting myoblast differentiation and myotube size. *Am. J. Physiol. Cell Physiol.* **2009**, *296*, C1258–C1270. [CrossRef] [PubMed]
54. Morissette, M.R.; Cook, S.A.; Buranasombati, C.; Rosenberg, M.A.; Rosenzweig, A. Myostatin inhibits IGF-I-induced myotube hypertrophy through Akt. *Am. J. Physiol. Cell Physiol.* **2009**, *297*, C1124–C1132. [CrossRef]
55. Philip, B.; Lu, Z.; Gao, Y. Regulation of GDF-8 signaling by the p38 MAPK. *Cell Signal.* **2005**, *17*, 365–375. [CrossRef]
56. Yang, W.; Chen, Y.; Zhang, Y.; Wang, X.; Yang, N.; Zhu, D. Extracellular signal-regulated kinase 1/2 mitogen-activated protein kinase pathway is involved in myostatin-regulated differentiation repression. *Cancer Res.* **2006**, *66*, 1320–1326. [CrossRef]
57. Lipina, C.; Kendall, H.; McPherron, A.C.; Taylor, P.M.; Hundal, H.S. Mechanisms involved in the enhancement of mammalian target of rapamycin signalling and hypertrophy in skeletal muscle of myostatin-deficient mice. *FEBS Lett.* **2010**, *584*, 2403–2408. [CrossRef]
58. Fang, L.; Chang, H.M.; Cheng, J.C.; Yu, Y.; Leung, P.C.; Sun, Y.P. Growth Differentiation Factor-8 Decreases StAR Expression Through ALK5-Mediated Smad3 and ERK1/2 Signaling Pathways in Luteinized Human Granulosa Cells. *Endocrinology* **2015**, *156*, 4684–4694. [CrossRef]
59. Pan, S.; Zhang, L.; Liu, Z.; Xing, H. Myostatin suppresses adipogenic differentiation and lipid accumulation by activating crosstalk between ERK1/2 and PKA signaling pathways in porcine subcutaneous preadipocytes. *J. Anim. Sci.* **2021**, *99*, skab287. [CrossRef]
60. Li, F.; Yang, H.; Duan, Y.; Yin, Y. Myostatin regulates preadipocyte differentiation and lipid metabolism of adipocyte via ERK1/2. *Cell Biol. Int.* **2011**, *35*, 1141–1146. [CrossRef]

MDPI
St. Alban-Anlage 66
4052 Basel
Switzerland
www.mdpi.com

Genes Editorial Office
E-mail: genes@mdpi.com
www.mdpi.com/journal/genes

www.ingramcontent.com/pod-product-compliance
Lightning Source LLC
Chambersburg PA
CBHW061448210926
51850CB00008B/1647
* 9 7 8 3 7 2 5 8 0 3 8 7 3 *